Nuclear Vigilance

The Creation of the Atomic Energy Detection
System (AEDS) and its Impact on Nuclear Arms
Control Goals during the Eisenhower Administration

James Michael Young, Ph.D.
Command Historian
Air Force Technical Applications Center

Air Force Technical Applications Center
United States Air Force
Patrick Space Force Base
Melbourne, Florida, 32925
May 2022

Opinions, conclusions, and recommendations expressed or implied within are solely those of the author and do not necessarily represent the views of the Air force History and Museums Program, the U.S. Air Force, the Department of Defense, or any other U.S. government agency.

Photographs on pages 326 and 327 appear here courtesy of Cornell University's Rare and Manuscript Collections.

Nuclear Vigilance: The Creation of the Atomic Energy Detection System (AEDS) and its Impact on Nuclear Arms Control Goals during the Eisenhower Administration

ISBN: 978-1-312-42453-1

9 781312 424531

The Author

The author, James Michael Young, is a historian for the Air Force Technical Applications Center (AFTAC). He joined AFTAC in 2011 after serving in the U.S. Army for 30 years. Young has written a number of AFTAC histories, and has published articles in *Air Power History* and on the web at defensepolicy.org and the avfforum.org. He holds a B.A. and M.A. from the Ohio State University, an M.S. from the Defense Intelligence College, and a Ph.D. from Antioch University.

Dedicated to the men and women of the Long Range Detection program for their numerous contributions and sacrifices in providing this vital treaty monitoring capability to the nation.

TABLE OF CONTENTS

INTRODUCTION

To fully appreciate the historical evolution of the Long Range Detection (LRD) mission and the creation and implementation of the United States Atomic Energy Detection System (USAEDS), it is necessary to understand the context of the early Cold War era.[1] Major political forces and scientific advancements in the initial years of the Cold War compelled the development of a capability to detect nuclear detonations at extended distances. Those developments occurred within a climate of severe mistrust between two superpowers: the United States and the Soviet Union.[2] Soon after the Second World War, the United States realized its monopoly on nuclear weapons would not last forever. Thus, the paramount question of the late 1940s was: When would the Soviet Union detonate its first nuclear weapon?

Against all predictions (and estimates greatly varied), the Soviets detonated their first nuclear device—which the Americans nicknamed Joe-1 as a sardonic reference to Joseph Stalin—on August 29, 1949, much earlier than most experts anticipated.[3] Fortunately, insightful leaders within the U.S. Government had turned to new scientific technologies to address this question several years earlier when it appeared unlikely that American human intelligence resources were capable of penetrating the secretive Soviet nuclear program. Enlisting some of the nation's foremost experts in the fields of radiochemistry, nuclear physics, acoustics, seismology, and meteorology, the Air Force quickly developed the mission to "detect atomic explosions anywhere in the world." In 1948, it created a special unit to do so. Commanded by an Air Force major general, the unit was designated the 1009th Special Weapons Squadron (SWS). The administrative arm of the unit, which housed the research and analysis elements, was known by its department name, "AFOAT-1," for Atomic Energy Office, Section One, under the Air Force Deputy Chief of Staff for Operations. As such, the Commander wore two hats: Commander of the 1009th SWS and Director of AFOAT-1 (Chapter 1).[4]

This unprecedented unit was born on the eve of a decade that is remembered, often nostalgically, as an era of new and exciting technological achievements. For the Department of Defense (DoD), however, it was a time of great uncertainty. With fascism destroyed, the emerging threat quickly shifted to the Soviet Union and communist expansionism. The senior leaders throughout the government recognized how innovative technology had brought an immediate halt to the Second World War. In formulating a new national security policy, therefore, they believed technology could likewise provide definitive solutions for America's defense in the nuclear era (Chapters 2 and 3).[5]

At first, understanding the scope of the LRD effort proved challenging. However, the detonation of Joe-1 in August 1949, jolted the nation into immediate action to stay ahead of the Soviets in the development and weaponization of nuclear science (Chapter 4). Unable to ascertain the state of nuclear developments within the Soviet Union by conventional means, the government heavily invested in the exploration of technological solutions. Consequently, the Air Force received prioritized funding to create a multidisciplinary system that could monitor and collect data on nuclear detonations originating in the Soviet Union. This was unexplored territory. The first few years involved extensive research and development (R&D) in order to understand the effects of a nuclear explosion and ways to collect evidence of the detonation. Fortunately, AFOAT-1 would learn to do so by conducting extensive experiments during U.S. nuclear tests (Chapters 5 and 6).[6]

Experimenting with new LRD concepts at the U.S. testing grounds in Nevada and in the Pacific proved vital in rapidly developing the AEDS. However, AFOAT-1 could not do it alone. Despite the strict secrecy enshrouding AFOAT-1's activities, the Air Force relied extensively on private and other government organizations to create the AEDS. Initially, the system relied heavily on the "A" technique—the aerial collection of nuclear debris. The Air Weather Service led the way in conducting this technique, with AFOAT-1 specialists embedded in the crews. After flying collection sorties, the AFOAT-1 crew members shipped the samples to several government and commercial laboratories for analysis (designated the "L" technique). The AEDS also began using seismic sites—the "B" technique—where seismographs recorded data that either indicated an earthquake or an explosion had taken place. The U.S. Coast and Geodetic Survey and several universities with valuable experience in seismology routinely collaborated with AFOAT-1. The fourth technique, which initially jumpstarted the AEDS, was the acoustic or "I" technique. The Army, with many years of experience, managed the acoustic sites in the AEDS throughout the decade. Rudimentarily, all four techniques were operational by the time of the second Soviet detonation on September 24, 1951.

As the 1950s progressed, international events elevated tensions between the two superpowers and spurred both countries to more rapidly advance their nuclear weapons programs. As those weapons tests evolved and became more frequent, so did AFOAT-1's knowledge and ability to refine various LRD techniques. On November 1, 1952, just three days before Dwight Eisenhower won the presidential election, the United States successfully tested the world's first thermonuclear device. Codenamed Shot Mike, the world's first hydrogen weapon obliterated the island of Elugelab in the Marshall Islands.[7] The 10.4 megaton blast also produced deadly radioactive fallout which soon travelled around the Earth (Chapter 7).[8] Thermonuclear or hydrogen weapons used fusion in addition to fission to generate the chain reaction of energy release. The fact that they were 1,000 times greater than the first generation of atomic bombs, such as those used against Japan in 1945, created a severe division within the nuclear science community as some viewed such weapons as immoral.[9] That division would later fuel the contentious debates over Eisenhower's attempts to formulate a nuclear disarmament treaty with the Soviet Union, and impact AFOAT-1's stewardship over the AEDS. The unfathomable destructive power horrified the old soldier now turned president. For the next eight years, Eisenhower would take ever-increasing actions to curb the spiraling arms race and to achieve some measure of nuclear arms control.[10]

During his first term in office, Eisenhower did little to advance his strong desire for nuclear disarmament. Other high-stakes priorities preoccupied the President, such as ending the Korean War and replacing Truman's foreign policy of containment with a policy called the "New Look" (soon to be associated with the term "massive retaliation"). His first term in office coincided with the beginning stage of the Cold War, when perfecting nuclear weapons and building up a stockpile for deterrence drove national security policy. However, once the public became anxious over the enormous destructive power of hydrogen bombs and the radiation dangers they imposed, Eisenhower became more engaged in seeking ways to curb the accelerating arms race.

In those early years of the 1950s, fear of communist expansionism pervaded American society. Although McCarthyism (a term describing Senator Joseph McCarthy's popular crusade to root out suspected communists in the government) waned by the end of Eisenhower's second term, strong anti-Soviet sentiment caused the President to walk a fine line between his attempts to significantly reduce defense spending and the perception that he was "soft" on communism (an allegation which had caused the Democratic Party to lose the 1952 election). As the Soviets advanced their nuclear testing program, the Atomic Energy Commission (AEC), and DoD in particular, advocated for more nuclear weapons. Such fears arose out of a lack of knowledge about the state of the Soviet nuclear program. Because American human intelligence assets had no success in acquiring information about Soviet nuclear research, all stakeholders involved in national security looked to technology to penetrate that veil of secrecy. In this regard, the AEDS became a panacea for doing so (Chapter 8).

On the eve of the 1956 presidential election, the fledgling AEDS (or rather, its capabilities) became a political football at the center of the election debates over the nuclear arms race and the dangers of nuclear radiation fallout. Throughout the summer of 1956, the Democratic nominee, Adlai Stevenson, II, made radiation fallout and a test ban the central themes of his campaign. In referring to the AEDS, he erroneously stressed that such technology could easily detect any Soviet violations of a test ban treaty. He once proclaimed that "you can't hide the explosion any more than you can hide an earthquake."[11] From that moment on, despite the security classifications of AFOAT-1's LRD work, politicians and scientists began citing the effectiveness or ineffectiveness of the AEDS (depending on which side of the arms control debate one fell) as the only means to detect any treaty violations within the borders of the Soviet Union.

After his re-election in 1956, Eisenhower strongly pursued nuclear disarmament. However, he met heavy resistance from the AEC, headed by Commissioner Lewis Strauss, a vehement anti-communist, and the DoD, especially the Air Force. Until late 1957, when the Soviets successfully launched an intercontinental ballistic missile (ICBM) and placed two satellites into Earth's orbit, Strauss had monopolized nuclear scientific advice reaching the President. From the beginning of his appointment in 1953, Strauss always succeeded in dissuading the President from making any serious concessions in negotiating with the Soviets. The Soviet scientific advances, however, suddenly threatened American nuclear superiority and prompted Eisenhower to consult other nuclear physicists for advice on nuclear policy. The result was the creation of the President's Scientific Advisory Committee (PSAC). For the first time in U.S. history, the Office of the President had a scientific advisor entitled Special

Assistant to the President for Science and Technology. Eisenhower appointed James Killian, president of MIT, into the new position. Killian, as a proponent of arms control, supervised the PSAC. In short order, he stacked the PSAC with pro-treaty scientists, including the future Nobel laureate Dr. Hans Bethe (Chapter 9).[12]

Bethe was an ardent pro-arms control advocate who sided with other prominent scientists such as Dr. J. Robert Oppenheimer and Dr. Isador I. Rabi to oppose the development of the hydrogen bomb. Soon after the detonation of the second Soviet test in September 1951, Bethe began working closely with AFOAT-1 to help evaluate samples collected from Soviet tests. Throughout the decade and beyond, he led AFOAT-1's Scientific Advisory Committee (known as the Bethe Panel within AFOAT-1), which reviewed significant LRD R&D issues. Consequently, he understood the capabilities and limitations of the AEDS better than any other scientist outside the U.S. Air Force. By late 1957, Bethe leveraged that knowledge in his strong advocacy for a test ban.

In 1958 and 1959, debates over a test ban treaty placed the AEDS at the center of the international treaty negotiations. In fact, AFOAT-1's technical director, Mr. Doyle Northrup, and his top seismologist, Dr. Carl Romney, would serve on the U.S. teams at the Geneva negotiations that eventually produced the Limited Test Ban Treaty of 1963. As we will see, their contributions helped provide the most important factors in shaping the technical discussions. More importantly, AFOAT-1's thorough assessment of the limitations of the AEDS (especially seismology) would ultimately derail President Eisenhower's dream of establishing an arms control agreement with the Soviet Union (Chapters 10 and 11).[13]

*

My first intent with this history is to detail how the AEDS was created and to explain how AFOAT-1 operationalized science to do so. Also, because of previous classification restrictions, a unit history of the 1009th Special Weapons Squadron has never been written. In actuality, the history of the 1009th is essentially a history of the creation of the AEDS. It was a unique unit specifically designed to learn, experiment, and innovate in order to create an unprecedented technical surveillance system which could monitor nuclear detonations within the borders of the Soviet Union.

My second and more important intent is to place the role and impact of AFOAT-1's scientists, especially Doyle Northrup and Carl Romney, into their actual historical context. Because classification levels over the AEDS and most AFOAT-1 activities remained in place throughout the Cold War and were only recently relaxed, their story has never been told. Their influence on American foreign policy was profound and best illustrates how non-political actors—in this case, Air Force scientists working in a highly classified organization—significantly influenced political events with far-reaching, international consequences. The narrative that follows focuses on how they did that and why their activities grew in importance as the decade unfolded.

As with any work of this nature, I am indebted to many people who helped make this writing possible. First, this project came to fruition because of the full encouragement and support of my previous and current bosses, AFTAC Directors of Staff Mr. Jim Whidden, II, and Dr. Kevin Muhs. I am indebted to my colleagues at AFTAC, Dr. Bill Johnson, Dr. Mark Woods, and Dr. Ward Dougherty for their review of the manuscript and feedback, particularly in understanding the sciences involved in creating the AEDS. Thanks also to my colleagues on the AFTAC staff, Ms. Susan A. Romano and Mr. David Charitat, for their reviews. Both offered superb edits and comments. I am especially appreciative of the support I received from Cornell University's Rare and Manuscript Collections. Ms. Julia Gardner, Head of Research Services, and her staff, particularly Ms. Hillary Dorsch Wong, were extremely helpful and accommodating. They went the extra mile in guiding me through the extensive papers of Dr. Hans Bethe.

As I strive to become a better historian and author, I am constantly inspired by the professional drive and accomplishments of my children, Dr. A J Mata, MD, MPH, an OB/GYN chief resident at Vanderbilt University Medical Center; and Dr. Adrian Young, PhD, an assistant professor in the Department of History at Denison University. Last, but certainly not least, I am eternally grateful to my wife and editor, Angelika Young. Her many hours of tolerating my frustrations with this project and my poor spelling habits have made her a hero in my eyes. Without her enduring love, I am not quite sure I could have ever completed this book.

The primary audience for this history are the men and women of the Air Force Technical Applications Center. Your heritage has remained unknown for far too long.

TECHNIQUES OF THE AEDS

"A" Airborne Particulate & Gas Sampling Tech. (1948-Present)

"B" Seismic Tech. (1948-Present)

"C" Ground-Based Particulate Sampling Tech. (1950-Present)

"D" Ground Based Gaseous Sampling Tech. (1950-Present)

"Db" Balloon Whole Air Sampling Tech. (1954-1967)

"E" Surface-Based Resonance Scatter Tech. (1963-1971)

"F" Geophysical Diagnostics-Tech. (1962-1975)

"H" Magnetotelluric (Earth Current) Tech. (1961-67; 71-73)

"I" Acoustic Tech. (1948-1975)

"J" VLF/HF EMP Tech. (1980-1988)

"K" Ionosphere Tech. (1967-1971)

"L" Laboratory Sample Analysis Tech. (1950-present)

"N" HF Narrowband Phase Anomaly Tech. (1967)

"O" Hydroacoustic Tech. (1967 -present)

"Q" Electromagnetic Pulse Tech. (1953-1975)

"R" Backscatter Radar Tech. (1961-1967)

"T" Satellite Tech. (1965-present)

"U" VLF Phase Anomaly Tech. (1961-67; 1971-73)

"V" Vertical Incidence Tech. (1962-1967)

"W" HF Propagation Anomaly Tech. (1968-1972)

"Z" Atmospheric Fluorescence Detection Tech. (1961-67; 1971)

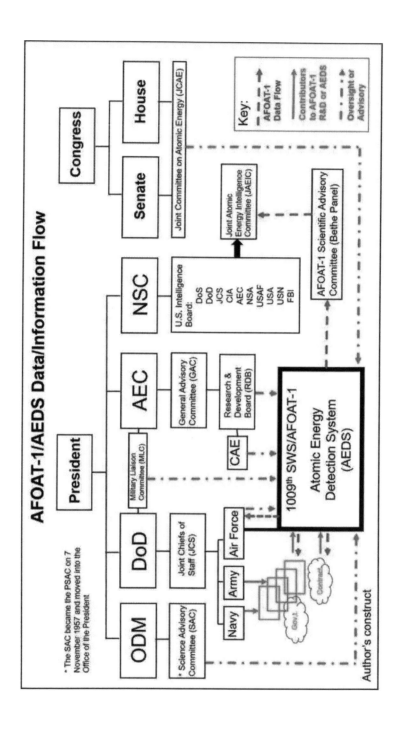

AFOAT-1/AEDS Data/Information Flow

* The SAC became the PSAC on 7 November 1957 and moved into the Office of the President

Key:
AFOAT-1 Data Flow
Contributors to AFOAT-1 R&D or AEDS
Oversight or Advisory

Author's construct

Doyle Northrup

In the history of the AEDS, Northrup was the most significant member of AFOAT-1 and AFTAC. A talented scientist with an astute awareness of political realities, Northrup adroitly led some of the nation's best scientists in the rapid creation of the AEDS. His early recognition of the importance of geophysical techniques proved critical in establishing all of the techniques that made the AEDS fully functional and reliable.

Photo source AFTAC archives

Notes for Introduction

[1] Until 1987, the system was simply called the Atomic Energy Detection System (AEDS). In 1987, the name was changed to the United States Atomic Energy Detection System (USAEDS) to denote it as a national, strategic system.

[2] For a thorough analysis of this mistrust see especially Deborah Weich Larson, *Anatomy of Mistrust: U.S.-Soviet Relations during the Cold War* (Ithaca, NY: Cornell University Press, 1997); and John Lewis Gaddis, *We Now Know: Rethinking Cold War History* (New York, NY: Oxford University Press, 1997). For Cold War context, particularly relative to technology developments see Richard Rhodes, *The Making of the Atomic Bomb* (New York, NY: Simon and Schuster, 1986); Martin J. Sherman, *A World Destroyed* (New York, NY: Random House, 1973); and Martin Walker, *The Cold War* (New York, NY: Henry Holt and Company, 1995).

[3] See especially Michael D. Gordon, *Red Cloud at Dawn* (New York, NY: Farrar, Straus, and Giroux, 2009). Dr. Anthony L. Turkevich coined the term "Joe" (as Northrup later verified). Turkevich was a radiochemist involved in the Manhattan Project where he was instrumental in estimated the amount of energy released in an explosion. He was also an early advocate of debris sampling.

[4] AFOAT-1 personnel primarily worked at the headquarters in Washington D.C. Personnel involved in field work, either in the detachments or at the test sites, were assigned to the 1009th SWS although many researchers worked in the field during the U.S. nuclear tests. However, it was essentially one single organization. To avoid unnecessary confusion and to reflect the preferred name used throughout the government and the Air Force, I am using AFOAT-1 as a reference to the entire organization.

[5] See especially David Halberstam, *The Fifties* (New York, NY: Random House, 1993).

[6] Though largely focused on the meteorological effects of fallout, one of the best books on the U.S. test shots is Richard L. Miller, *Under the Cloud: The Decades of Nuclear Testing* (Woodlands, TX: Two-Sixty Press, 1991).

[7] For all U.S. nuclear tests, each detonation—called a "shot"—was given a code name. Note that an entire series of shots fell under an "operation" that also had a codename. I have used all

capital letters for operations and have capitalized the first letter of the shot codenames.

[8] See especially Richard Rhodes, *Dark Sun: The Making of the Hydrogen Bomb* (New York NY: Touchstone, 1995).

[9] See especially Gregg Herken, *Brotherhood of the Bomb: The Tangled Lives and Loyalties of Robert Oppenheimer, Ernest Lawrence, and Edward Teller* (New York, NY: Henry Holt and Company, 2002).

[10] For general background see Stephen E. Ambrose, *Eisenhower: The President* (New York, NY: Simon and Schuster, 1984); and Robert A. Divine, *Eisenhower and the Cold War* (New York, NY: Oxford University Press, 1981).

[11] Ibid.

[12] See especially Gregg Herken, *Cardinal Choices: Presidential Science Advising from the Atomic Bomb to SDI* (Stanford, CT: Stanford University Press, 2000); also Harold Karan Jacobson and Eric Stein, *Diplomats, Scientists, and Politicians* (Ann Arbor, Ml: University of Michigan Press, 1966); and Thomas C. Reed and Danny B. Stillman, *The Nuclear Express: A Political History of the Bomb and its Proliferation* (Minneapolis, MN: Zenith Press, 2009).

[13] See Romney's memoirs. Carl Romney, *Recollections* (Bloomington, IN: Author House, 2012).

CHAPTER 1

Mission Origins and Growing Pains

The story of Long Range Detection (LRD) and the creation of the Atomic Energy Detection System (AEDS) actually began during the Second World War with the Manhattan Project. As the United States raced against time to develop a nuclear weapon during the war, those involved in the project feared the Germans were pursuing a similar course of research. After all, many of the world's leading physicists were Germans. The overall director of the Manhattan Project, Brigadier General Leslie Groves, took these concerns so seriously that he approached General George Marshall in 1943 to obtain responsibility for atomic intelligence.[1] With Marshall's approval, Groves established a Foreign Intelligence Section within his headquarters. Although Groves was careful to not duplicate any existing capabilities of other intelligence organizations, he created specially-trained teams of civilians and military personnel who followed the advancing Allied forces across Europe in 1944 and 1945. Codenamed ALSOS, the operation looked for any evidence of atomic research.[2] Thus, from the earliest days, concern for foreign nuclear weapons production constituted a high priority for U.S. decision-makers. LRD was unique, though, as it offered an alternative to "human intelligence" as a means of gathering information in favor of leveraging technology to ascertain foreign detonation events at extreme distances. Understanding how to do so, however, was not apparent at first.[3]

On August 6th and 9th, 1945, the United States dropped atomic bombs on the Japanese cities of Hiroshima and Nagasaki, resulting in the deaths of more than 200,000 Japanese citizens. Literally overnight, those atomic bombs instituted a sea change in the nature of warfare. Soon thereafter, American foreign policy itself experienced a similar radical change, as the nature of foreign relations came to rest on the bedrock of what Cold War historians have labeled atomic or nuclear diplomacy.

Coming out of the Second World War, President Harry S. Truman's foreign policy rapidly evolved as the United States' wartime alliance with the Soviet Union quickly disintegrated. On April 23, 1945, less than two weeks after President Roosevelt's death, Soviet experts briefed the new president on Soviet expansionism that, in retrospect, led Truman to shift American foreign policy away from Roosevelt's trust of the Soviet Union as a wartime-ally to one of a potential adversary. Almost immediately, suspicions about Soviet expansionism quickly led to concrete actions. By October, the Joint Chiefs of Staff (JCS) had drafted contingency plans targeting twenty Soviet cities with atomic bombs.[4]

By late 1945, with the division of Europe remaining unsettled, a series of Russian activities escalated tensions as Romania, Bulgaria, and Poland all established pro-Soviet governments. In December, the Russians seized control over northern Iran, and threatened encroachment into Turkey by placing 200,000 troops in Bulgaria and 12 combat divisions on Turkey's eastern border. By January 5, 1946, Truman had had enough. In a letter drafted that day, Truman wrote: "Unless Russia is faced with an iron fist and strong language, another war is in the making. Only one language do they understand—'how many divisions have you?'"[5]

Seven weeks later, on February 22, 1946, George Kennan, the charge d'affaires in the American embassy in Moscow, sent an 8,000-word telegram that immediately shook up the State Department and soon thereafter formed the basis of Truman's foreign policy for the next six years. Famously known in Cold War history as the "Long Telegram," Kennan used frank language to depict a Soviet Union that would leverage its huge conventional forces and communist ideological fervor to expand control over Europe, the Middle East, and other areas of the world where emerging nations could be vulnerable to communist ideology. He concluded his correspondence by stating "in foreign countries Communists will, as a rule, work towards destruction of all forms of personal independence—economic, political, and moral."[6]

On March 5, 1946, two weeks after Kennan's telegram, former British Prime Minister Winston Churchill added to Truman's hardening views when he visited the president in Missouri. While there, Churchill delivered his famous "iron curtain" speech. His remarks clearly underscored Kennan's warnings by citing recent examples of communist expansionism occurring across central and southern Europe. Toward the end of his speech, Churchill delivered eloquent words, which would endure over time:

> A shadow has fallen upon the scenes so lately lighted by the Allied victory. Nobody knows what Soviet Russia and its Communist international organization intends to do in the immediate future, or what are the limits, if any, to their expansive and proselytizing tendencies. . . . From Stettin in the Baltic to Trieste in the Adriatic, an iron curtain has descended across the Continent.[7]

As 1946 ended and 1947 began, one of the harshest winters on record in Europe seriously thwarted economic and infrastructural recovery efforts. Faced with millions of starving people, especially in Germany, Allied occupation forces quickly increased aid and assistance levels. As the winter subsided in March, the Truman administration believed struggling countries, those devastated by the war or seeking independence from colonialism, were ripe for communist subjugation.

Truman's response was a new policy statement underscoring his evolving foreign policy of "containment." Delivered in a speech on March 12, 1947, before a joint session of Congress, the president called for U.S. intervention in the civil war in Greece. The political situation in Greece had deteriorated as communist insurgent forces made gains in the face of deteriorating economic conditions. Concurrently, Turkey's weak government faced Soviet pressure to share control of the strategic Dardanelles Straits. When Great Britain announced on February 21, 1947, that Britain could no longer provide financial aid to both countries, Truman saw little choice but to argue for American intervention. As he told Congress, "it must be the policy of the United States to support free peoples who are resisting attempted subjugation by armed minorities or by outside pressures."[8] Undersecretary of State Dean Acheson also told Congress that the fall of Greece and Turkey would lead to similar overthrows by communist insurgencies "like a row of falling dominos." Both Acheson's domino theory and Truman's doctrine of foreign assistance and intervention would henceforth provide the basis of U.S. foreign policy until the end of the Cold War decades later.[9]

In essence, Truman and Acheson ended the period of Roosevelt's wartime cooperative relationship with the Soviet Union by formulating a doctrine of containment (i.e., containing Soviet conventional forces). What started as concern over the conventional threat to national security quickly transformed into a doctrine based on countering ideological threats and, later, on defending against nuclear threats. Certainly, early Cold War events hastened that process. Between mid-1947 and mid-1950, a number of national events and international confrontations served to harden the Cold War and usher in nuclear diplomacy as a critical tool in formulating American foreign policy.

Three months after Truman's speech, on June 5, 1947, Secretary of State George C. Marshall addressed the graduating class of Harvard University. In his speech, Marshall called for a large program of economic assistance to Europe. He did so in response to the severe winter and, more importantly, as a pillar of the Truman Doctrine to counter communist expansionism. Congress approved the Marshall Plan, officially called the Economic Cooperation Act, in March 1948. Russia immediately disavowed the plan as an attack by capitalists designed to subjugate the people of Europe.[10]

In the United States, Cold War entrenchment grew as widespread anti-communist fervor reached unprecedented heights. On November 24, 1947, the U.S. House of Representatives Committee on Un-American Activities (HUAC) found 10 Hollywood film industry employees in contempt of Congress and banned them from working at the major studios. They were sentenced to a year in jail. Efforts to blacklist suspected communists grew over the next few years and expanded well into the early 1950s. In the minds of many Americans this "Red Scare" served to solidify the ideological confrontation between the United States and Russia.[11]

As communist political parties with strong Soviet support and influence continued to take control over Eastern European countries, the Soviet Union initiated a blockade of Berlin on June 24, 1948. The intent was to force the western allied powers to abandon their sectors of the divided city. The Soviet Union, in the face of the Marshall Plan, viewed Berlin, located deep in their occupation zone, as a significant obstacle to the solidification of a communist Eastern Europe. Such fears were firmly grounded in the "German Question": How should the wartime allies ensure that Germany would remain incapable of ever threatening her neighbors again? When the western allies introduced the Deutsch Mark as West Germany's official currency on June 20, 1948, Russia responded with the blockade four days later. The blockade would last for 13 months. During that time, the United States and Great Britain successfully sustained the city with a massive airlift.

As the blockade was underway, the Soviets established a separate municipal government in East Berlin on November 30, 1948. Cold War tensions further increased when the North Atlantic Treaty Organization (NATO) formed in April 1949, followed by the creation of the Federal Republic of Germany several weeks later. Within days, the Soviet Union approved a new constitution for East Germany, and on October 7, 1949, East Germany formally became the German Democratic

Republic. The German Question was no longer in question. The division of Europe now looked permanent.

The events occurring between March 1947, with the articulation of the Truman Doctrine and the Soviet Union's first test of a nuclear device in late August 1949, served to harden U.S. Cold War foreign policy. Interestingly, until his departure from office in 1952, Truman never threatened the Soviet Union with nuclear weapons. Paradoxically, the U.S. nuclear monopoly had no deterrent effect on Soviet behavior. There were practical reasons as to why the American monopoly of atomic bombs did not translate into nuclear diplomacy during his administration. First, the nuclear arsenal was small and capable bombers were relatively few. Although military planners drafted contingency plans prior to 1950 targeting Soviet cities with nuclear weapons, they recognized the U.S. could not win a preemptive war nor weaken Soviet power under such a scenario. Second, perhaps because of the psychological impact of recently using two bombs on Japan, the U.S. government "never transformed their atomic monopoly into an effective instrument of peacetime coercion." Consequently, Stalin pursued aggressive policies and activities as though nuclear weapons did not exist. He simply refused to be intimidated and, by his actions, embarked upon a strategy of what Pulitzer Prize winning historian Martin Sherwin called "reverse atomic diplomacy."[12] As a result, the Truman administration found itself open to calls for international controls over nuclear arms.

As historians now recognize, in the initial years of his administration, Truman sought to limit and control the proliferation of nuclear weapons. However, ardent anticommunists among conservative politicians and the armed services strongly contested the president's efforts to do so. In the late 1940s, however, many scientists challenged the conservative views and the Cold War political order.[13] They also participated in the heated debates about whether all matters of atomic energy should fall under military or civilian control (with most favoring the latter). Historians have thoroughly chronicled that contestation between General Groves, who saw such oversight as an extension of the Manhattan Project, and Congress, which strongly believed in civilian stewardship over nuclear energy and weapons. The Atomic Energy Act of August 1, 1946 resolved that debate in favor of civilian control and thus established the U.S. Atomic Energy Commission (AEC). Although they welcomed this as a favorable outcome, the scientists were disappointed that the new law did not address controls or limitations over the development of nuclear energy for military purposes.[14]

Still, scientists and other proponents of civilian control continued to argue that peace could best be maintained through a foreign policy of international cooperation in the development of nuclear energy. In June 1946, they were hopeful when the Baruch Plan was presented to the United Nations. Named after Bernard Baruch, an advisor to President Truman, the plan outlined a UN-controlled program of oversight to govern the development and use of atomic energy.[15] A proposed UN organization—the United Nations Atomic Energy Commission (UNAEC)—subordinate to the Security Council, would manage any nuclear facility capable of producing nuclear weapons. UNAEC would also be authorized to inspect any research facility and to punish any violators with UN imposed sanctions. The lynchpins of the proposal were on-site inspections and the future destruction of the U.S. nuclear arsenal. However, as the aforementioned context of the early Cold War years reveal, those control stipulations proved unacceptable as mistrust between the U.S. and the USSR grew. Throughout the 1950s, Eisenhower would never lose hope that some form of verifiable arms control, similar to the Baruch Plan, was possible. In the interim, however, America's defense demanded a capability to monitor a Soviet nuclear weapons program. It is within this context that the nation required a system that could monitor nuclear tests from great distances. Thus, the Atomic Energy Detection System was born.

*

As the seeds of the Cold War sprouted on the heels of the Second World War, U. S. government officials realized the United States would not permanently retain a nuclear monopoly. To surveille the Soviet Union, they quickly recognized the need for a monitoring program grounded in technological innovation rather than human intelligence. Unfortunately, establishing such a detection capability immediately after the war was simply not feasible. America's foremost nuclear scientists, the very people needed to create the AEDS, were still involved with the Manhattan Project. Most of those scientists and engineers were eager to return to their former pre-war technical pursuits, and many opposed further experimentation of nuclear physics for weapons development. Consequently, the early nuclear establishment in the United States remained fairly unorganized until July 1946, when the Atomic Energy Commission (AEC) gained stewardship over all nuclear matters. Even then, however, it would take time to work through bureaucratic obstacles and to determine specific roles and missions.[16]

In October 1946, Lieutenant General Hoyt S. Vandenberg, the Director of the Central Intelligence Group (CIG), wrote to General Leslie Groves, recently the Director of the Manhattan Project, soliciting recommendations on how to set up a program to obtain information concerning the Soviet Union's possible possession of a nuclear bomb.[17] Because General Vandenberg's request occurred as the programs and responsibilities of the Manhattan Project transferred to the new AEC, Vandenberg did not receive a reply from Groves. Following the transition, one of the AEC's new commissioners, Admiral Lewis L. Strauss, conducted several conferences with General Vandenberg, urging the appointment of an interservice committee to study the problem of long range detection.[18]

On March 14, 1947, one week after the announcement of the Truman Doctrine, Vandenberg wrote a letter to the Departments of War and Navy, the AEC, and the Joint Research and Development Board (JRDB) to express his concern for the need of an LRD capability. On June 6, 1946, the Secretary of War and the Secretary of the Navy had organized the JRDB to oversee the coordination of research programs between the departments. The JRDB replaced the Joint Committee on New Weapons and Equipment, which had fallen under the auspices of the Joint Chiefs of Staff. Vandenberg's letter underscored the urgency of an LRD program and stated it should be a high priority. In his letter, General Vandenberg proposed a coordinated effort by the addressed organizations to formulate an overall plan for developing LRD equipment and operations. He stated the CIG would sponsor the first meeting and recommended specific individuals to serve on an LRD committee.[19]

On May 14, 1947, the Secretary of War, Robert P. Patterson, approved CIG sponsorship of a LRD committee and appointed Colonel Benjamin G. Holzman as his representative.[20] Holzman was then serving on the Air Staff as the Research and Development Officer for Atmospheric Sciences. He had recently served as the staff weather officer in the first U.S. nuclear test series, Operation CROSSROADS.[21] One week later, on May 21, 1947, representatives from the War Department, the Navy Department, the AEC and CIG held their first meeting in room 7117 of the North Interior Building, Washington D.C. The 10 attendees appeared to include the right people to steer the birth of LRD. In addition to Holzman, Colonel Lyle E. Seeman was present to lead off the meeting as chairman of the committee. Seeman had been Groves' associate director of the Manhattan Project's Los Alamos laboratory-wide system. While the real challenge of this initial meeting was to address operational responsibilities for LRD, prominent Navy chemist Edward S. Gilfillan's summary of current capabilities of detection equipment used at CROSSROADS actually initiated serious misunderstandings that would persist for years to come; namely, that existing equipment was adequate for the job at hand. Gilfillan stated "the sensitivity of the currently used instruments is sufficient to detect explosions anywhere on Earth." As time would tell, this was a gross overstatement.[22]

Halfway through the two-and-a-half-hour meeting, physicist I. Creasy, a former member of the Manhattan Project, finally turned the discussion toward the immediate problem, namely identifying an organization for the LRD mission. Gilfillan's summary certainly highlighted the work of various stakeholders in LRD. As CROSSROADS had demonstrated, the Army Air Forces (AAF) conducted aerial sampling, the Army employed acoustic and seismic equipment, the Navy used its hydroacoustic stations, and a number of governmental and civilian organizations conducted experiments as well. Creasy noted the difficult task was to determine overall responsibility for LRD in light of these many functional responsibilities. Navy Captain Horacio Rivero, Jr. concurred, stating that the problem of the committee was "to find out what should be done and who is to do it."[23] The committee recognized the challenge and agreed they first had to determine the requirements for LRD. All decided to form a subcommittee to do so; Holzman, Gilfillan, and Major F. A. Valente (an engineer at Hanford during the Manhattan Project), agreed to serve.[24]

William T. Golden, an assistant to the AEC Commissioner Lewis Strauss, though remaining silent during the day's discussions, evidently disagreed with Gilfillan's statement that existing technological capabilities were adequate in detecting explosions anywhere in the world. The day after the meeting, Golden wrote a memo to Strauss outlining his observations of the May 21, 1947 meeting. Golden informed Strauss that the meeting was inconclusive. He noted while the existing techniques (acoustic, seismic, and radiological) looked promising when used collaboratively, they were currently inadequate. His greater concern, however, was organization. Bluntly, he stated "primary responsibility for the establishment and operation of a detection system must be placed with a single organization. Divided responsibility or operation by committee invites another Pearl Harbor." Golden recognized the need of interagency contributions to an LRD system but argued why each of the principles could not effectively exercise overall command. He concluded only the CIG was in a position to do so. Golden urged Strauss to discuss the issue with Admiral Roscoe H. Hillenkoeter, now the Director of the CIG. He emphasized time was short given the committee's anticipation of a fully operating system in two years from a current starting date—"four years after Hiroshima."[25]

The subcommittee met again on June 6, 1947 to conclude work on identifying requirements and specifications for an LRD program. Their report was short, succinct, and served to encapsulate the main points discussed with the committee on May 21st. They advocated for an immediate implementation of a worldwide "system of systems" (i.e., aerial sampling aircraft, laboratory analyses, and seismic and acoustic stations) that could "determine the time and place of all large explosions which occur anywhere on the earth, and to establish beyond all doubt whether any of them are atomic in nature." In short, they recommended a significant expansion of all existing methods and equipment that, in their view, were "available, actually or potentially, and possess adequate sensitivity." They also urged the construction of a "Control Central" to coordinate all collections and operations. In echoing Gilfillan's primary concern, the subcommittee emphasized the immediate need to fly aerial sampling missions in order to establish a database of the existing radiological background existing in the atmosphere as a result of CROSSROADS or from natural occurrences. Although the subcommittee members recognized the importance of multiple stakeholders in LRD, the report made no specific recommendations on an overall responsible authority.[26]

Three weeks later, on June 30, 1947, Hillenkoetter submitted his final report to Patterson, with copies to the Secretary of the Navy and the AEC.[27] Hillenkoetter's three-page report reiterated the subcommittee's narrative, largely verbatim. He did, however, strongly press for the immediate collection of atmospheric samples to establish a background baseline. He added such operations would not require a network infrastructure—viewed by all as taking two years to become fully operational—and thus could commence immediately. The most critical aspects of his report dealt with the selection of a central authority. He informed Patterson that a great amount of effort went into formulating a recommendation for an overall authority. While he noted there was much debate over civilian versus military control, all parties eventually agreed it should be a military organization. He concluded his report by stating:

> [I]t is my responsibility as Director of Central Intelligence that the conclusions of the Committee be accepted and implemented forthwith by appropriate directive to the Army Air Forces for overall responsibility, supported by requests to other interested agencies for necessary cooperation and assistance to carry out the program.[28]

Throughout July and August, debates over Hillenkoetter's recommendations ensued through various correspondences and conferences. On July 24th, Groves, now head of the joint (Army-Navy) Armed Forces Special Weapons Project (AFSWP), reluctantly concurred with the recommendations, but with some caveats. First, he strongly argued the AAF was ill-prepared to garner the requisite expertise to staff the envisioned LRD system and that it would be very challenging to do so. Second, he recommended his AFSWP should receive all data collected by the LRD program for analysis and evaluation.[29] A week later, on July 30th, Lieutenant General

Lewis H. Brereton, Chairman of the Military Liaison Committee (MLC), backed Groves' position. Brereton stated the AFSWP was best suited to analyze and evaluate collected samples, and unlike the AAF, such a mission would incur little increase in staff. Brereton also wanted to see significant involvement of the CIG and the AEC in any LRD program.[30]

Quickly becoming aware of the dissenting opinions, Hillenkoetter solicited Lieutenant General Curt LeMay's views and requested his response within 24 hours.[31] LeMay, now serving as the Deputy Chief of Staff for Research and Development, replied the following day with a well-written justification for the AAF to direct the nation's LRD program. In his reclama to the views of Groves and Brereton, LeMay voiced several arguments. First, he countered the assumption that an LRD mission was simply an atomic problem related to nuclear explosions. He explained "the observations and analyses of data involved in this program is a geophysical problem and mainly a meteorological one." Therefore, the involvement of the Air Weather Service (AWS) would be paramount. The AWS would play the most critical role because seismic and sonic detections would not yield information on the specifics of "atomic bursts," only the direction and location of the detonations. Second, LeMay argued the AWS had already conducted research in detecting hurricanes at remote distances and that a small network of recording stations already existed. He assured the War Department "the knowledge to handle the collection and analysis of seismic data is already contained in the meteorological services."[32]

LeMay further argued, in detail, that the AWS was already in place to conduct the LRD mission and possessed operational knowledge to address the challenge of aerial sampling. The AWS, he stated, had an extensive number of B-29 weather reconnaissance planes already flying over designated aerial routes where all upper-air parameters were quantitatively documented. Those upper-air parameters constituted a background database vital to the analyses of the samples. After a weather reconnaissance run, radiological chemists could routinely analyze the samples to determine the degree and composition of radioactivity. The chemists could determine from the decay curves the time at which a possible atomic explosion occurred. From the chemistry of the fission products, the analysis could reveal information on the efficiency and character of the explosions. LeMay also argued that the AWS played a key role because LRD required meteorologists to determine the trajectory and the source of the sample of air containing high radioactive content. Meteorologists could accurately direct the air reconnaissance necessary to verify the significance of the original high radioactive sample.[33]

The most impressive aspect of LeMay's letter was his extensive knowledge of the difficulties involved in long range detection. As head of AAF R&D, he had direct involvement in the CROSSROADS experiments the previous year. While he stressed the necessity of having seismic and acoustic techniques, Groves stated it was "highly uncertain" whether those techniques could "furnish any positive identification of nuclear explosions at remote distances." LeMay noted the Air Materiel Command (AMC) had conducted research on detecting remote explosions and detection techniques were still in the research stage, and "practical application [was] uncertain." LeMay was also aware of the geographical challenges involved in the application of seismology, especially in distinguishing between explosions and earthquakes.[34]

On September 5, 1947, the new Secretary of War, Kenneth Royall, concurred with the recommendations of the LRD Committee. In his concurrence letter to Hillenkoetter, Royall assured the director of the CIG that all of the other LRD stakeholders were critical to the program. "In order for the program to be a success, the full cooperation and assistance of many other interested agencies will be necessary. The Army Air Forces will affect the necessary coordination."[35]

A week later, on September 15th, Admiral James V. Forrestal, Secretary of the Navy, followed with a similar endorsement. Forrestal, while clearly supporting Hillenkoetter's recommendation, appeared to take a central position between the LeMay and Groves-Brereton arguments. He qualified his concurrence by stating the AAF should have overall responsibility but only with the understanding that "in the analysis and evaluation phases of the program maximum use will be made of the resources and facilities of the [AFSWP and the AEC] and other appropriate agencies."[36]

These debates were important for several reasons. First, the wrestling over roles and responsibilities served as a microcosm of the extensive inter-service rivalry then occurring at the top levels. In a small way, the debates over the LRD mission were an extension of those arguments and threatened to delay the establishment of an effective monitoring system, especially in regard to funding. Second, the Cold War was quickly escalating. The uncertainty over when the Soviet Union would end the U.S. nuclear monopoly complicated President Truman's efforts to formulate a post-war foreign policy. An effective AEDS would certainly assist in ascertaining the initiation and subsequent development of a Russian nuclear program. Finally, with LRD, the nation was venturing into unknown areas of research. As the two detonations conducted in the 1946 CROSSROADS tests revealed (discussed in Chapter 2), nuclear science, with all of its potentialities, was in its infancy. As such, there were many interested parties, both military and civilian, already involved. They would need to unite their evolving efforts if the U.S. was to establish a viable AEDS in the shortest time frame possible.

After reviewing all of the viewpoints and studying recommendations on the proposed programs, the Chief of Staff of the Army, General Dwight D. Eisenhower, assigned overall responsibility for long range detection to the AAF. His order came only two days before the birth of the U.S. Air Force. On September 16, 1947, Eisenhower signed the basic directive with wording that was broad and all-inclusive. It charged the commanding general of the AAF with "overall responsibility for detecting atomic explosions anywhere in the world. This responsibility is to include the collection, analysis, and evaluation of the required scientific data and appropriate dissemination of the resulting intelligence."[37]

Following Eisenhower's directive, the Air Force took no significant action until October 27, 1947. On that date, the Air Staff delegated responsibility for preparing the overall LRD plan to the Deputy Chief of Staff for Operations (DCS/O), Lieutenant General Lauris Norstad. General Vandenberg, as the Vice Chief of Staff, directed AMC to initiate necessary research and for the AWS to prepare for the operational collection of air samples.[38] Vandenberg emphasized haste in developing the necessary means of long range detection so that LRD stakeholders could conduct experiments during the 1948 nuclear tests in the Pacific, codenamed SANDSTONE.[39] Norstad submitted the general plan to Vandenberg on November 13, 1947, with the recommendation that the Special Weapons Group (SWG) within the Deputy Chief of Staff for Materiel (DCS/M) gain responsibility for LRD. Vandenberg immediately accepted Norstad's recommendation and formally assigned responsibility to the DCS/M the following day.[40] The chief of the SWG was well-respected Major General William E. Kepner. Shortly before his assignment with the SWG, Kepner served as the deputy commander of Joint Task Force 1 for the 1946 nuclear test series Operation CROSSROADS.[41] Norstad also submitted recommendations to a number of agencies expected to participate in LRD to include the CIA, the Chief of Naval Research, the commanding general of AMC, representatives from the Army, the chairman of the Research and Development Board (RDB), the AEC, and the AFSWP. Kepner called for a meeting of these participants for November 17, 1947.[42] In preparing to meet his new responsibilities, General Kepner included representatives of more than 25 agencies in his November 17th conference. Attendees agreed the Department of the Navy, the AWS, and other operational agencies would gather technical information and data within their capabilities for analysis and evaluation before submitting any final recommendations to the SWG.

During the first week of December, the service secretaries circulated several memos acknowledging their subordinate organizations' support of Air Force stewardship over the LRD program. On December 4th, Kepner notified AMC that the entire program now had a "1A priority."[43] Over the next month, the LRD program rapidly developed as the Office of Naval Research, the Army Signal Corps, and the Air Force AWS tackled their assigned LRD responsibilities. The SWG quickly approved the joint research program under the code name WHITESMITH.

During the first half of December, the various stakeholders of the LRD program also exchanged several memoranda pledging cooperation. While widely-reported inter-service rivalries were being played out at the highest levels, the LRD participants were immediately concerned with planning for the upcoming SANDSTONE test series. Although the task force exercising overall responsibility of the operation fell to Lieutenant General John E. Hull, Commanding General, U.S. Army Pacific, the nuclear tests in the Marshall Islands required a large naval contingent. Interestingly, given the competitive exchanges concerning roles and responsibilities during the previous weeks, the Navy pledged its full support of LRD for SANDSTONE in a memo to the Department of the Air Force on December 16, 1947. This memo was a follow-up to a joint memo from the Navy and Air Force secretaries (John L Sullivan and W. Stuart Symington, Jr.) to the Secretary of Defense (now Forrestal) on December 10th, reassuring Forrestal that the Air Force governed the LRD program. While speculative, Symington may have been especially sensitive to the fact that the former secretary of the Navy was now the secretary of the newly formed Department of Defense. On December 8th, in a brief memo to Air Force Chief of Staff General Carl Spaatz, Symington alerted Spaatz about the joint memo and stated "it would seem to me of greatest importance that in this connect [LRD program] we go ahead and do what we say we are going to do."[44] However, Sullivan's December 16th memo to Symington, while pledging support on one hand, continued to express concern about the Navy's role in the LRD program on the other. "I want to assure you of the desire of the Navy to cooperate in this important project to the maximum extent practicable *within the limitations imposed by other equally important commitments of this Department.*"[45]

As SANDSTONE approached, even the AEC, while initially concerned the Air Force could not manage the entire LRD program, pledged its support to the Air Force Chief of Staff on December 10, 1947: "As you are aware, the Commission considers this project [LRD] of the utmost urgency. For that reason, the Commission is prepared to render every possible assistance in order that no further delays may be experienced in placing the important program into operation."[46]

In early December 1947, Lieutenant General Howard A. Craig, the DCS/M, organized a LRD division within Kepner's SWG, known as Air Force Materiel Special Weapons-One (AFMSW-1). Its mission was to conduct R&D in order to design a functional LRD system. Several days later, on December 14th, Major General Alfred F. Hegenberger was appointed chief of the LRD division. Like Kepner, Hegenberger was well-respected throughout the military flying community as a navigation pioneer who was considered the "father" of instrument flying. As a navigator and pilot, he and Lieutenant Lester J. Maitland had flown the first flight from California to Hawaii in June 1927 in a Fokker

C-2. In 1932, after developing a blind flight system, Hegenberger accomplished the first official solo blind flight. During the Second World War, Hegenberger had commanded 10th Air Force.[47]

Three days after Hegenberger's appointment, Kepner hired prominent geophysicist Dr. Ellis Adolph Johnson as the technical director of AFMSW-1. In the 1930s, Johnson had become an expert on terrestrial magnetism and worked as a section chairman in the Department of Terrestrial Magnetism at the Carnegie Institution in Washington, D.C. Shortly before World War II, Johnson became the Associate Director of Research for the Naval Ordnance Laboratory (NOL), working at Pearl Harbor on mines and countermining techniques. In 1942, he accepted a commission as commander in the Navy Reserve and deployed to the Pacific theater to serve on the staff of the Commander in Chief, Pacific. In that assignment, he developed and promoted the aerial mining campaign against Japan for the Navy and the Army Air Forces. After the war, Kepner found Johnson working at the Carnegie Institution and convinced him to join the LRD program.[48] Johnson's appointment proved critical at this early stage of LRD. In the days and weeks ahead, Johnson recruited superbly talented scientists to join AFMSW-1. Notably, they included nuclear physicist Dr. William D. Urry, seismologist Dr. J. Allan Crocker, and, most importantly, Mr. Doyle Northrup (Johnson's future successor). Northrup had worked for Johnson at Pearl Harbor in 1941. In this new role, Johnson charged Northrup with exploring new types of effects for LRD.[49]

From January to April 1948, Johnson prepared a broad outline of a two-year program of intensive R&D in the seismic, acoustic, and nuclear fields. His first task was to formulate an R&D budget. He quickly surveyed various educational, commercial, and government facilities, as well as numerous reports to estimate a two-year budget of 46 million dollars (2021 = $520 million). At that time, the priority of effort was the formulation of a comprehensive research program, in connection with current U.S. nuclear detonations, to test the use of scientific techniques for LRD. Participants included personnel from the Air Force, the Army, the Navy, the Coast Guard, and the Geodetic Survey.[50]

On April 1, 1948, HQ USAF activated the 51st Air Force Base Unit as a field extension of the Air Force Chief of Staff. Thus, the 51st became the military support organization for AFMSW-1. This new organization, located at Gravelly Point in Washington, D.C., was formed to carry out the experimental tests on various activities in coordination with several agencies participating in the LRD program. Command of the new unit was assigned to Hegenberger, as an additional duty.

On July 1, 1948, the Air Staff transferred the SWG from the DCS/M to the DCS/O, and re-designated the SWG as the "Office of the Assistant Deputy Chief of Staff, Operations for Atomic Energy" (abbreviated "Office for Atomic Energy" or AFOAT). The 51st Air Force Base Unit was now considered to be under AFOAT, although the initial reassignment directive did not specifically cover the transfer.[51] Initially, this cumbersome reporting channel caused confusion. Consequently, the Air Staff issued a letter dated August 20, 1948, to clarify the relationship: "The 51st AF Base Unit, as an exempted activity of the Chief of Staff USAF, is authorized the usage of the symbol AFOAT-1 (as office symbol) which reflects the relationship of the 51st AF Base Unit to the Office of the Assistant Deputy Chief of Staff of Operations for Atomic Energy." The 51st Air Force Base Unit was authorized 23 officers, one warrant officer, 63 enlisted men, and 45 civilians.[52]

On August 6, 1948, Major General David M. Schlatter, Assistant Deputy Chief of Staff, Operations for Atomic Energy, relieved Kepner as commander of AFOAT (until recently, the SWG). Then, on August 28, 1948, the 51st AF Base Unit was re-designated 1009th Special Weapons Squadron

(SWS) with the assumed office symbol of AFOAT-1. On that date, Major General Hegenberger became the unit's first commander. In that position, Hegenberger wore two hats: as commander of a "numbered" Air Force unit and as director of an extensive staff. Henceforth, commanders would continue in that dual role. This clear division reflected the two distinct elements of LRD: analyses and R&D on one side (AFOAT-1), and operations on the other (1009th). Both groups shared the same administrative staff. That same month, however, Hegenberger lost Johnson as his technical director. Johnson had become disenchanted with the prolonged budget fights (see Chapter 2) and had tendered his resignation to the Secretary of the Air Force on July 1, 1948, effective August 4th. Although Johnson served the LRD mission for only eight months, he accomplished two great tasks which would serve AFOAT-1 well in the years ahead. Johnson had continued to recruit an excellent team of scientists which now included Gilfillan, Jr., George H. Shortley, and Donald H. Rock. More importantly, Johnson developed an extensive R&D plan. This plan identified scientific experiments that would lead to the engineering of LRD instruments. Upon Johnson's departure, Shortley became acting technical director. However, he too, departed in late September to return to the Ohio State University. At that time, Doyle Northrup officially became the technical director.[53]

It is important to remember that the rapidly expanding Cold War and extensive anti-communist sentiment in Congress during this period of time dramatically shaped political support for the LRD efforts. As a result, almost anything dealing with nuclear matters was given the highest priority. In the summer of 1948, pressures to expand LRD capabilities were the result of high-level interest in testing new detection methods against U.S. nuclear tests. Those pressures came directly from the RDB. The RDB, authorized under Section 214 of the National Security Act of 1947, had replaced the JRDB on September 30, 1947. Chaired by Dr.

Vannevar Bush, the RDB organizational structure closely resembled the JRDB framework, which included a number of specialty committees. The RDB's Committee on Atomic Energy (CAE) was now directly responsible for providing oversight of LRD research (see AFOAT-1/AEDS Data/Information Flow graphic in the Introduction). The CAE consisted of nine members: six officers who comprised the Military Advisory Committee (two of whom were General Groves and Rear Admiral William "Deke" Parsons), and three civilians.

The civilians, some of the most highly respected leaders within the small nuclear research community at that time, were James B. Conant (chair), J. Robert Oppenheimer (former Director of the Manhattan Project), and Clifford W. Greenwalt (Vice President of DuPont). In January 1948, the CAE unanimously pronounced that "we, on the Research and Development Board, deem it very unlikely that Russia will be in a position to test an atomic bomb as early as 1950, or within several years of that date."[54] In sum, LRD efforts were to focus on U.S. nuclear tests to develop technological advancements that could soon result in an effective worldwide monitoring system.

The birth of LRD had been painful. The 17 months of starts and stops that characterized the establishment of an initial organizational framework for LRD were directly related to a major reorganization of the entire U.S. military establishment then underway. When the war ended in August 1945, the U.S. found itself as the leading world power. Senior government leaders fully understood the U.S. could never return to its pre-war inclination toward isolationism. Throughout 1946, as tensions mounted with the Soviet Union, America's military posture seemed uncertain as the Truman administration wrestled with formulating a policy of containment while executing a massive demobilization of American forces. Indeed, the U.S. armed forces had downsized from 12 million personnel during 1945 to less than two million by June 1947.[55]

In short time, it was clear to senior military and political leaders that the existing national security system, which had served the war well, was now inadequate for the new world order. Marshall and Eisenhower recognized the need for extensive reforms and wanted strong centralized control of the national and theater levels of authority. However, Admiral James V. Forrestal, then Secretary of the Navy, favored a less centralized system largely resembling the Second World War model in which each service exercised strong authorities. Truman and much of Congress hoped to stave off a postwar recession by creating an efficient and economical military establishment. By early 1947, inter-service rivalries had reached contentious levels as each branch vied for dwindling resources, compounded by an unsettled foreign policy designed to deter a large Soviet conventional force.[56]

The first attempt at reform occurred on July 26, 1947, when Truman signed the National Defense Act of 1947 into law, a solution largely favoring the Forrestal viewpoint. Most importantly for LRD, the Air Force became a separate service under the act in September, two days after the Eisenhower directive for LRD. Now, all three services became executive departments headed by civilian secretaries, each of whom had direct access to the President. In addition, the secretaries joined the heads of several other government organizations on the new National Security Council (NSC), which was also created under the act. Unfortunately, the new reforms did little to resolve interservice conflicts. The Secretary of Defense provided general direction to the services but lacked real influence to force the departments to cooperate. The real power resided with each service's chief of staff. Collectively, they formed the Joint Chiefs of Staff (JCS), a statutory body in the Office of the Secretary of Defense that essentially formed a loosely federated National Military Establishment.

As the President signed the act into law, LeMay was drafting his correspondence to the War Department staff to argue that the AAF/USAF should govern the LRD mission. When the new USAF formally received the LRD mission in September, furious debates over roles and missions continued to plague the "reformed" military establishment. Unfortunately, it would take Congress two more years to finally correct the flaws in the 1947 act with an amendment— the National Security bill that Truman signed into law on August 10, 1949 (19 days before the Russians detonated their first nuclear device). The new legislation strengthened the powers of the defense secretary as the armed services became military departments within the Department of Defense. The JCS also received a chairman under the bill who reported to the Secretary of Defense. Together, they were able to force cooperation between the armed services. For AFOAT-1 and LRD, however, such turmoil directly impacted the R&D budget and seriously hindered the creation of the AEDS (discussed in Chapter 3). Against this backdrop, LRD underwent an awkward organizational startup.

Although strong proponents such as LeMay and Vandenberg ensured a future for LRD within the USAF, priorities of LRD within the national security strategy remained uncertain. Through the spring and summer of 1949, the Joint Chiefs failed to clarify the national priority of LRD and delayed important funding decisions required to move the mission forward. Hegenberger, as the commander of AFOAT-1, dealt with the effects of inter-service rivalries within the Pentagon but shielded his scientists from the bureaucratic inertia. In fact, their research efforts were moving quickly ahead as they built upon the rudimentary experiments conducted in the 1946 Operation CROSSROADS, and as they applied Johnson's R&D roadmap to Operation SANDSTONE in 1948. They were eager to learn how to detect.

Major General
Alfred F. Hegenberger

First commander of the 1009th SWS and Director of AFOAT-1. Considered the "father of instrument flying." As a navigator, he and Lieutenant Lester J. Maitland flew the first trans-Pacific flight from California to Hawaii in June 1927. In 1932, after developing a blind flight system, Hegenberger accomplished the first official solo blind flight. During the Second World War, Hegenberger commanded the 10th Air Force. His close relationship and partnership with Doyle Northrup set the standard for talented LRD leadership in the years ahead. *Photo source: AFTAC archives*

Hegenberger and Johnson (1947)

Extremely qualified to initially lead the LRD mission, Johnson, as the first technical director of AFOAT-1, developed the extensive R&D plans that would successfully lead to an effective AEDS in the years ahead. Becoming frustrated with budget issues, he left AFOAT-1 within a year of his arrival.

Photo source: AFTAC archives

General Hoyt S. Vandenberg

In 1946, as Director of Central Intelligence, Vandenberg sought advice from Leslie Groves on how to create a program to monitor Russian nuclear activities. In late 1947, and throughout 1948 and 1949, he was instrumental in shepherding AFOAT-1 through contentious mission and budgetary battles. Later, as Air Force Chief of Staff, he made LRD a top priority.

Photo courtesy of the United States Air Force.

Dr. J. Robert Oppenheimer

Oppenheimer was a theoretical physicist who headed the Los Alamos Laboratory for the Manhattan Project. Highly respected among scientist and government officials after the war, Oppenheimer served on a number of committees that impacted the development of LRD and the AEDS. He was the initial chairman of the CAE and led the verification meeting for the Joe-1 analysis. In 1954, his security clearance was revoked after highly publicized hearings viewed his pre-war political views as supportive of communism.

Photo courtesy of the Department of Energy

Major General Leslie Groves

Director of the wartime Manhattan Project. In 1947, as Commander of the Armed Forces Special Weapons Project (AFSWP), he argued that the AFSWP should control nuclear weapons as an extension of the Manhattan Project. An adversary of Lewis Strauss, he lost the fight over who would administer control with the AEC gaining that responsibility.

Photo courtesy of Los Alamos National laboratory

General Curtis E. LeMay

Best known as Comm ander, Strategic Air Command (1948-1957) and USAF Chief of Staff (1961-1965). In mid-1947, as a lieutenant general responsible for all R&D within the Army Air Force, LeMay strongly advocated for the (soon to be) USAF to receive the LRD mission. More than any other military officer at that time, LeMay had an impressive grasp of the challenges involved in creating the AEDS.

Photo courtesy of the United States Air Force

Lieutenant General William E. Kepner

Played a key role in initially organizing the LRD mission. He served as Deputy Commander, Joint Task Force One at Operation CROSSROADS, where he observed the first LRD experiments. During WWI, he saw extensive combat as an infantry commander and later achieved fame as a pioneer balloonist. During the Second World War, he commanded the Eighth Fighter Command and the Ninth Air Force.

Photo courtesy of the United States Air Force

Vice Admiral Roscoe Hillenkoetter

A Naval intelligence officer for most of his career, Hillenkoetter was wounded in the Japanese attack on Pearl Harbor in 1941. After the war, he served as the first director of the CIA. He was fired from that position after the CIA failed to anticipate Joe-1 and the attack that initiated the Korean War.

Photo courtesy of the Library of Congress

Lewis L. Strauss

Chairman of the AEC from 1953 to 1958. Until 1957, Strauss essentially became the sole source of advice to President Eisenhower on nuclear policy. Strongly anti-Russian, Strauss became an obstacle to the president's goal of an arms control treaty with the Soviet Union. Alongside Teller and Lawrence, he advocated for thermonuclear weapons. Also, his strong dislike of Oppenheimer contributed to the famous physicist's downfall. *Photo courtesy of the National Archives*

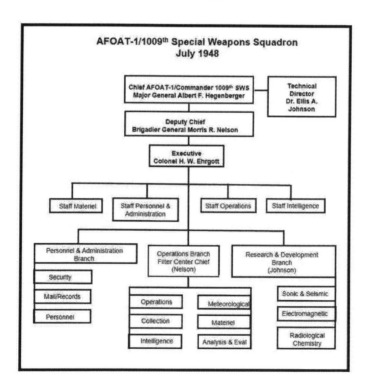

Notes to Chapter 1

[1] Leslie M. Groves, *Now it can be Told* (NY: Harper, 1962), 185. "In the fall of 1943, General Marshall asked me, through Styer, whether there was any reason why I could not take over all foreign intelligence in our area of interest."

[2] Ibid., 4; see especially Samuel A. Goudsmit, *Alsos* (Newbury, NY: AlP Press, 1996).

[3] During the 1950s, almost any information gathered on the Soviet Union was viewed as "intelligence." LRD, however, was viewed as a special, scientific form of information gathering. Data and information are not intelligence. Though often confused, especially in the early years of the Cold War, intelligence is actually a complex process. In general, that process is divided into three phases: collection, analysis, and dissemination. Intelligence organizations employ skilled analysts. Their job is to analyze data and information, from a myriad of sources (both classified and unclassified), to produce a "product" that is useful to decision makers or other analysts. Once those finished analytical products are disseminated they are considered intelligence at that point. In the 1950s, LRD involved the invention of unique technologies to detect detonations originating within the Soviet Union. To protect that work, those R&D and testing programs were highly classified. As such (especially when they collected data from a Soviet nuclear detonation), they were viewed as a highly classified means of surveillance. Over time, however, such protection proved unnecessary, and in 1976, the AEDS became an unclassified system.

[4] Martin Walker, *Cold War*, 19, 26.

[5] As quoted in Ibid., 33-37.

[6] Telegram, George Kennan to George Marshall, February 22, 1946, Harry S. Truman Administration File, Elsey Papers, at https://www.trumanlibrary.org, accessed October 26, 2016. Also cited in Walker, *Cold War*, 40.

[7] Entire transcript of his speech is at http://www.winstonchurchill.org. Accessed October 26, 2016.

[8] "Address of the President to Congress, Recommending Assistance to Greece and Turkey," March 12, 1947, Elsey Papers, at https://www.trumanlibrary.org/whistlestop/study collections/doctnne/large/documents/index.php?documentid=5-9&pagenumber=1, last accessed November 1, 2016.

[9] Walker, *Cold War*, 49; and Robert R. Bowe and Richard H. Immerman, *Waging Peace: How Eisenhower Shaped an Enduring Cold War Strategy* (New York, NY: Oxford University Press, 1998), 12.

[10] There are many books on the Marshall Plan. The most recent, however, is especially good. See Ben Steil, *The Marshall Plan: Dawn of the Cold War* (New York, NY: Simon and Schuster, 2018).

[11] Halberstam, *The Fifties*, 49-59; also Ambrose, *Eisenhower*, 136-139.

[12] Martin J. Sherwin, *A World Destroyed: Hiroshima and the Origins of the Nuclear Arms Race* (New York, NY: Vintage Books, 1987), 238.

[13] Jessica Wang, *American Science in an Age of Anxiety: Scientists, Anticommunism, and the Cold War* (Chapel Hill, NC: University of North Carolina Press, 1999), 7.

[14] William Lawren, *The General and the Bomb: A Biography of General Leslie R. Groves, Director of the Manhattan Project* (New York, NY: Dodd, Mead, and Company), 1988.

[15] For the text of the Baruch Plan that was presented to the United Nation's Atomic Energy Commission on June 14, 1946, see Philip L. Cantelon (Ed.), Richard G. Hewlett (Ed.), and Robert C. Williams (Ed.), *The American Atom: A Documentary History of Nuclear Policies from the Discovery of Fission to the Present* (Philadelphia, PA: University of Pennsylvania, 1991), 91-96.

[16] Alice L. Buck, *A History of the Atomic Energy Commission* (Washington DC: U.S. Department of Energy, 1983).

[17] President Harry S. Truman created the Central Intelligence Group (CIG) under the direction of a Director of Central Intelligence by presidential directive on January 22, 1946, and this group was transformed into the Central Intelligence Agency (CIA) by implementation of the National Security Act of 1947. Vandenberg held that position from June 10, 1946, to May 1, 1947, at which time he became Vice Chief of Staff of the Air Force.

[18] Phillip S. Meilinger, *Hoyt S. Vandenberg: The Life of a General* (Bloomington, IN: Indiana University Press, 1989), 150.

[19] Letter, CIG to Army, Navy, AEC, and JRDB, recommending LRD committee, March 14, 1947, AFTAC archives.

[20] Memo from Seeman to Holzman, reference Patterson appointing Holzman as his personal representative. AFTAC archives.

[21] Letter, CIG, to members of the committee, May 14, 1947. AFTAC archives. His appointment underscores the important recognition of meteorology during the inception phase of LRD. Holzman was recognized as one of the top scientists (PhD, meteorology) in the AAF. He played a critical role in weather operations for the invasion of Europe in 1944 and would lead weather operations in 1948 for Operation SANDSTONE. He would go on to hold high level R&D positions, retiring as a brigadier general in 1961 to work as Chief Meteorologist for American Airlines. Throughout the 1950s, he became an important ally of AFOAT-1.

[22] "Minutes of the Long Range Detection Committee," May 21, 1947; and Agenda, "Long Range Detection," May 13, 1947, AFTAC archives.

[23] Rivero was an influential member of the committee. At that time, he served as a technical assistant on the Staff of Commander Joint Task Force One for Operation Crossroads, and was on the Staff of Commander, Joint Task Force Seven during the atomic weapons tests in Eniwetok in 1948. He would later become the first Latino four-star admiral in the U. S. Navy, and ambassador to Spain from 1972-1974.

[24] Ibid.

[25] Memo from William T. Golden to Lewis L. Strauss, Subject: "Long Range Detection Committee (Monitoring)," May 22, 1947, AFTAC archives.

[26] "Report of the Subcommittee to the Long Range Detection Committee," June 6, 1947, AFTAC archives.

[27] In replacing Vandenberg, Hillenkoeter assumed the position on May 1, 1947. Patterson was in his last month as Secretary of War and would be replaced by Kenneth C. Royall.

[28] Rear Admiral Roscoe H. Hillenkoetter, Memorandum for the Secretary of War, "Long Range Detection of Atomic Explosions," June 30, 1947, AFTAC archives.

[29] Major General Leslie Groves, Memorandum for the Chief of Staff, "Long Range Detection of Atomic Explosions," July 24, 1947, AFTAC archives.

[30] Lieutenant General Lewis H. Brereton, Memorandum for Chief of Staff, "Long Range Detection of Atomic Explosions," July 30, 1947, AFTAC archives.

[31] R. F. Ennis, Memorandum for the Deputy Chief of Air Staff for Research and Development, "Long Range Detection of Atomic Explosions," July 31, 1947, AFTAC archives.

[32] Curtis E. LeMay, Memorandum for the War Department General Staff, "Long Range Detection of Atomic Explosions," August 1, 1947, AFTAC archives.

[33] Ibid.

[34] Ibid.

[35] Letter from Kenneth C. Royall, Secretary of War to Rear Admiral R. H. Hillenkoetter, Director, Central Intelligence Group, September 5, 1947, AFTAC archives.

[36] Memo from Secretary of the Navy (Forrestal) and the Director of Central Intelligence (Hillenkoeter), September 15, 1947, AFTAC archives. Two days later, Forrestal became Secretary of Defense and was replaced by John L. Sullivan.

[37] Dwight D. Eisenhower, Memorandum for the Commanding General, Army Air Forces, "Long Range Detection of Atomic Explosions," September 16, 1947, AFTAC archives.

[38] On November 7th, AMC replied to the Chief of Staff acknowledging the tasking. AMC delegated responsibility to the Electronic Sub-Division of its Engineering Division under the direction of Brigadier General Alden R. Crawford. Importantly, and somewhat prescient, Crawford emphasized the need to understand the current scope of geophysical research in order to avoid duplication of efforts with so many parties already involved in LRD.

[39] Hoyt S. Vandenberg, Memorandum for the Deputy Chief of Staff, Operations, "Long Range Detection of Atomic Explosions," October 27, 1947, AFTAC archives. In reiterating Eisenhower's verbiage, HQ, USAF followed up with its official directive to Major General William E. Kepner, Chief, Armed Forces Special Weapons Project on November 12th. Office of the Chief of Staff, USAF, Memorandum to Commanding General, AMC; Chairman, Atomic Energy Commission; and Chief, Armed Forces Special Weapons Project. "Long Range Detection of Atomic Explosions," 12 November 1947, AFTAC archives.

[40] Memo from Vandenberg to DCS/M, November 14, 1947, AFTAC archives.

[41] Kepner had seen extensive combat during the First World War and became the Air Service's balloon pioneer and expert

during the 1920s. During the Second World War, he commanded the 8th Fighter Command and the 9th Air Force.

[42] Note that the RDB had recently replaced the JRDB. Office of the Chief of Staff, USAF, Memorandum to Commanding General, AMC; Chairman, Atomic Energy Commission; and Chief, Armed Forces Special Weapons Project, "Long Range Detection of Atomic Explosions," November 12, 1947, AFTAC archives.

[43] See for example, Memo from A. S. Barrows, Acting Secretary of the Air Force to Secretary Sullivan reference LRD, December 2, 1947; memo from Major General Aurand, Director of Research and Development to Chief of Signal Office, December 2, 1947, AFTAC archives; memo from Lieutenant General H. A. Craig, DCS/M to Major General Aurand, Director of Research and Development, December 2, 1947; and memo from Kepner to CG, AMC, December 4, 1947. "Minutes of the Long Range Detection Committee," May 21, 1947; and Agenda, "Long Range Detection," May 13, 1947, AFTAC archives.

[44] Memo from Symington to Spaatz, Subject "Status Report on Program of Long Range Detection of Atomic Explosions," January 8, 1948, AFTAC archives. Symington was the first Secretary of the Air Force.

[45] Italics added. Memo from Sullivan to Symington reference full support of Navy, December 16, 1947, AFTAC archives. Note that John L. Sullivan replaced Forrestal as Secretary of the Navy on September 17, 1947 when Forrestal became the first Secretary of Defense.

[46] Memo from Assistant Vice Chief of Staff to DCS/M reference memo for Mr. Symington to Mr. Forrestal reiterating full support to LRD, December 10, 1947, AFTAC archives; memo from Sullivan to Symington reference full support of Navy, especially in regard to SANDSTONE, December 16, 1947, AFTAC archives; memo from AEC to Chief of Staff, USAF, reference "Long Range Detection of Atomic Explosions," December 10, 1947, AFTAC archives.

[47] On January 15, 1948, AFMSW-1 became the official office symbol for Hegenberger's division. Memo from DCS/M to All Personnel, SWG, January 15, 1948, AFTAC archives. For Hegenberger's accomplishments, see Robert F. Hegenberger, "The Bird of Paradise: The Significance of the 1927 Hawaiian Flight," *Air Power History* (Summer 1991), 6-18.

[48] Johnson worked across all of the armed services throughout the 1940s and 1950s, and became one of the nation's top operational research professionals. His R&D work on using the AAF to deploy mines near the end of the war was well known to LeMay. As head of AAF R&D in 1946, LeMay may have had a role in getting Kepner and Johnson together. After leaving the LRD program, Johnson served as the Director of Organizational Research of the U.S. Army. See Charles R. Shrader, *History of Operations Research in the United States Army: 1942-1946* (Washington, D.C.: Office of the Deputy Under Secretary of the Army for Operations Research, 2000), 164-165.

[49] Transcripts, Doyle Northrup oral history interview, USAF Oral History Program, Interview #K239.0512-685, July 24, 1973, AFTAC archives. During the Japanese attack on December 7th, both scientists helped in the rescue efforts. Soon thereafter, they used their experimental acoustic equipment to locate sunken Japanese mini-subs in Pearl Harbor. After finding one and bringing it to shore, Northrup (small in stature) climbed into the sub and helped remove the bodies and equipment.

[50] Memorandum, Ellis A. Johnson to Hegenberger, "Preliminary Estimate of Long Range Detection Problem," January 15, 1948, AFTAC archives.

[51] Ibid. Air Force Letter 20-35 did not specifically cover the transfer. This was accomplished retroactively by a letter dated September 10, 1948, AFTAC archives.

[52] Ibid.

[53] *History of Long Range Detection: 1947-1953*, author unidentified AFTAC archives.

[54] Ibid. Conant held a PhD in chemistry and was President of Harvard University at the time. He would later serve as the first U.S. ambassador to the Federal Republic of Germany in 1955.

[55] Walker, *Cold War*, Chapter Two. See also Gaddis, *We Now Know*, Chapter One.

[56] Meilinger, *Vandenberg*, Chapter Six.

CHAPTER 2

Learning to Detect

Scientists involved in the first U.S. nuclear test—the detonation of TRINITY on July 16, 1945—had actually stumbled upon the rudiments of LRD, albeit indirectly. Aware that particulates could travel long distances following a detonation and concerned that people living near the test site could conceivably bring legal action against the government for contamination, the TRINITY participants placed detection instruments spanning distances out to 500 miles from El Paso to Denver. For TRINITY, these instruments were primarily seismographs and radiation survey meters.[1] Although the seismographs failed to detect any earth waves, observers speculated that blast-generated particles could remain in the atmosphere for extended periods of time. Their hunches were confirmed when a B-29, modified with a crude cylinder lined with tissue paper, flew sorties on August 10th and 15th, and returned with radioactive particulates. Test results looked promising, and researchers eagerly awaited opportunities for future experiments.[2]

In late 1945, General Groves headed the planning for a series of tests in the Pacific anticipated for late spring 1946. In the planning leading up to the test series codenamed CROSSROADS, organizers from the Army and the Navy included experiments specifically designed to evaluate LRD techniques, with an eventual monitoring system in mind. Both the Army and Navy provided personnel to staff the Remote Measurements Section of CROSSROADS. Leading this section was Commander George Vaux, who reported directly to Rear Admiral William "Deke" Parsons, the technical director and deputy commander of the test series. Parsons had participated in the Manhattan Project as head of the Ordnance Division at Los Alamos. Major General Kepner, not yet named to head up the Special Weapons Group (see Chapter 1), served as the commander of all air operations at

CROSSROADS. Reporting to him was Brigadier General Roger M. Ramey, who held responsibility for all AAF activities during the operation. Like Kepner, Ramey had transferred from the cavalry to the Air Corps early in his career. He was decorated for his attempts to save aircraft at Pearl Harbor in 1941, and again with the Distinguished Service Cross for his actions as a B-17 command pilot in leading an attack on a Japanese base at Rabaul, New Britain, in 1943. The following year, in July 1947, Ramey would play an important role in one of LRD's most important classified projects codenamed Project MOGUL.[3]

Project MOGUL was an adjunct to the air activities embedded in CROSSROADS. The AAF utilized the nuclear tests for its independent investigation of an experimental LRD technique. The AAF originally designed MOGUL, which later extended well beyond CROSSROADS, to continue the cooperative wartime relationship between civilian research institutions and the military, and to help maintain America's technological superiority. In time, MOGUL expanded under contracts with leading scientific institutions, such as New York University, Woods Hole Oceanographic Institution, Columbia University, and the University of California at Los Angeles.

MOGUL was conceived as a program to build upon the Navy's wartime research of sound transmissions and oceanic propagations. In his work with the Navy in 1944, geophysicist Dr. W. Maurice Ewing discovered a horizontal sound channel existing at a great depth in the oceans that permitted sound to travel at enormous distances without signal degradation or detection from above or below. The AAF became interested in the idea when Ewing wrote to General Carl Spaatz in October 1945, speculating that a similar type of channel existed somewhere in the Earth's atmosphere. After recognizing that jets or rockets passing the axis of such a stratospheric channel could be detected and triangulated thousands of miles away, Spaatz agreed to sponsor further research.

Spaatz placed responsibility for this R&D program under the leadership of Colonel Roscoe C. Wilson. Almost immediately, as the AAF planned for CROSSROADS, other projects fell under the MOGUL program as well.[4] At that time, Wilson was one of the AAF's most capable experts on the military application of nuclear science. Early in his career, he worked on the prototype of the future B-17 bomber and taught science at West Point. Following several other engineering assignments, Wilson was appointed as the AAF project officer to the Manhattan Project's Engineering Division in June 1943. In this post, he played a significant role in the construction of the first atomic bomb and was instrumental in choosing Alamogordo, New Mexico, as the test site for TRINITY.

After combat duty in the Pacific in 1945, Wilson joined the team that surveyed the effects of the atomic weapons used at Hiroshima and Nagasaki. In July 1946, the Air Force assigned Wilson to the Office of the Deputy Chief of Staff for Research and Development, at that time headed by Lieutenant General Curtis LeMay. In this capacity, he oversaw the AAF's portion of Program 8—Remote Measurements. Program 8 was just one of several experimental programs of CROSSROADS. The primary purpose of the series was to study the effects of nuclear weapons on ships, equipment, and materiel within a fleet of more than 90 surplus U.S. and captured enemy vessels positioned in the Bikini Lagoon. Indeed, the remote measurements personnel were only a small fraction of the 42,000 men aboard the 150 ships that made up Joint Task Force 1 (JTF 1). Still, the radiation measurements experiment would gather important data that the AAF/USAF would soon operationalize for LRD.[5]

Returning to Washington, Wilson served in several staff positions at the Pentagon, and in 1947 he was designated as the Deputy Chief of AFSWP. In that position, he functioned as the Air Force representative to the Military Liaison Committee (MLC) responsible for coordinating activities between the Defense Department and the AEC. Among other duties, Wilson also served as a member of the Committee on Atomic Energy (CAE), a subordinate committee of the RDB. The AEC would soon exercise significant influence over LRD R&D (see AFOAT-1/AEDS Data/Information Flow graphic in the Introduction). From the time of AFOAT-1's inception until the late 1950s, Wilson greatly but quietly helped advance AFOAT-1's R&D programs and experiments.[6]

On July 1, 1946, an AAF B-29 named *Dave's Dream* initiated Shot Able with an air drop at 520 feet over a fleet of obsolete ships and submarines. However, the crew dropped the bomb approximately 2,000 feet off the designated target. Consequently, the detonation destroyed fewer ships than the planners anticipated. Despite the error, the experiments yielded a large quantity of data. The researchers participating in the LRD experimental programs conducted a total of 21 projects. Some encompassed remote measurement experiments at sites around the world designed to record the effects produced by the detonations. In addition to Ewing's stratospheric channeling experiment, others included measurements of tides, wave action, atmospheric reflectivity, atmospheric pressure, atmospheric ionization, radioactivity, and long-range radio waves.[7]

Particulate collection became the center piece of the LRD experiments. Modified bombers demonstrated the feasibility of collecting samples at long distances for radiation measurements. Another LRD experiment called for the collection and measurement of air samples to determine the efficiency of the detonations. To do that, the AAF flew unmanned drone aircraft (modified B-17s and F6Fs) in and around the detonation cloud to obtain radioactive samples.

For Shot Able, drone controllers operating out of their drone control radio jeeps sent four B-17 drones into the air from Enewetak. The five B-17 airborne control aircraft seamlessly took control of the drones to position them at various altitudes between 6,000 and 30,000 feet just minutes before the detonation. With generals Kepner and Ramey circling the test area in their command B-29, the drone controllers placed their drones on automatic pilot and sent the unmanned aircraft through the center of the cloud at 24,000 and 30,000 feet. The controllers quickly flew around the cloud to the other side and regained control of the drones.

B-17 drones were outfitted with special filter boxes mounted in place of their top turrets. Inside each was a filter paper designed to collect particulates. A large, inflatable rubber bag located in the bomb bay captured 90 cubic feet of air. Upon command from the controller aircraft, the filter box opened as the drone entered the cloud. It closed automatically after 30 seconds. Upon landing, to release the filter unit, ground personnel attempted to avoid contamination by pulling on a lanyard that ran alongside the outside of the fuselage to the handle fixed in the door of the plane. One short pull on the lanyard brought the entire apparatus tumbling to the ground. Los Alamos personnel then collected the bag and filter paper for immediate shipment aboard a C-54 from Kwajalein, Marshall Islands, for analysis.[8]

The AAF also flew modified B-29s at various distances from Bikini to collect drifting air and particulate samples. Routes included flights over Guam, Okinawa, Hawaii, Washington, Arizona, Florida, and Panama at an altitude of 25,000 feet. The filter units aboard those planes housed special oil-bathed filters specifically designed for particulate collection. These daily five- to eight-hour flights ended with the shipment of the filter papers to the Corps of Engineers in Berkeley, California, where they were processed and forwarded to a Standard Oil Laboratory for analysis.[9]

Testing an underwater detonation, CROSSROADS organizers initiated Shot Baker on July 25, 1946, with a nuclear device suspended at 90 feet beneath an auxiliary craft anchored in the midst of the target fleet. As the mushroom cloud formed, expanded, and rose, it pulled up "an immense stem of saltwater."[10] For this test, two of the four drones flew directly over the target ship at the moment of detonation. Flying at 6,000 and 16,000 feet, the blast threw the aircraft several hundred feet but, though damaged, they landed safely. The other two drones flew through the cloud at 7,000 and 11,000 feet respectively. Observers noted at the time that Baker produced more fallout than Able—valuable data that would soon advance LRD.[11]

In sum, the results of the TRINITY and CROSSROADS tests (as well as Hiroshima and Nagasaki) gave impetus to the development of a national LRD capability. The AAF's performance at CROSSROADS was impressive, and provided Vandenberg and LeMay with strong arguments for their advocacy of AAF stewardship over LRD in the ongoing debates over roles and missions. In time, the AAF/Air Force soon became the dominant U.S. armed service to execute the operational arm of the nation's national security strategy. This, of course, grew in scope as the Cold War escalated, and national security rested largely on the Air Force's extensive ability to deploy nuclear weapons over long distances with intercontinental ballistic missiles (ICBMs) and long range bombers. However, on the heels of CROSSROADS, all of that was yet to come. Of immediate concern were the questions of how far the Soviet Union had advanced with nuclear research, and how close it was to the development of a nuclear weapon?

In addressing those questions, the CROSSROADS experiments proved invaluable in the development of several LRD techniques, particularly airborne sampling and seismic recordings. While the potential was apparent, the operationalization of effective detection techniques was far from realization. Indeed, CROSSROADS generated more questions about LRD than it provided answers.

Unfortunately, press reports and articles published in scientific journals after the CROSSROADS tests implied LRD was easy to implement. The writings reported that monitoring stations located thousands of miles from the test sites had detected the blast effects. These writings also proffered that the airborne collection of particulates in radioactive fallout made LRD easy. In reality, however, the opposite was true. Analyses of particulates might reveal the presence of a detonation and some data on a bomb's characteristics, but it could not yield any information on where the detonation took place. Seismology held the potential for location identification, but the state of that technology at the time could not distinguish between explosions and natural earthquakes. Most likely, the combination of these popular beliefs and the opinion that the Soviets were incapable of producing an atomic weapon in the near future combined to influence the move to cut LRD R&D funding in 1948 and 1949 (discussed in Chapter 3). In the interim period between CROSSROADS (July 1946) and the next test series codenamed SANDSTONE (May 1948), those decision makers closest to LRD were occupied with the aforementioned organizational activities, especially after the establishment of the U.S. Air Force as a separate armed service in September 1947. The scientists involved, however, advanced the science with data in hand as they prepared for SANDSTONE.[12]

The first to articulate the details of the technical challenges was Dr. Ellis A. Johnson, the first technical director of AFMSW-1 and AFOAT-1. On January 15, 1948, Johnson issued Technical Directive No.1, which was a preliminary estimate of the overall LRD problem. In his memo, he outlined four physical means of detection: (1) measurement of seismic waves in the earth's crust; (2) measurement of sonic waves and pulsations in the earth's atmosphere; (3) radiological and chemical measurements of pollution of the earth's atmosphere and crust by fission; and (4) electromagnetic effects produced by fission.

The inherent difficulties in each of the four methods were discouraging. Seismically, the problem differed from conventional seismology. The frequency content of the wave was unknown, and the absorption of different frequencies within the earth was unknown for each of the various layers. Sonic propagation in the air depended upon temperature and pressure gradients of the atmosphere and upon meteorology. Also, the noise background was based on many factors. The radiological and chemical methods for accurate analyses seemed to offer even greater difficulties.

The relationship of the physical, radioactive, and chemical activity of the fission products, together with the interrelationship of these factors with meteorology, had to be understood before scientists could produce effective detection technologies. Scientists also had a poor understanding of the entire problem of atmospheric diffusion. Such knowledge was critical because the dilution of the fission products in the air would determine the lower limit of detectability.[13]

The first step in the solution of the multitudinous problems of LRD was the formulation of a research program for SANDSTONE. Hegenberger designated this program as Operation FITZWILLIAM.[14] The initial planners anticipated experiments to measure air ionization, air current to determine the presence of the conducting cloud, and the pulsation of the atmosphere to contrast with the measurement of the sonic waves produced by the explosion. They also wanted to sample gaseous fission products, and rain and snow from weather stations in the path of the contaminated clouds. In Johnson's view, SANDSTONE was essentially a "beginner's course" in LRD. CROSSROADS had revealed that, in terms of requirements, LRD overlapped six or seven sciences already being developed across various military and industrial agencies. To save time, the LRD researchers solicited assistance from organizations such as the Coast and Geodetic Survey, which had one hundred years of experience in seismic operations, and the Evans Signal Laboratory of the Army Signal Corps, which had been working for many years with various acoustic and sound-ranging devices. The assistance of these organizations made it possible to conduct experiments in all envisioned elements of LRD during

Operation SANDSTONE. To manage the radiological portions of FITZWILLIAM, Johnson relied heavily on TracerLab, Inc. (after CROSSROADS, TracerLab became AFOAT-1's sole contract radiochemistry laboratory).[15]

The AEC, with support from Los Alamos Scientific Laboratory (LASL), sponsored the three-shot SANDSTONE test series in the late spring of 1948. Their goal was to proof-test new weapon designs. When President Truman approved the preliminary SANDSTONE test program on June 27, 1947, the United States possessed only 13 nuclear weapons in its stockpile. Therefore, the SANDSTONE participants hoped greater efficiencies would allow the United States to more rapidly expand its nuclear stockpile in order to stay far ahead of the anticipated Soviet nuclear weapons program.[16]

Planning for SANDSTONE first began on October 18, 1947, when the various agencies and organizations involved formed Joint Task Force 7 (JTF-7). For planning purposes and correspondence, the organizers initially assigned the code name SWITCHMAN to the project. The AEC exercised overall responsibility for the shots. In addition to the four armed services, a number of other government agencies were included as well as several civilian organizations under contract. U.S. Army Lieutenant General John E. Hull commanded JTF-7 which was subordinate to the Joint Chiefs of Staff (JCS). Within JTF-7, the test director of Task Group (TG) 7.1 was Captain James S. Russell, USN, head of the Weapons Branch, Division of Military Application, AEC. He was responsible for the direction of all technical test activities and policies, and all military experiments were under his direction. General Kepner, as the Chief, Special Weapons Group, served as the deputy commander of JTF-7, and was also responsible for all air operations, meteorological services, inter-island air transportation, air-sea rescue, and aerial photography. At the same time, the Air Force again called upon General Ramey to command SANDSTONE's Air Task Group 7.4. This task group included a total of 1711 USAF personnel, 481 of whom operated subgroup 7.4.2., responsible for the airborne sampling operations and seismic testing.[17]

During Operation SANDSTONE, JTF 7 detonated three nuclear fission devices at the AEC's Pacific Proving Ground (PPG) in the Eniwetak Atoll. The AEC selected Eniwetak for the test series because it was large enough for the three nuclear detonations, and the steady westerly trade winds would carry fallout from the shots out over the open ocean. Organizers codenamed the three shots, all detonated from 200-foot towers, as shots X-ray, Yoke, and Zebra. Shot X-ray, detonated on Enjebi Island on April 15, 1948 produced a yield of 37 kilotons (kt.). Shot Yoke was detonated on Aomon Island on May 1st with a yield of 49kt. Finally, Shot Zebra was detonated on May 14th on Runit Island and yielded 18kt.[18]

For SANDSTONE, the Air Force provided 24 modified B-17s from the 1st Experimental Guided Missiles Group (EGMG) at Eglin Field, Florida, under the command of Colonel John R. Kilgore. However, only 8 drones and 8 controllers flew the cloud sampling mission. The sampling drones were configured as they were for CROSSROADS. For Shot X-ray, two filter-equipped WB-29s deployed to track the debris in order to verify the predicted path of the radioactive cloud. To track this drifting debris, the WB-29s flew 12-hour shifts for five days after the detonation.[19]

Similarly, for Shot Zebra, the Air Force flew two WB-29s to track the radioactive cloud over an extended distance. Eight B-17 drones were directed through the cloud for sampling operations, with each aircraft making three passes. After their runs, the drones and samples were noticeably more radioactive than on the previous two shots. In fact, as the five Los Alamos personnel removed the wire-mesh filter-paper holders from the filter papers, three of the men received serious beta radiation burns on their hands from high gamma exposures. They immediately showered and flew back to LASL on the two C-54 aircraft. Even before arriving at LASL their hands had become red, swollen, and itching. At LASL they were hospitalized, and their badly burned hands later required skin grafts. In these early days of airborne sampling, filter handling and decontamination procedures were rather crude. In time, concerns over safety and radiation hazards would have a profound effect on LRD and, more importantly, on domestic and foreign policy as fallout became a worldwide concern.[20]

In addition to the extensive operations for airborne sampling, the Air Force conducted six experiments specifically associated with other potential LRD methods. However, these FITZWILLIAM experiments were permitted with the understanding that the experiments would not interfere with the primary objectives of SANDSTONE. Because the sites for the experiments were far removed from Eniwetak, the LRD experiments posed no radiation hazards and were termed "special operations." Of the six experiments, the Air Force conducted four, three unilaterally and one (magnetic effects) in partnership with the Naval Ordnance Laboratory.[21]

The data and reports Johnson and his team of researchers received from the initial SANDSTONE detonation—Shot X-ray—were disappointing. Though preliminary, the X-ray reports released on April 27th, revealed that the existing network of sensors, which would soon become the AEDS, were not as sensitive as many believed. New automatic recording counters (both ground and airborne), designed to detect atomic explosions at various distances, largely failed. Ground-based air filter units provided very little samples, leading Johnson to question the volume of air necessary to provide adequate samples, or to suspect the viability of the filter papers. While airborne particulate filters produced positive results to detect an explosion within a few hours out to a distance of 200 miles, Dr. William Urry, Johnson's Chief of the Radiological Division, expressed concern that a Soviet test conducted within 10 days of a U.S. nuclear test would produce "evasive" results; meaning nuclear debris would intermix, making analysis difficult. Finally, Northrup's report to Johnson on seismic detection indicated no seismic stations beyond 1,000 miles detected the 37-kiloton detonation.[22]

The following day, April 28, 1948, Johnson consolidated the Urry and Norhrup reports in a memorandum to General Hegenberger. In his memo, Johnson summarized the key findings and emphasized that currently, the airborne collection method was the only LRD technique that could ensure a positive detection of a Soviet test. Even so, he cautioned that an underground or underwater test could be evasive, and adequate coverage would require extensive resources. Johnson estimated they would need five B-29 air groups for continuous airborne sampling operations. He utilized his report to strongly advocate for an extensive R&D program that would emphasize geophysical LRD methods (i.e., sonic, acoustic, and seismic). However, because the government saw little value in geophysics R&D, his advocacy would go unheeded and frustrate Northrup for a decade. As he noted:

> The negative results obtained on sonic and seismic equipment during the present results are not a significant measure of our ultimate ability to utilize these methods for detecting either underground, underwater or air bursts of atomic bombs. The evidence from the CROSSROADS and SANDSTONE tests together make it probable that with further extensive research, sonic and seismic measurements can be established to provide positive evidence.[23]

Within FITZWILLIAM, three of the four Air Force LRD experiments were rudimentary and produced little or no significant results. These were magnetic effects, remote detection by ground observation, and the lunar technique.[24] In regard to the latter, the Air Force experimented with this bizarre method during Shot Yoke. In this experiment, researchers attempted to photograph atmospheric nuclear test detonation light flashes as they were reflected off the darkened surface of the moon. This technique required the presence of a "new" moon; that is, the phase of the moon occurring when the moon is at conjunction and the nearside is totally unilluminated by the sun. Observers at the CROSSROADS detonations in 1946 reported such a flash. However, LRD researchers soon concluded this method had limited value because of the necessity for suitable positioning of the moon and the need for clear atmospheric visibility. Due to these physical constraints and the lack of positive results in SANDSTONE, the researchers quickly abandoned this novel concept.[25]

In contrast, the fourth experiment—remote detection by airborne filters—showed promise, albeit with some significant shortcomings. In general, the second and third shots (Yoke and Zebra) confirmed the first reports after Shot X-ray. The main purpose of collecting atomic bomb debris was to present irrefutable evidence of a detonation by isolating key fission products and verifying them to be the source of anomalous activity in the atmosphere or hydrosphere. However, the radiochemical analyses of the long-range samples collected in SANDSTONE were disappointing. First, only ultra-micro quantities of samples were available, which made accurate quantitative analysis difficult. Second, the dissolution of the filter paper containing the samples presented many difficulties, as Urry first speculated. Researchers quickly surmised that they had to improve the ultra-micro methods for detecting fissionable material to permit a lower limit of observation. Fortunately, an operational method for collecting better samples at long distances suddenly arose.

Well into SANDSTONE, researchers believed rising detonation clouds were too radioactive for manned aircraft to penetrate or to collect an adequate amount of particulates for any substantial analysis. Hence, they used only drone aircraft. The crews of the WB-29 weather reconnaissance aircraft were valuable in long distance sorties to verify radioactive cloud travel in the predicted direction but were limited to extended stand-off distances. As they had in CROSSROADS, the WB-29s tracked the drifting effluent clouds for five days. However, during the final detonation of Operation SANDSTONE (Shot Zebra), Major Paul H. Fackler, the commander of the 514th Reconnaissance Squadron (VLR) Weather, "accidentally" flew his aircraft through a protruding finger of the expanding cloud, thus violating the ten-mile standoff directive for the B-29s. Upon landing, onboard radiation monitoring equipment showed the exposure levels were well within an acceptable range for crew safety. Also, contrary to the skepticism of many researchers, Fackler's filter papers returned rich in particulate samples. Fackler's demonstration greatly impacted the future of LRD, as the next test series scheduled for 1951 would reveal.[26]

Fackler's flight path that day proved WB-29s could obtain good samples at closer distances to the cloud with little or no risk to the crew. His "mistake" demonstrated that the WB-29 could provide greater precision in sampling size and within more accurately identified locations than drones. Evaluators, however, still favored the use of drone aircraft but argued the Air Force had to improve the drone control equipment. More importantly, they noted the filter boxes required a re-design given the radiation burns suffered by the three ground personnel. Finally, despite Fackler's demonstration, they recommended the Air Force establish a permanent drone aircraft unit in order to train and retain qualified personnel.

Overall, the FITZWILLIAM experiments, combined with knowledge gained from previous tests, showed promise and helped refine plans for additional research. The good news was that seismic stations within 400 miles of ground zero reported the blast, and the station at San Diego detected sonic waves transmitted through the temperature inversion layer deep in the ocean. Researchers believed at the time that both methods would be important in the future for detecting underground or underwater explosions, but only of secondary importance in confirming other types of detonations. While Johnson recognized these were only small steps forward, he was disappointed that the tests did little to promote geophysical research. Despite his personal views, he felt the pressures to field an operational LRD detection system, even if that system initially only included airborne sampling.[27]

On the same day as the last SANDSTONE detonation (Shot Zebra on May 14th), Johnson submitted a proposal to Hegenberger for placing the "interim net" (i.e., the AEDS) into operation. While admitting the FITZWILLIAM results were "imperfectly evaluated" and seismic and sonic techniques were not yet adequate, Johnson strongly argued for the immediate implementation of airborne sampling, which would ensure "100% coverage of any atomic air burst in Russia."[27] To emplace such a net would require six B-29 squadrons for a total of 83 aircraft added to the AWS. Full coverage would require 10 daily flights covering a number of routes: from Alaska to England by way of Greenland, from England to India, from India to the Philippine Islands, from the Philippines to Japan, and from Japan to Alaska. Johnson even proposed placing filters on selected commercial aircraft flying international routes.[28]

Even though Johnson was satisfied with radiological capabilities, he remained especially frustrated by the fact that, going into SANDSTONE, the seismic R&D program was seriously behind schedule. Although seismic readings were recorded at distances of 700 miles at TRINITY and 5,000 miles at CROSSROADS, the existing seismographs exposed severe limitations resulting in poor data gathered from those operations and the SANDSTONE shots. New seismographs would not become available until the 1951 GREENHOUSE series of tests. Like seismic R&D, the work accomplished in the acoustic program was only 20 percent of that originally planned due to delays in obtaining approval for the various projects. Acoustic measurements revealed the inadequacy of existing instruments and techniques. The microbarographs (instruments that record atmospheric pressure changes) lacked sufficient sensitivity, low frequency response, stability, and an adequate recording and analyzing system. In addition, the U.S. tests revealed the need for a systematic study of the nature of noise background and the development of noise-reducing techniques.

These significant shortcomings led Northrup to form the Acoustic Technical Working Committee soon after SANDSTONE. In their first meeting held on July 7, 1948, Gilfillan set the agenda. This Navy chemist had been the Technical Director of the Joint Operations CROSSROADS Committee of DoD that initially conducted the scientific survey of the Bikini Atoll and, therefore, was well aware of the challenges ahead. Prior to SANDSTONE, researchers outlined a plan during the FITZWILLIAM preparations to operate 33 AWS and Office of Naval Research (ONR) radiological stations in an interim research and surveillance system. However, with fresh SANDSTONE data in hand, their new task was to determine the operational requirements for a successful surveillance net. They reviewed a summary of the data obtained from CROSSROADS and SANDSTONE, and discussed the relative merits of the acoustic systems (i.e., the ground reception of low and high frequencies as well as balloon reception). The committee was under great pressure to make an assessment almost immediately, either during that meeting or by August 1st. They all agreed to stress the urgency for R&D, and to note that the "scope of [R&D] in acoustics is limited only by the extreme urgency for installation of the final surveillance network at the earliest possible date."[29] They also acknowledged the need to solicit assistance from several government agencies and contractors. The committee estimated a total budget of $8.8 million (2021 = $96.7 million). The primary problem they were trying to address was instrument fidelity in detecting acoustic signals at longer distances. In a reference to the FITZWILLIAM data, they noted "the failure to receive definite signals beyond 1,000 miles should not be interpreted as reason for abandoning acoustic methods of detection."[30]

That first week in July, when the Gilfillan meeting emphasized the urgency of rapid R&D, Johnson had become very frustrated with several obstacles standing in the way of organizing a high-priority R&D program. Despite the tremendous efforts to develop close, collaborative relationships among the various services and agencies to recognize Air Force stewardship over LRD and a successful FITZWILLIAM program, the most serious resistance to collaboration came from within the Air Force itself. On July 2nd, just five days prior to the important Gilfillan meeting, Johnson submitted a scathing 29-page memo to Hegenberger condemning AMC for not supporting the LRD program. The DCS/M had originally directed AMC to provide R&D for the LRD program on October 24, 1947. Johnson had become quite angry that AMC had failed in every way possible to advance LRD. Johnson accused AMC of extensive incompetence and the "deliberate sabotage" of LRD R&D by not processing critical contracts with commercial partners, especially General Electric and TracerLab, Johnson's sole partner for radiological sciences. He added that AMC's negligence seriously jeopardized his relationship with the few key scientists who could possibly conduct the required research. For more than a dozen pages, Johnson detailed a number of serious examples of ineptitude, the most serious being AMC's failures in the recent FITZWILLIAM experiments. "This record is a clear indication that AMC was not only non-cooperative, but jeopardized the best interests of the Air Force by its negligence in connection with the FITZWILLIAM program." Johnson wanted AFMSW-1 (soon to be AFOAT-1) to have total control over a proposed dedicated lab and LRD R&D. "An emergency research program cannot be carried out unless both responsibilities and authority are directly provided to the organization within the Air Force which has primary responsibility."[31]

A second major obstacle confronting Johnson was the battle Vandenberg was beginning to fight over funding (see Chapter 3). Johnson's memo highlighted his frustration with the mixed signals he was receiving from the Air Staff. Although the Air Force had given the LRD program "1A priority" on December 4, 1947 (Kepner's directive to AMC), funding was being held up pending a priority ranking from the JCS. Johnson, operating under the impression that fielding a detection network was vital to U.S. national security, stressed the urgency in his July 2, 1948, memo by underscoring the existing capabilities and limitations of the rudimentary system currently in place. In a separate 26-page memo submitted as a supplement to the AMC criticism memo (also on July 2nd), Johnson issued a dire warning to emphasize the urgency.[32]

> It is believed that the Russians may possibly now have sufficient material for one bomb, and that they will make a test, under conditions of strictest secrecy, probably far below the ground in a mine shaft, before committing themselves to the economic drain incident to stockpiling. It is felt, however, that they will not commit atomic bombs and planes to an attack without first observing an air burst under service conditions, but that an attack might follow such a successful demonstration within a few hours. It is, therefore, felt to be imperative that surveillance be initiated as soon as possible and that a complete surveillance program be in operation on or before 1 January 1950.[33]

He informed Hegenberger that the operation of the Interim Net (i.e., routine airborne surveillance) and LRD R&D were interwoven and, together, were essential to moving LRD forward as quickly as possible. Cognizant of Vandenberg's budgetary struggles, Johnson stated R&D should focus on radiological and geophysical methods now while the other sciences could wait. He provided Hegenberger an assessment indicating the airborne collection and analysis technique was already operationally sound. "We can now detect any burst in seven days by radioactive measurements at 10,000 miles and in two hours at distances of 1,500 miles by sonic means." Pessimistically, he countered "we will never be able to positively identify an underground burst as an atomic event." In retrospect, given his drive to build a strong geophysical R&D arm, Johnson's negative assessment was meant to shock decision makers.[34]

While Johnson had suggested the means by which R&D could feasibly prioritize the R&D expense, his assessment of the costs associated with fielding a fully effective detection system by mid-1950 did little to assist Hegenberger and Vandenberg in their budget battles. For fiscal year 1949 (July 1948 to June 1949), Johnson estimated a budget of $30 million (2021 = $331 million). If that investment was made, Johnson stated, the envisioned "routine surveillance system providing evaluated information *for any type of an atomic burst* can be provided by 1950." He concluded his report by emphasizing that even with such a level of funding, success could only be achieved if AFMSW-1 was provided technical supervision of a dedicated lab, full authority over all R&D, and the authority to initiate and supervise prime contracts.[35]

Johnson's recommendation in May 1948 to begin flying airborne sampling operations simply reflected his and his colleagues' awareness that decision and policymakers outside of AFOAT-1 greatly preferred radiochemistry over geophysical methods. In confronting their skepticism of the plan, Johnson pushed geophysical R&D very hard throughout the summer of 1948. Vandenberg, fully supporting Johnson, resisted heavy pressures to reduce the Air Force budget and forwarded the full LRD budget request of $30 million. Still awaiting a JCS decision on setting a national priority for LRD (see Chapter 3), Vandenberg was determined to have a fully functional and effective AEDS by mid-1950. However, he would continue to lead the fight without his LRD technical director. Soon after his scathing 29- and 26-page memos to Hegenberger on July 2, 1948, Johnson tendered his resignation. Contentious confrontations over budgets throughout his eight-month tenure had taken their toll on the talented scientist.

On October 12, 1948, Northrup, now the AFOAT-1 Technical Director, led an important meeting of the Acoustic Technical Working Committee. He stated the committee was willing to extend the FY49 allocations out beyond the fiscal years in anticipation of more budget cuts but they would not curtail the existing R&D plan. The crux of the problem was the expense of large quantities of high explosives for testing. At this meeting, Captain Hallan from AFOAT-1 reported that the current plan called for 4600 tons of explosives for seismic and 44,000 tons for acoustic tests. The total expense was 13 million dollars, obviously a huge portion of the entire LRD budget.[36]

In trying to reduce this expense, the committee evaluated all possible explosives that would adequately but cheaply support the tests. They explored the use of methane, liquid oxygen, dynamite, TNT, and aluminum nitrate. TNT, they concluded, was the most practicable given the Army's large stockpile. In debating test requirements, Dr. Albert Focke of the Navy Electronics Laboratory (NEL), stated "long range detection requires very large shots . . . and likely, 1,000 tons [of TNT] is not enough." Northrup entertained a number of recommendations upon receiving each laboratory test plan. In the end, he reduced each plan by 25 to 50 percent. In sum, the committee faced the harsh reality of fiscal constraints but agreed to continue the current R&D plan, which would now include only smaller, shorter-range tests. By late afternoon, the committee voiced concern that their way ahead, without large-scale testing, did not really address the long-range detection problem. Dr. James A. Peoples of AMC overstated the obvious: "[I]t might prove necessary in order to establish a performance of LRD, we should have to explode some very large charges, possibly A-bombs, and that we do not know whether 10,000 tons or 50,000 tons of ordinary explosives would meet the requirements."[37]

While Northrup was struggling to advance the geophysical R&D program throughout the late summer and early fall of 1948, with little supportive data from the latest experiments, other researchers provided decision makers with detailed evaluation reports on the radiochemical analyses of the airborne collections in SANDSTONE. The aircraft had detected radiological debris at a distance of 12,000 miles, and had successfully traced the cloud for 18,000 miles over the course of three weeks. There were strong indications solid fissionable material from altitudes below 20,000 feet would suffer serious loss of concentration due to rain or snow. At long distances from ground zero, solid fissionable material had settled. Consequently, the use of ground radiological filters or automatic Geiger counters did not appear promising in comparison to particulate collection at high altitudes by aircraft or balloons. Measurements of the ionosphere, and magnetic or electromagnetic measurements offered little hope of a practical surveillance method. Some researchers believed measurements of atmospheric electric currents or conductivity would offer a good method of detection, but they had yet to complete analysis of those particular measurements.[38]

By the time AFOAT-1 came into existence in August 1948, planners had given adequate thought to the challenges of LRD, especially after analyzing the results of CROSSROADS and SANDSTONE. Although the experiments conducted in the two U.S. test series were moving LRD forward, the researchers clearly understood that to catch the first evidence of a Russian atomic test, they needed instruments with a sensitivity far in excess of anything used in those tests. Indeed, the ultimate goal of the new mission was to detect the first Soviet atomic bomb. Strategically, the concern was that the Soviets would develop atomic weapons in secrecy and subsequently produce an adequate stockpile of weapons to challenge American nuclear supremacy. In truth, the experts themselves were in disagreement as to when the Soviets would test their first device. Estimates ranged anywhere from two to 20 years, although planners viewed a mid-1950 date as a real possibility. LRD planners anticipated that any rudimentary detection system would not be operational any earlier than mid-1950, and the system would require at least six months preparatory time prior to that date in order to achieve a satisfactory level of efficiency.[39]

From the beginning, LRD researchers had envisioned three types of physical means to detect a remote but large atomic explosion. These were seismic and acoustic observations, and the radiochemical analysis of post-detonation particulates. However, because of the extensive use of airborne samplers in CROSSROADS and SANDSTONE, the real emphasis was on radiochemical analyses. Initially, the primary challenge was that the sample had to be both large enough to permit accurate measurement of the half-lives of the fission products and young enough to permit identification of the short-lived products. Conducting this type of analysis at CROSSROADS and SANDSTONE was relatively easy because the analysts already knew the time and place of the explosions. Also, the weather was good, and the air mass movements were sufficiently well established to permit aircraft to follow the cloud for great distances.[40]

In the case of a Soviet detonation, there was no reason to expect such favorable conditions. The Soviet Union possessed potential geographical test locations that could make LRD difficult (e.g., their Arctic territories). Any bomb debris released there might be caught in the general circulation of the atmosphere of this region, for which the western powers possessed little or no meteorological information. The cloud might continue to circulate in the local air masses of that area until it settled or was scavenged out by precipitation. Planners also feared the Soviets could detonate their first weapon on the western slopes of the Himalayas where the atomic cloud might be greatly reduced in intensity by precipitation, which was known at times to go as high as 30,000 to 35,000 feet. Moreover, an atomic explosion of low-level intensity in Central or Eastern Siberia, where low-level winds or light debris from such a cloud might fail to reach upper-level winds, and possibly be lost within the boundaries of the Soviet Union. Researchers also noted the Soviets could resort to an underground or underwater blast, making the collection of fission particles virtually impossible.[41]

From the point of view of the LRD researchers trying to thread their way through all of those obstacles to a successful detection system, it was evident there were three very important requirements for fulfillment of the LRD mission: (1) to determine the best time and place for the collection of adequate samples; (2) to detect and collect a sample of the bomb debris; and (3) to establish the atomic nature of the sample. The first requirement was largely dependent upon geophysical techniques. Pressure waves transmitted through the atmosphere and seismic waves transmitted through the Earth's crust to suitably located geophysical instruments would establish the time and place of the explosion; and if the blast should occur underwater, the Navy's SOFAR (i.e., hydrophones) system would accomplish the same end.[42] The LRD personnel considered the development of these techniques as essential to determining the best time and place for the collection of samples. They believed highly sensitive air conductivity instruments and Geiger counters, compensating for cosmic ray background, would lead to the most active part of the atomic cloud. Airborne and ground filters as well as precipitation collection equipment were considered for collection of particulate products. Rare gas collection equipment, ground or airborne, was also needed to collect gaseous products. By 1948, radiochemical techniques used in CROSSROADS and SANDSTONE seemed very promising as a means of establishing the atomic nature of an explosion and its sample.

Despite all of these initial challenges, the Air Force aggressively implemented its long-range surveillance mission. On June 1, 1948, two weeks after Shot Zebra in Operation SANDSTONE, the AWS began to systematically fly filter-equipped WB-29s as routine missions between Alaska and Japan. This was the birth of the Atomic Energy Detection System initially referred to as the "Interim Research Network" or simply the "Interim Net." It began with 22 northern hemisphere ground stations and 24 AWS units for aerial filtering. The ground stations were equipped with automatic recording counters, ground filter units (GFUs), and wraparound Geiger counters. Four of the 24 AWS units were stations basing the WB-29s used for filtering operations. Those bases were located in Alaska, Bermuda, California, and Guam. Because the newly formed AFOAT-1 lacked experience in command and control, the Assistant for Atomic Energy, DCS/O, delegated the responsibility for operational control of the entire Interim Net to the Military Air Transportation Service (MATS) on October 26, 1948. He stipulated, however, that MATS would coordinate all major plans and procedures with AFOAT-1 to ensure operational efficiencies would not hinder AFOAT-1's R&D efforts. This reorganization also included a new Filter Center run by AFOAT-1's collection and meteorological sections.

In November 1948, the Air Force completed an upgrade to Shemya AFB in the Aleutian Islands. These improvements permitted the 375th Weather Reconnaissance Squadron to locate there from Ladd AFB to conduct airborne sampling operations, flying seven sorties every 48 hours. With this move, the immediate interim net surveillance requirements were fulfilled.

AFOAT-1 handed over operational control of the Filter Center on January 25, 1949, and formally transferred 16 airmen to AWS. By that time, there were 20 stations and 55 filter-equipped RB-29 AWS aircraft flying routine sampling missions from Guam to the North Pole. In total, the Interim

Net now employed 1,342 personnel in direct and secondary support roles.[44]

The year 1948 was a fruitful year for LRD research. Throughout the year, a large body of scientists extensively studied the data sets from CROSSROADS and SANDSTONE to examine all aspects of the LRD problem, with special emphasis placed on the detection of nuclear debris in the atmosphere and at great distances. In sum, by the end of 1948, the United States was well positioned to develop the initial LRD techniques of acoustic, seismic, and airborne sampling in order to establish an effective, fully functional AEDS network by mid-1950. The only thing standing in the way was the issue of adequate funding.

Shot Baker at Operations Crossroads - July 25, 1946

The second of 194 U.S. nuclear tests conducted during the 1940s and 1950s. From the beginning, radiochemical analysis of detonation debris would prove to be the most reliable LRD technique. Unfortunately, emphasis on this AEDS technique delayed important R&D for several geophysical techniques. *Photo courtesy of the Department of Defense*

Crew removes debris sample from B-17 drones at operation SANDSTONE in 1948.

Photo courtesy of the Harry S. Truman Library

**Lieutenant General
Roscoe C. Wilson**

The most knowledgeable general officer on nuclear science in the late 1940s. Wilson served on the critical oversight organizations responsible for the creation and implementation of the AEDS. Quietly, behind the scenes, he served as a tremendous champion of AFOAT-1 for more than a decade.

Photo courtesy of the United States Air Force

Oppenheimer and Groves at the Trinity site - September 1945
Photo courtesy of Los Alamos National Laboratory

Notes to Chapter 2

[1] The terms seismograph and seismometer are frequently used interchangeably. By convention, a "seismometer" is the sensor which consists of an inertial mass, a spring based suspension system, and a frame which holds everything together and is fixed to the Earth (usually on a massive pier). The "seismograph" is the combination of the sensor and the recording system. That seismic data was recorded on paper (often photographic) mounted on a rotating drum. The paper printout was called a seismogram. Courtesy Dr. Mark Woods, AFTAC.

[2] J.W. Blair, D.H. Frisch, and S. Katcoff, "Detection of Nuclear-Explosion Dust in the Atmosphere," October 2, 1945, https://fas.org/spg/othergov/doe/lanl/docs1/00423503.pdf, last accessed February 3, 2019.

[3] Leland B. Taylor, *History of Air Force Atomic Cloud Sampling* (Air Force Systems Command, January 1963), 3. The Joint Chiefs of Staff formed Task Force 1 to conduct CROSSROADS. Because the Navy comprised 90 percent of the task force, command of JTF-1 went to Vice Admiral William H.P. Blandy.

[4] *The Roswell Report: Fact versus Fiction in the New Mexico Desert* (Headquarters, United States Air Force, 1995), 2-3. Ewing was the co-inventor of the Press-Ewing seismometer that was initially used in the AEDS.

[5] L. Berkhouse, S.E. Davis, F.R. Gladeck, J.H. Hallowell, C.B. Jones, E.J. Martin, F.W. McMullan, and M.J. Osborne, *Final Report of Operation Crossroads- 1946* (Washington, DC: Defense Nuclear Agency, May 1984), 1-2, 63, 78-81. For Wilson biography see http://us.af.mil/About-us/Biographies/ Display/Article/105188/lieutenant-general-roscoe-c-wilson/.

[6] Not much is known about the details of Wilson's work and support. However, it is clear he played a huge role behind the scenes in AFOAT-1's birth and early progress.

[7] L. Berkhouse, et. AI., *Operation Crossroads*. See also Miller, *Under the Cloud*, 76-78.

[8] Ibid. The Navy also collected air samples with a rudimentary single filter paper mounted under the left wing of the F6F drones.

[9] Paul Kruger to Leslie Groves, "Remote Air Sampling," September 18, 1946, LANL Archives. The USAF did not possess its

own radiochemistry lab at this point but would soon partner with Tracerlab, Inc. as its sole radiochemistry lab for testing debris collection samples (see Chapter 4).

[10] Miller, *Under the Cloud*, 78-79.

[11] A water burst causes more condensation which then produces more local fallout. An airburst causes more far-field fallout. Berkhouse, et. Al., *Operation Crossroads*, 1-2.

[12] Carl Romney, *Detecting the Bomb: The Role of Seismology in the Cold War* (Washington, DC: New Academia Publishing, 2009), Chapter One.

[13] Memo from Johnson to Hegenberger, Subject: "Preliminary Estimate of Long Range Detection Problem," January 15, 1948, AFTAC archives.

[14] Note that after the next test series, LRD experimental programs would not adopt an operational code word.

[15] Technical memo #2, Johnson to Hegenberger, "Subject: Research Projects for FITZWILLIAM," January 21, 1948, AFTAC archives; Technical memo #3, Johnson to Hegenberger, Subject: "Status of Personnel and Training for the Radiological Measurements of FITZWILLIAM," February 18, 1948, AFTAC archives; See also *Report of Operation Fitzwilliam* (Washington, D.C.: Department of The Air Force, Headquarters United States Air Force, October 1952). DTIC accession number ADA598183.

[16] J.E. Hull, Report to the Joint Chiefs of Staff, "Atomic Weapons Tests, Enewetak Atoll, Operation Sandstone, 1948," June 16, 1948, DTIC accession number: AD-8316210.

[17] Ibid., L.H . Berkhouse, J.H. Hallowell, F.W. McMullan, S.E. Davis, C.B. Jones, M.J. Osborn, F.R. Gladeck, E.J. Martin, and W.E. Rogers, "Operation SANDSTONE: 1948," Technical Report (Washington: Defense Nuclear Agency, December 19, 1983), 42.

[18] Ibid., 1.

[19] Ibid., 105-107, 153.

[20] Ibid., 114-117.

[21] Ibid., 90. Also Hull, 37. The non-Air Force experiments were the U.S. Coast and Geodetic Survey's seismic observations and the Navy's hydroacoustic observations. These experiments yielded promising results and would soon expand as strong LRD techniques.

[22] Technical Memo #9 from Urry to Johnson, Subject: "Evaluation of Results of Long Range Detection of SANDSTONE by Radiological Methods to April 24, 1948 Inclusive," April 27, 1948, AFTAC archives. Also, Technical Memo #10 from Northrup to Johnson, Subject: "Evaluation of Sonic, Seismic and Other Methods of Long Range Detection of SANDSTONE Test X-ray," April 27, 1948, AFTAC archives.

[23] Technical Memo #12 from Johnson to Hegenberger, Subject: "Preliminary Results in Long Range Detection of Atomic Bombs" April 28, 1948, AFTAC archives.

[24] The magnetic effects experiments involved the use of magnetometers to conduct aerial surveys in the vicinity of the blasts. The ground observations experiment used an electron multiplier type of photo tube. Scientists believed that under ideal conditions light variations would be of sufficient magnitude to be detected with this instrument. USAF, "Report of Operation FITZWILLIAM," Vol. 1, p. 7, AFTAC archives.

[25] Technical Memo #12.

[26] See the Air Weather Reconnaissance Association website at http://www.awra.us/qalleryfeb07. html, last accessed September 11, 2018. Fackler, later promoted to colonel, would serve as AFOAT-1's Chief of Operations in 1958 and 1959. [27] Technical memo #15 from Johnson to Hegenberger, Subject: "Proposals for Radiological Surveillance and Interim Net," May 14, 1948, AFTAC archives. The term initially used to describe the first effort at establishing the AEDS.

[27] Ibid., Note: this assurance was the only sentence in his report that he underlined.

[28] Ibid.

[29] Meeting minutes of the Acoustic Technical Working Committee, July 1948, AFTAC archives.

[30] Ibid.

[31] Emphasis is Johnson's. Technical Memo #16, Johnson to Hegenberger, July 2, 1948, AFTAC archives. At the same time, AMC was deficient in supplying spare parts for the Interim Net. Consequently, AWS and AFOAT-1 worked around AMC by ordering duplicate parts and cannibalizing existing equipment. Also, Automatic Recording Counters at four stations had to be withdrawn to service other priority stations.

[32] Johnson's wording in several of his technical memos reveal his belief and fear that a Russian surprise attack was a real possibility. In this regard, he echoed the Air Force's strong opposition to Eisenhower's quest for an arms control agreement.

[33] Technical memo #17 from Johnson to Hegenberger, Subject: "Tentative Research Programs for Long Range Detection," July 2, 1948, AFTAC archives. Note that all but one page was a very detailed budget for R&D that encompassed all LRD stakeholders, both military and civilian, and their individual research projects.

[34] Technical Memo #16, Johnson to Hegenberger, July 2, 1948, AFTAC archives.

[35] Ibid, emphasis added.

[36] Meeting minutes of the Acoustic Technical Working Committee, October 12, 1948, AFTAC archives. Hallan's first name is unknown.

[37] Ibid.

[38] *History of Long Range Detection: 1947-1953*, Chapter II.

[39] Ibid.

[40] Ibid.

[41] Ibid.

[42] SOFAR (Sound Fixing and Ranging), was a system (using hydrophones) in which the sound waves of an underwater explosion were detected and located by three or more listening stations.

CHAPTER 3

How Important Is LRD?

R&D, of course, was the real expense associated with establishing an effective AEDS. From July 1948 to August 1949, Hegenberger, Johnson, and Northrup were embroiled in a series of intense budget battles. Receiving a strong sense of urgency from senior policymakers and his own service, Hegenberger had pressed Air Force Chief of Staff General Vandenberg, and Secretary of the Air Force W. Stuart Symington to encourage the Secretary of Defense, James V. Forrestal, to articulate the priority of LRD. In his July 20, 1948 memo to Vandenberg and Symington, Hegenberger stressed the need for Forrestal to broaden the mission and to emphasize the priority of LRD. Unbeknown to Hegenberger, Symington was already applying pressure to do so. In a July 14, 1948 memo to the chairman of the MLC, Symington clearly solicited a restatement of the LRD mission and its (assumed) high priority. "The long range detection mission [should] be restated so as to . . . emphasize the priority of the mission by stating its purpose, i.e., to provide *earliest possible warning of possession* by another power of a successful atomic bomb. . . ." Symington also emphasized that the restated mission and priority should come from the Secretary of Defense, and the Air Force would have complete authority over wide-ranging resources both within the Department of Defense and any necessary civilian resources.[1] However, major international events and serious inter-service jurisdictional and budgetary fights were stalling the momentum of Hegenberger's LRD program.

Vandenberg and Symington had their hands full. Three weeks earlier, on June 24, 1948, the Soviets blockaded Berlin. Tensions had built over the preceding months directly related to the "German Question." That is, how would the four occupying powers reunite Germany and ensure that its new form of government would be incapable of aggression in the future? Unfortunately, as the Cold War escalated throughout 1946 and 1947, a viable agreement between the former allies slipped beyond the realm of possibility when Soviet ideology took hold over Eastern Europe and as Western economic recovery plans progressed. In fact, only four days prior to the Berlin blockade, France, Britain, and the United States introduced the Deutsche Mark as West Germany's official currency—an act the Soviets viewed as solidifying a permanent division of Germany. Now Air Force Chief of Staff (Carl Spaatz had retired on April 29, 1948), Vandenberg oversaw the massive airlift into West Berlin, a time-consuming command responsibility. The blockade would endure for another eleven months, and when it ended on July 29, 1949, the Air Force had hauled four and one-half million pounds of food and fuel. Despite the demands of the airlift, Vandenberg remained tightly engaged with AFOAT-1 and the need to articulate the highest priority for LRD within the JCS.[2]

Simultaneously, Vandenberg and Symington battled their Navy and Army counterparts over roles, missions, and budgets. Although they all had collaborated quite effectively for the FITZWILLIAM experiments in SANDSTONE, the current fights— the most contentious in the history of the joint chiefs—reflected fears derived from extreme uncertainties over the possibility of war with the Soviet Union. Truman's policy of containment, which called for a mixture of conventional forces and the unspoken threat of American nuclear weapons, placed unattainable demands on the service chiefs as they attempted to formulate a feasible wartime contingency plan. When the Truman Doctrine was announced in May 1947, the AAF had reduced manpower strength from 2.25 million personnel to 300,000, and combat ready air groups had fallen from 218 to 2.[3] Simultaneously, even in the face of Soviet aggression over Berlin, Truman held firm to his austerity program. As the blockade was underway, Truman placed a limit on the defense budget of 14.4 billion dollars

(2021 = \$172 billion)—less than half of what the services claimed they needed as a *minimum*. The conflict became uglier. As Vandenberg's biographer Phillip Meilinger noted,

> What made the clashes so violent was the frustration, almost desperation, caused by budget cuts that left each service feeling it was ill-prepared to carry out its mission. Soldiers, sailors, and airmen fought bitterly because they believed they were fighting for their institutional lives.[4]

As Vandenberg observed the Air Force becoming the primary armed service under an evolving foreign policy undergirded by nuclear diplomacy, he found it difficult to compromise, especially when (not if) the Soviets acquired nuclear weapons. Indeed, U.S. dominance was beginning to emerge. Only three weeks after the Soviets blockaded Berlin (and two days before Hegenberger sent his July 20th memo to Vandenberg and Symington), Vandenberg deployed two Air Force groups comprised of sixty B-29s, now capable of dropping nuclear bombs, to their new bases in the United Kingdom.[5]

Not deterred by the naysayers of LRD nor waiting for Forrestal to act on his recommendations, Hegenberger pressed forward with a robust R&D plan and FY 49 budget proposal on August 13, 1948.[6] Although Hegenberger and Vandenberg expected a reduction in funding any day, they both agreed the Air Force should confirm the planned \$30 million LRD program budget "in principle," pending confirmation by the RDB and the MLC.[7] The 79-page plan constituted the first major R&D plan encompassing all of the major scientific techniques. The plan emphasized the urgency in fielding an effective AEDS by January 1, 1950. The stated goal was to perfect the techniques and instrumentation used in FITZWILLIAM, with the understanding that unfavorable conditions would exist in a real-world Soviet detonation. The plan assessed that the radiological techniques were in good shape. Seismic technology, however, required broad fundamental research in wave propagation before instruments could be added to the AEDS. Sonic technology also demanded extensive research in the propagation of subsonic waves through the atmosphere. For the sonic program, testing would require a large quantity of explosives. The goal was to produce effective instruments for the surveillance network. Finally, meteorology required wide-ranging studies of the mechanics of turbulent diffusion in the atmosphere and those of movements of air masses. Each of the plan's research programs outlined numerous sub-programs, the agency of primary responsibility, and projected budgets.

Impressively, the plan also included a prognosis of results. There was full confidence in the nuclear program (i.e., nuclear debris collection and analysis). For acoustics, researchers had confidence out to certain ranges: certainty of detection within 1,000 miles but questionable for ranges between 1,000 and 2,500 miles. The goal of the acoustics program was to develop instruments reaching well beyond 2,500 miles. However, the acoustics scientists appraised only a 25 percent chance of doing so.

For the seismic program, the plan acknowledged it was currently impossible to distinguish underground detonations from earthquakes. They assessed a 50 percent chance of discovering the characteristics enabling them to do so. The subdivision of the $30 million was: nuclear ($4.67 million), seismic ($6.35 million), acoustics ($9 million), and administrative ($6.38 million). Most impressive of the plan was the scope of participation by every stakeholder of LRD. Clearly, many civilian and military organizations had committed their resources and scientists to the urgency of establishing an effective AEDS.[8]

The August 13th documents also contained several recommended letters for Forrestal to sign. These letters specifically directed the Army and the Navy to support the Air Force's LRD program as a high priority. Two other letters were drafts for the Secretary of Defense articulating the same message to the MLC and the RDB.[9]

Five days later, however, on August 19, 1948, the CAE passed a unanimous motion restricting the plan and its associated budget. Attacking what Johnson had always considered essential R&D, the CAE wanted a scaled-down version of seismic research to only determine "whether seismic disturbances can be differentiated from those of earthquakes." If so, then a second research program could proceed. The CAE believed a delay of a year would be acceptable in establishing a seismic surveillance network. For acoustics research, the CAE all but abandoned any R&D program. "[S]ince the sonic method shows so little technical probability of success at long ranges and is so ambiguous and has so few virtues in comparison with the nuclear method of detecting air bursts, it does not at the present time justify a major effort." The CAE then forwarded the resolution letter to Forrestal and the RDB for "necessary action."[10]

On August 27, 1948 (one day before the activation of the 1009th SWS/AFOAT-1), Symington approved the LRD budget for FY49 and directed Vandenberg to implement those phases of the LRD program requiring immediate action. It was now absolutely clear that the LRD program occupied a high priority in Air Force planning. As a result, AFOAT-1 personnel began initiating certain research projects deemed feasible, albeit within the parameters of an expected budget cut, and at the same time began preparations to modify the seismic and acoustic research programs as the CAE recommended. They formulated three programs for the operation of a routine surveillance system: radiochemistry (debris sampling), acoustic, and seismic. Although the AWS had already initiated airborne sampling operations to fly the Alaska–Japan route in June 1948, the first program — radiochemistry — called for a fully operational airborne detection system by January 1, 1950. The second, emphasizing an acoustic research program, specified a target date of

January 1, 1951. The third, which called for the positioning of a worldwide system of seismic stations, anticipated an operational date of October 1, 1952.[11] When Hegenberger submitted the three plans to the CAE on September 23, 1948, he also requested the committee defer a detailed fiscal review until the JCS issued its guidance on the importance and priority of LRD. The CAE concurred.

Simultaneously, Hegenberger asked for an impartial panel comprised of nuclear, acoustic, and seismic experts to evaluate the proposed LRD program and to furnish continuing advice on their subjects. The CAE conditionally approved the requests and passed them on to the RDB. On September 28, 1948, the RDB proposed to the JCS a modified LRD program to serve as an interim guide pending further evaluation. The RDB stated an atomic energy detection system would be effective by January 1950, but the seismic network would be delayed for another year and only a minimum of acoustic work would be carried on until its desirability and feasibility was established. The RDB urged the JCS to provide, as soon as possible, guidance concerning the military value of the LRD program the Air Force envisioned.

While Hegenberger awaited the JCS decision (not rendered until six months later), he continued to push the LRD program forward. In mid-October 1948, the RDB established an LRD Ad Hoc Scientific Panel to consider long range detection problems. The chairman was the renowned physics researcher Dr. Alfred L. Loomis, who was assisted by three experts: Charles P. Boner of the University of Texas (nuclear), L. Don Leet of Harvard University (acoustic), and Joseph C. Boyce of MIT (seismic). The "Loomis Panel," as it was generally called, met in New York on October 27, 1948, where AFOAT-1 representatives presented a modified LRD plan.

Two days later, the RDB's Committee on Geophysics and Geography reviewed the modified seismic and acoustic programs, and resolved that there appeared to be an adequate scientific approach to the problem of LRD with these technologies. In their opinion, an R&D program was justified given the purported high priority. The committee recommended the Air Force begin the implementation of a comprehensive R&D program to determine the feasibility of utilizing seismic and acoustic techniques in an LRD surveillance system. AFOAT-1 now had complete authority to do so. On October 22nd, as the Loomis Panel began its work, Hegenberger finally received total control over all R&D. Johnson's criticism of AMC's poor management of LRD R&D was now vindicated. Unfortunately, Johnson was not present to celebrate, having departed AFOAT-1 the previous month, fed up with budget constraints.[12]

On December 4, 1948, the Loomis Panel formally submitted its report to the CAE. Importantly, the panel noted that identifying the exact date of a Soviet detonation was less important than gaining knowledge from nuclear research. While this view was certainly correct for the development of the AEDS, it served to reduce the sense of urgency to accelerate the LRD program. In fact, the panel presented two plans, anticipating a JCS decision of either a high or a medium priority for LRD. The only difference was the total cost of implementation. Both plans emphasized airborne collection over seismic and acoustic research.[13]

On December 17, 1948, the CAE passed a resolution accepting the panel's report with the provision that the panel could reconvene based on the impending JCS priority decision. The chairman of the CAE, James Conant, received authorization to accept, at his discretion, such changes in the handling of specific LRD projects as were approved by Dr. Loomis. In essence, the *ad hoc* panel limited the LRD R&D program to basic theoretical and experimentation investigations of relatively low cost. In their view, a slower pace with reduced costs would not impede a July 1952 LRD operational date.

Consequently, the CAE/RDB approval of the panel's report resulted in the cancellation of many of the LRD acoustic and seismic research projects. A dramatically reduced budget drove those cancellations. Although Hegenberger had unilaterally lowered the $30 million budget to $22.6 million, the panel's report further reduced program funds to $11 million. In January 1949, the program began the calendar year with a funding of $10.6 million. Hegenberger hoped the reduced program could at least provide the instrumentation and expertise for test explosion programs scheduled for 1950 and 1951. Budget planning for FY51 did not look much better. In February 1949, Conant approved a budget of only $13.7 million for FY50 (2021 = $158 million).

Facing the reality of fiscal constraints and the power of the CAE, Northrup reluctantly accepted most of the Loomis recommendations. On December 20th, he gently refuted some of their conclusions and recommended several modifications to the LRD R&D programs. Having nothing to lose, he also recommended some modest increases in funding for several seismic and acoustic sub-programs, hoping to gain some traction in those sciences.[14]

Air Force senior leaders did not readily accept the budget reductions. On December 31, 1948, Vandenberg informed the JCS that the RDB's actions made it impossible for AFOAT-1 to effect a fully operational AEDS by mid-1950. Vandenberg once again requested the JCS confirm the importance of an accelerated R&D program for LRD in order to place the detection system into operation no later than mid-1951, the possible date of a Russian nuclear test.

Seeing that Vandenberg's memo failed to solicit a prompt response from the service chiefs, Hegenberger accepted the budgetary restrictions as an interim decision. On February 2, 1949, he modified the LRD program to reflect three phases lasting through FY51 (i.e., ending June 30, 1952). Phase A, the remaining seven months of FY49 (1950), would operate on a $12 million budget. Phase B aligned with FY50 (July 1950–

June 1951) and would make the Long Range Experimental Propagation Study its priority, with funds not exceeding $14 million. Phase C placed emphases on the design, procurement, testing, and installation of final detection instrumentation. However, Hegenberger offered no budget estimate for FY51.[15]

Meanwhile, Hegenberger's primary radiochemistry laboratory, Tracerlab, Inc., informed him that further analysis of the SANDSTONE filter papers revealed that most of the radioactive particles discovered consisted of iron from the vaporized steel towers, which had housed the weapons. This discovery cast serious doubt on the efficiency of radiochemical methods of detection, which meant both the acoustic and seismic methods might be relatively much more important than many had believed. Hegenberger immediately requested a reevaluation of the Able shot data from CROSSROADS as a means of comparison.

On February 23, 1949, Hegenberger informed the RDB of Tracerlab's findings and their caution about placing too much emphasis on the nuclear method alone. This, he argued, was why the AEDS must include geophysical detection techniques. After reviewing the information available, the RDB requested further technical data and resubmitted the entire LRD problem to the CAE and the Committee on Geophysics and Geography. At the same time, the RDB again requested the JCS to issue its guidance on the strategic importance of LRD.

On March 7, 1949, the Air Force succeeded in persuading the RDB to convene a new joint committee of experts to review the LRD problem in light of the strategic importance of LRD as well as its technical aspects. The committee met on March 17th to consider the technical differences of opinion between the Air Force and the RDB. The discussion focused on whether or not the Air Force could adequately advance its geophysical R&D program within an experimental study of wave propagation in the atmosphere and lithosphere. The attendees identified five required scientific objectives for the proposed R&D program:

1. To determine the criteria for distinguishing between a natural and an artificial source of seismic waves.

2. To study the effect of range on the reliability of these criteria in determining the nature and location of the source.

3. To study the physical properties of the medium for long range propagation of acoustic waves.

4. To determine the effect of the physical properties of the medium for long range propagation of acoustic waves upon the reliability of acoustic transmission.

5. To evaluate the capabilities of competitive seismic and acoustic instrumentation developments currently underway.[16]

Unfortunately, the attendees reached no conclusions but agreed to reconvene on April 7, 1949. They once again urged the JCS to set a priority for LRD and this time stated the RDB required a decision before the April meeting in order to meet the adjusted 1952 operational timeline for a fully functional AEDS.

Vandenberg was growing impatient. On March 23, 1949, he again reminded his fellow service chiefs that the Air Force had prepared a program to accomplish its LRD mission but had been unable to proceed in an orderly fashion because the JCS could not agree on the urgency for an LRD program. He also told his colleagues there was significant disagreement within the RDB concerning the technical merits of the seismic and acoustic portions of the LRD Program. He then advocated for an accelerated R&D program to begin at once.

Northrup immediately provided Vandenberg with new data and a capabilities estimate to reinforce Vandenberg's efforts to push his fellow chiefs into action. Within 24 hours, he produced a concise eight-page report underscoring the urgency of a strong geophysical R&D program. He first summarized a CIA estimate, backed by the Joint Atomic

Energy Intelligence Committee (JAEIC) and the Joint Intelligence Committee (JIC) of the Joint Chiefs of Staff, who concurred with AFOAT-1's prediction on when the Soviets would test their first nuclear device. That estimate predicted a possible mid-1950 date, with concurrence from British and Canadian atomic energy experts. Northrup then described the pristine, ideal conditions under which the previous U.S. tests had operated and warned that no one should expect any of those conditions to be present in a Soviet test: "It cannot be too strongly emphasized that in the event of a foreign atomic test, these factors will not be known except by techniques and instrumentation provided by the surveillance system." Finally, he cautioned that the oversight committees were wrong in their assessments: "This office is informed of an erroneous and highly optimistic estimate within reviewing agencies of present capability to detect a foreign air burst. Any such estimates have no basis in fact." This memo was impactful because it clearly aligned urgent mission requirements with capabilities (or lack thereof).[17]

While it is unclear what impact Northrup's memo precipitated a reply from the JCS, the chiefs forwarded their stated position only four days later (on March 28, 1949) to the RDB. Their response hinted at the severe limitations they faced under Truman's austerity program. While they admitted the state of the Soviet nuclear weapons program could possibly be inaccurately assessed with existing technical capabilities, they believed that a time lag of one to two months between an atomic explosion and its discovery was strategically acceptable. The chiefs also stated that while knowing the location of such an explosion was desirable, it was not essential beyond verifying the detonation had occurred in the USSR or the Soviet sphere of influence. The JCS statement was not as strong as AFOAT-1 had hoped. Still, it was sufficient to justify the full-scale implementation of seismic and acoustic R&D as well as the test explosion program.

The JCS decision altered the views of the RDB. Meeting the day after the decision, on March 29, 1949, the RDB stated the LRD program must have the requirement to determine the approximate time and place of foreign atomic explosions. They also admitted that seismic, acoustic, and underwater methods appeared to be the only means of obtaining the desired information. In order to further LRD, they decided to cancel the meeting of the Joint Committee of April 7th and to establish a joint panel to review the whole problem. Dr. Charles P. Boner, who had served on the Loomis Panel, headed the new committee, which was immediately known as the Boner Panel.

The belated JCS approval on the LRD program, however, impeded the original plans for R&D, and by the spring of 1949 it was evident to Hegenberger that the program could not utilize the available $12 million in the short time remaining within the current fiscal year. Consequently, AFOAT-1 returned $3.5 million to the budgetary reserve and retained $8.5 million for the LRD program. At the same time, Hegenberger was notified that the FY51 budget was approved for only $12 million. Consequently, he was once again forced to re-prioritize the LRD R&D program and delay some high priority research.

On June 3, 1949, Hegenberger re-issued the objectives of the revamped R&D program. Strategically, the program would focus on how to determine the optimum location and time for the collection of samples of explosion products; to detect and collect samples of explosion products for analysis; and to establish the atomic nature of the sample. To achieve these capabilities, Hegenberger listed eight technical and scientific objectives. These were:

> 1. To develop instruments and techniques for recording pressure waves transmitted over long distances through the atmosphere, hydrosphere, and the lithosphere and investigate natural disturbances in these media.

2. To study the long range propagation of pressure waves through the atmosphere, hydrosphere and lithosphere and ascertain the limitations on accuracy, reliability and discrimination of the recording instruments.

3. To determine the exact physical and chemical properties of the particulate matter disseminated by the explosion.

4. To determine the mechanism of transport of explosion products through the atmosphere and hydrosphere.

5. To develop techniques and equipment for detecting and collecting suitable samples of explosion products from the atmosphere and hydrosphere.

6. To develop techniques for concentration and analysis of explosion products.

7. To proof test all instruments and techniques developed for the realization of the technical objectives.

8. To supervise all the work done in connection with the technical objectives.[18]

At the end of FY49 (June 1950), R&D on these objectives remained in a preliminary stage. Researchers had developed new or improved instruments and techniques to record pressure waves, but further progress depended on the proposed test explosion programs. Previous RDB actions had limited the study of long-range propagation of pressure waves in favor of studying explosion sites and explosive materials. In the nuclear program, researchers had begun to determine the properties of particulate matter and the mechanism of transport of explosion products through air and water. They had already made some progress developing techniques and equipment for collecting and analyzing samples of explosion products from the atmosphere and hydrosphere. However, this work remained in the laboratory research and experimental stage. The full attainment of the scientific objectives required an atomic test without the possibility of extraneous materials such as vaporized shot towers.

Consequently, AFOAT-1 prepared its FY50 budget with more than 50 percent of its funding obligations scheduled for nuclear, seismic, and acoustic test explosions as indicated by the following table (which includes the FY49 obligations by way of comparison):

PROGRAM	FY 1949	FY 1950
Geophysical Instruments and Techniques	3,194,000	965,000
Long Range Wave Propagation	74,700	*4,802,900
Nuclear Researches	2,873,500	1,790,000
Total Research Cost	6,142,200	7,557,900
Research Contingencies	333,400	350,000
Administration and Overhead	794,370	695,000
FITZWILLIAM Expenditures	1,230,030	* * *
Improvement of Instrumentation FY 1951 Tests	* * *	800,000
Total with Additional Costs	8,500,000	9,402,900

* Note the huge shift in funding allocation within geophysics R&D toward Long Range Wave Propagation. That was the key to solving detection at very long ranges.

19

DoD-wide opposition to LRD remained stubborn through the spring of 1949, despite the JCS memorandum of March 28th articulating the high priority of establishing an AEDS. At its meeting of June 1, 1949, the CAE reviewed the AFOAT-1 budget for FY50 ($9.4 million) and recommended the RDB withhold $5.4 million slated for geophysical research until such time as the Boner Panel could pass judgment on the program. At the same time, the CAE proposed that the $20 million total cost of the entire LRD program could be better spent in other fields of R&D. Vandenberg strongly objected to both recommendations. Again, the future of LRD geophysical R&D programs was thrown into a state of uncertainty.[20]

On July 12, 1949, the Air Force members of the RDB presented their formal non-concurrence of the CAE recommendations. They noted only a thorough study of pressure wave propagation at great distances would lead to the development of accurate and reliable instruments capable of forming an effective AEDS. Hegenberger had submitted his aforementioned proposal to accomplish that task to the CAE in August 1948, but the RDB had neither approved such a program nor advised the Air Force as to how the mission could be accomplished without such an expensive study. The generals on the committee argued an atomic detonation was a unique event, and accurate detection therefore required unique detection capabilities. They reminded the committee the Air Force was charged with developing a viable system that could reliably detect the first Soviet test. Time was running out, they stated, because some experts had predicted the Soviets could test as early as mid-1950. In their conclusion, the generals stated "[t]his Board, by not taking positive and timely review action in this case, is in the untenable position of hampering, or even preventing, the accomplishment of an urgent operational mission of great importance to the national security." The RDB conceded to some extent and authorized the Air Force to proceed "conditionally with the advanced studies on LRD" pending the final recommendations of the Boner Panel.[21]

By mid-August 1949 (unbeknownst to anyone that the Russians were only two weeks away from testing), Northrup had all but dismissed any hopes that an effective acoustic network would be in place before an anticipated Soviet test (conceivably within a year or so). On August 15th, he convened a meeting of the Acoustic Technical Advisory Group. During this meeting he reviewed the current views of the RDB, CAE, and the Bonner Panel. He vented, unnecessarily given the expertise of those present, that an effective AEDS had to include geophysical techniques. Only those techniques, he said, could provide near real-time confirmation of a detonation, and that the blast occurred at a given time and place. Acoustic methods "now appear promising" were best for air bursts, and seismic methods were best for underwater or underground detonations. Northrup acknowledged that in light of the delays and huge cutbacks, their limited R&D efforts were now forced to work in tandem with the limited operational systems currently in place.[22]

By this time, Hegenberger had retired from the Air Force (August 31st) and was replaced by his deputy, Brigadier General Morris Nelson.[23] Nelson had been with Hegenberger since January 1948, prior to the activation of the 1009th SWS/AFOAT-1, and was well versed in all aspects of Johnson's and Northrup's struggles with getting LRD off the ground. Nelson graduated from West Point in 1926 and became a pilot in the Air Service. During the Second World War, he first served in Washington as President of the Air Defense Board and Chief of Air Defense. With an assignment to the Mediterranean theater, he commanded a bomber wing from August 1944 until the end of hostilities. Returning to Washington, Nelson became Assistant Chief of Staff for Personnel and then commanding general of the Panama Air Defense Wing (later re-designated Sixth Fighter Wing) before joining Hegenberger to create the LRD program.[24]

Throughout the summer of 1949, the Boner Panel reviewed the latest version of the Air Force LRD program and then submitted a devastating report to the RDB on September 1, 1949. In short, the panel recommended "the Joint Chiefs of Staff be requested to reevaluate the necessity for detecting a foreign atomic explosion by instrumental means, with a view to cancelling the research and development program on LRD." The panel recommended the Eisenhower directive charging the Air Force with the overall responsibility for LRD be modified to recognize the limited capability of detection by instrumental means. Somewhat contradictorily, they urged the JCS to expedite the creation of an AEDS. In their view, however, the panel believed an AEDS was possible with current equipment and techniques. In terms of budgetary support, the panel recommended a ceiling of $4.8 million for LRD R&D, and those monies should come out of the Air Force's overall R&D budget.[25]

The effect of the Boner Report was immediate. DoD's austerity program was still in full force, and on September 14th, the Secretary of Defense's management committee decided to limit the LRD program to the maximum recommended by the Boner Panel. A memorandum to this effect was forwarded to the Air Force on September 16th, decreasing the FY49 program by $1.7 million and the FY50 program by $4.9 million.[26] No one involved in the budget decisions during those first two weeks of September knew that the Soviet Union had just detonated its first atomic weapon. On September 1st, less than 48 hours after the release of the Boner Panel report, USAF AWS crews and AFOAT-1 personnel collected proof of that historic event. In his previous memo to Hegenberger on February 2nd, Northrup had been quite prescient in predicting the effects of a sudden, unexpected Soviet test.

> The electrifying news that Russians had unquestionably exploded a successful atomic bomb would knife through the less essential activities and prod this country into ·a rate of preparedness for atomic war greater than the present rate by at least an order of magnitude."[27]

Now, seven months later, he was proven right. Everything about LRD was about to change—literally overnight.

Dr. James B. Conant

Chemist and former president of Harvard University. Conant chaired the powerful Atomic Energy Committee (AEC) which exercised extensive oversight of LRD activities. Conant was strongly opposed to the development of the hydrogen bomb. A close associate of Vannevar Bush, Conant advised Truman on the use of the atomic bomb and was present at the Trinity test in July 1945.

Photo source: public domain - GICOG, Information Bulletin, the monthly magazine of the Office of U.S. High Commissioner for Germany. Issue March 1953.

Alfred L. Loomis

Played a key role in the development of radar and the atomic bomb. In 1948, he chaired the Loomis Panel which concluded that the exact date of a Soviet detonation was less important than obtaining knowledge from nuclear research. Albeit short-lived, the Loomis Report sought to diminish the sense of urgency to accelerate the LRD program.

Photo courtesy of the Department of Energy

Dr. Charles P. Boner

Well-respected physicist at the University of Texas who served as the first Technical Director, Office of Government-Sponsored Research. In 1949, after serving on the Loomis Panel, he chaired the Boner Panel that recommended canceling the LRD R&D program. Ironically, the report was immediately discarded as it was released during the Joe-1 alert.

Photo courtesy of the University of Texas

Major General Morris Nelson

Initially Hegenberger's deputy, he assumed command of the 1009th SWS/AFOAT-1 just days before Joe-1. Nelson was especially frank with Vandenberg and Oppenheimer at the Joe-1 verification meeting in regard to the limitations of the young AEDS and AFOAT-1's erroneous estimate that the Soviets were probably years away in developing an atomic bomb.

Photo source: AFTAC archives

W. Stuart Symington

Symington was the first Secretary of the Air Force (1947-1950). In 1948, he strongly pressured the Secretary of Defense (Forrestal), to clearly restate the highest priority for the LRD mission. Occurring at a time of intense budget battles and prior to Joe-1, it took almost a year for the JCS to designate it as such. After the surprise of Joe-1, Symington resigned in protest in 1950 over inadequate funding for LRD.

Photo courtesy of the United States Air Force

James V. Forrestal

Served as the Secretary of the Navy from 1944 until 1947, at which time he became the first U. S. Secretary of Defense. During the initial days of organizing the LRD roles and missions, as Secretary of the Navy, Forrestal qualified his support for the AAF's assumption of the LRD mission. The AAF should do so with the understanding that "the analysis and evaluation phases of the program will make maximum use of resources and facilities of other appropriate agencies."

Photo courtesy of the National Archives

Notes to Chapter 3

[1] Emphasis added. Memo from Hegenberger to Vandenberg and Symington, Subject: "Long Range Detection of Atomic Explosions," July 20, 1948, AFTAC archives. Memo from Symington to MLC, Subject: ""Long Range Detection of Atomic Explosions," July 14, 1948, AFTAC archives.

[2] Curtis E. LeMay and Kantor MacKinlay, *Mission with LeMay: My Story* (New York: Doubleday and Company, Inc., 1965), 416.

[3] Meilinger, *Vandenberg*, 125.

[4] Ibid., 139.

[5] Walker, *Cold War*, 57.

[6] Brigadier General Nelson, Summary sheet and memo, "Proposed Research & Development Program for Long Range Detection ," August 13, 1948, AFTAC archives. Note that Nelson served as Hegenberger's deputy until he assumed command of the 1009th when Hegenberger retired in August 1949.

[7] Six months earlier, on February 11th, Ellis had estimated a total LRD budget of $46 million (2021 = $529 million) for the first two years. In their minds, the hopeful $30 million estimate was already a reduction of their desired budget by 35 percent.

[8] Technical memo #28, "United States Air Force Fiscal Year 1949 Program for Long Range Detection," Dr. Ellis Johnson, August 13, 1948, AFTAC archives.

[9] Major General D. W. Schlatter, Deputy Chief of Staff, Operations for Atomic Energy, "Proposed Research & Development Program for Long Range Detection." August 27, 1948. Note that $30 million was an enormous sum in 1948 (2021 = $343 million).

[10] CAE resolution, August 19, 1948, AFTAC archives.

[11] Memo from Symington to Vandenberg, [no subject], August 27, 1948, AFTAC archives.

[12] Memo DCS/M to CG, AMC, Subject: "Responsibility for Long Range Detection of Atomic Bombs, October 22, 1948, AFTAC archives.

[13] *History of Long Range Detection: 1947-1953*, author unidentified (Washington, D.C: The Air Force Office of Atomic Energy- One, Headquarters, United States Air Force, 1953), 28.

[14] Technical memo #30, from Northrup to Hegenberger, Subject: "Recommendations of Ad Hoc Panel of the Committee on Atomic Energy," December 20, 1948, AFTAC archives.

[15] Technical memo #31, from Northrup to Hegenberger, Subject: "U.S. Air Force Headquarters FY 1949 and FY 1950 Research and Development Program for Detection of an Atomic Bomb," February 2, 1949, AFTAC archives.

[16] Tech memo #32, from Northrup to Hegenberger. Subject: "Technical Difference of Opinion between Air Force and RDB on LRD Program," March 17, 1949, AFTAC archives.

[17] Emphasis is Northrup's. Technical memo #33, from Northrup to Hegenberger, Subject: "Material Requested by RDB in Substantiation of the Air Force Program for Long Range Detection (Tech. Memo. No. 31)," March 24, 1949, AFTAC archives.

[18] Technical memo #34, from Northrup to Hegenberger, Subject: "Research Required to Reach the Objectives of the Long Range Detection Program," June 3, 1949, AFTAC archives.

[19] Technical memo #35, from Northrup to Hegenberger, Subject: "Implementation in FY/49 of Research on Long Range Detection," June 6, 1949, AFTAC archives.

[20] DoD-wide objections to LRD funding continued well into 1950, even after the detection of Joe-1. On August 1, 1950, the Military Liaison Committee to the Secretary of Defense submitted its annual report for fiscal year 1950. Objecting to the high costs of geophysical R&D, they stated that "the maintenance of a detection system designed primarily for determining the time and place of nuclear explosions does not seem warranted." Annual Report of the Military Liaison Committee to the Secretary of Defense for the Fiscal Year 1950," August 1, 1950, AFTAC archives.

[21] *History of Long Range Detection: 1947-1953,* 32-33. AFTAC archives.

[22] Meeting minutes of the Acoustic Technical Advisory Group, September 8, 1949, AFTAC archives

[23] Ironically, Hegenberger retired just hours after Joe-1.

[24] Biography, Major General Morris R. Nelson, http://www.af.mil, last accessed June 4, 2018.

[25] *History of Long Range Detection: 1947-1953,* 33-34. AFTAC archives.

[26] Ibid.

[27] Technical memo #31, February 2, 1949, AFTAC archives.

CHAPTER 4

Joe-1 and the Birth of the AEDS

In May 1949, the AWS established a routine weather reconnaissance track, codenamed "Loon Charlie," from Eielson AFB, Alaska, to Yokota, Airbase, Japan. The 375th Reconnaissance Squadron (RS), Very Long Range, Weather, was just one of many reconnaissance squadrons flying such long-range missions since the first AWS B-29 had flown a weather reconnaissance flight over the North Pole on March 17, 1947 (code named "Ptarmigan"). The 375th RS originally flew the Loon Charlie missions from Shemya AFB in the Aleutian Islands but relocated to Eielson AFB on May 16, 1949. In supporting the fledgling AEDS, the squadron flew filter-equipped RB-29s like those that had participated in the CROSSROADS and SANDSTONE operations. These flights were long, cold, and uncomfortable. The mission flew a 3,600-mile track to Yokota and would then pick up a "turnaround" crew to return to Eielson. The return journey to Alaska was referred to as a "reverse Loon Charlie" mission.[1]

On August 24, 1949, First Lieutenant Robert C. Johnson, the Aircraft Commander at Yokota with Crew 5A of the 375th RS received Special Order No. 113, directing him to conduct a reverse Loon Charlie mission. In addition to Johnson, Crew 5A consisted of First Lieutenant Lawrence E. Paul, Pilot; First Lieutenants Charles E. Massey and Gene A. Culbertson, Navigators; First Lieutenant Robert L. Lulofs, Weather Observer; Master Sergeant James K. Boswood, Engineer; Master Sergeant Thomas R. Richardson, Radar Observer; Staff Sergeant Richard D. Boyce, Crew Chief; Staff Sergeant Steve T. Yapuncich, Radio Operator; Corporal Lauren G. Gackstetter, Dropsonde Operator; and Private First Class William H. Kelly, Radio Operator. Johnson and his crew attempted to fly the mission but experienced two engine failures, first on the primary aircraft and then with a standby aircraft. Although the second failure occurred shortly after takeoff, Lieutenant Johnson was forced to fly to Misawa Airbase, Japan, due to an inoperative ground-controlled approach radar at Yokota. Delayed by several days, Crew 5A finally received an RB-29 (tail number 44-62214) which had conducted a Loon Charlie mission on September 1st.[2]

At that time, Johnson and his crew of 10 were but 11 airmen out of a total of approximately 1,500 AWS personnel dedicated to the operation of the Interim Net (i .e. the AEDS), and their aircraft was one of more than 60 filter-equipped planes available to conduct LRD missions. On the heels of the CROSSROAD tests, the Air Force rapidly built the Interim Net by developing the capability to fly the periphery of the Soviet Union to routinely monitor air masses originating from that continent. They did so with increasing efficiency. With more than 150 air masses moving out of the Soviet Union each month, the AWS committed to flying seven flights every 48 hours. In November 1948, they were capable of intercepting 62 percent of those air masses. By mid-1949, they were consistently intercepting a minimum of 90 percent.[3]

On September 3, 1949, Lieutenant Johnson and Crew 5A took off from Misawa at 0424 Greenwich Mean Time (GMT) to fly the fourteen-hour mission back to Eielson. Flying at 18,000 feet, the crew experienced nothing unusual during the flight. As MSgt Richardson recalled:

> We flew our normal route which roughly paralleled the east coast of Siberia flying straight across to recover at Eielson AFB, Alaska. We landed, turned the filter paper in at debriefing to the processors, and they ran it through their analysis. The next day, someone from operations came and asked for a pick-up crew to fly another mission that night. We flew to Hawaii and again turned the filter papers in at the other end.[4]

What Richardson did not know at that time was that Crew 5A had just detected the Soviet Union's first atomic detonation.

Crew 5A's collection triggered the 112th alert for AFOAT-1 in 13 months. Because all of the previous alerts proved to be false alarms, no one expected anything different from this mission. Many suspected the threshold for triggering an alert—50 counts per minute (CPM), that is, the measurement of ionizing radiation on the filter paper—was set too low given the inaccuracy of the equipment at that time.[5] Still, AFOAT-1 personnel followed proper procedures. They ran the first filter paper through the wrap-around counter and observed a readout of 85 cpm, which initiated the alert. They then quickly processed a second paper that revealed 153 cpm, which they immediately sent to Tracerlab, Inc. at Berkeley, California. That sample would remain in transit for three days.[6]

As night approached on September 3rd, AFOAT-1 personnel plotted the half-life of the decay curve and reported the counting rate each hour from their analysis of the initial sample set. The following day, 24 hours after the collection of the first sample, the half-life of the decay curve was established at about 72 hours. Since the half-life of natural radioactivity on air samples had been running about 10.5 hours, this was the first indication that artificial fission products may have been introduced into the atmosphere. Throughout Sunday and Monday of the Labor Day weekend (September 4th and 5th), AFOAT-1 personnel anxiously followed the bi-hourly decay measurements on the sample. They also requested more airborne sampling sorties.[7]

By this time, AFOAT-1 realized that no other stations nor geophysical techniques in the Interim Net had detected the suspected event. It was soon apparent none of the seismographs or microbarographs in the surveillance network had recorded anything, most likely due to "the remoteness of the present station locations" and the limited capabilities of those inadequate instruments already scheduled for upgrades.[8]

On Wednesday, September 7th, AFOAT-1's George Olmstead, Northrup's deputy as Assistant Technical Director, contacted the Army's Evans Signal Laboratory located near Fort Monmouth, New Jersey, requesting an immediate, thorough review of all acoustic data recorded between August 22nd and September 4th. In speaking with Army Captain John M. Brittain, Olmstead was assured "the records will be reviewed with extreme care in order to pick up anything of significance that may have been missed in the field analysis." Brittain also added that he had just received a report from the station at Fort Lewis, Washington (one of seven acoustic stations in the net), that a suspicious signal lasting three seconds arrived but was only "apparent in the presence of considerable noise."[9] Olmstead called Brittain the following day and learned that most of the acoustic stations had already dispatched their records by courier or air mail. Brittain again reassured Olmstead "that the best trained people at Evans would be used in the continued search of records."[10]

On September 6th and 7th, the ground filter unit located at Fort Randall, Alaska, in the Aleutian Islands, detected particulates. These samples, when compared to other samples by analyzing two fission products with significantly different half-lives, could help determine the date of origin of the detonation within a 10 percent error rate. During this time, AFOAT-1 also sent samples to the Naval Research Laboratory (NRL) and the Los Alamos Scientific Laboratory (LASL) as well as to the United Kingdom.[11] They recognized time was critical. AFOAT-1 understood that the short-lived materials in the samples had to reach the lab within 10 days of the detonation in order to produce the best analysis.[12] The British laboratory at Harwell, though unable to determine the nature of the atomic explosion, reported that the event most likely resulted from "the fast fission of plutonium." While LASL was not prepared to conduct a quantitative analysis for fission products, they did carry out chemical analysis of two airborne filter papers they had received from TracerLab. "The results qualitatively indicate fission products of recent origin" that were certainly less than one month old. The NRL tested a sample from AFOAT-1 and verified "positive evidence of recent explosion from an A-bomb." They delivered an oral preliminary report to Nelson on September 14th, followed by a complete written report two days later.[13]

In the months preceding Joe-1, the NRL had developed another technique to collect radioactive samples. Tested during Operation SANDSTONE and later codenamed Project RAIN BARREL, the NRL experimented with gamma-ray detectors and filter papers attached to pumps on tanks of collected rainfall. In April 1949, they set up a station on Kodiak Island in Alaska, with a building covered by a 1,000-square-foot corrugated aluminum roof that collected the run-off into the tanks. Rain collected between September 6th and 12th triggered strong signals on the gamma-ray detectors. Subsequent analysis "yielded tens of thousands of counts per minute of the major fission product isotopes." Belatedly, they forwarded their report to AFOAT-1 on September 22nd, noting that NRL had tested four isotopes registering over 50,000 counts per minute. However, they could only conclude the detonation could not have taken place prior to August 24th.[14]

Key to the follow-on operations after the initial collection were the predictions of the routes of the air masses flowing out of the Soviet Union. The AFOAT-1 meteorologists were initially concerned about making projections because many weather reports from the USSR were unreliable. Also, they had not received some upper air data from Alaska due to a teletype failure. Fortunately, during the initial days of Joe-1, very little movement or change in wind patterns occurred in the area of interest. The direction of the wind remained constant at 18,000 feet, the altitude of the collections. The meteorologists understood from the U.S. tests in operations CROSSROADS and SANDSTONE that such stationary flow patterns generally gave reliable trajectories.[15]

Based on the initial track of Lieutenant Johnson's historic flight, the meteorologists conducted a backtracking analysis. In looking at the time and location of the hottest sample paper, they narrowed down the general search area to a 700-mile leg. They then subdivided the 700-mile leg into three segments. Based on this information, two additional sorties flew the most likely segment on September 5th. The sortie over Yokota collected samples of 1,000 cpm which gave the analysts a high degree of confidence that the path from the test site was located just north of the Himalaya Mountains. That trajectory was located along a zone of strong horizontal wind sheers (thus keeping the small particles elevated) which helped to explain why effluents were found on so many subsequent flights. Their backtracking efforts resulted in the production of three models with assumed detonation dates occurring between August 27th and 29th. They posited that their 29th model was the most likely (as time and additional data analyses would confirm).[16]

On September 7th, AFOAT-1 received the initial TracerLab report, which positively determined the presence of fission products. Three days later, the lab provided a detailed report offering proof of a Soviet nuclear test. Of all four analytical reports AFOAT-1 received, this report was by far the most thorough, and came closest to estimating the actual time and date of the Soviet test. The quality of analysis came as no surprise to Northrup. He had developed a large degree of confidence in TracerLab since first forming a relationship in February 1948, as planning for FITZWILLIAM was underway. Coming out of SANDSTONE, TracerLab had done a superb job in analyzing the radiochemical experiments. In essence, because AFOAT-1 did not have a lab of its own, TracerLab had essentially become an operational arm of the Interim Net. Their expertise in conducting fission product analysis was due to the number of world-class scientists the company employed. Many of those experts were hired as the lab recruited Dr. Lloyd R. Zumwalt from the AEC's Oak Ridge National Laboratory to establish a West Coast facility at Berkeley, California. By the time of Joe-1, Tracerlab had expanded to more than 250 employees, and almost half of Tracerlab's business was devoted to AFOAT-1. Consequently, when Northrup received Zumwalt's analysis on September 10th, he began compiling a comprehensive report that would eventually inform President Truman of the Soviet test.[17]

The September 10th Tracerlab report was based on nine samples Zumwalt received on September 6th. Eight of the nine samples were dissolved and quantitatively analyzed for activities of seven isotopes. The lab used the ninth sample for decay measurement. They estimated the date of the detonation from the unseparated fission product decay and from the isotopic counting rate ratio decay. The latter measurements disproved the event as a reactor accident—a speculation many shared at that time. This meant the bomb was a plutonium device similar to the American bomb dropped on Nagasaki in 1945.[18]

Immediately upon receiving the Tracerlab report on September 7th, AFOAT-1 and AWS went into action, flying as many sorties as they could muster. In total, between September 3rd and 16th, AFOAT-1 vectored 93 sorties covering an area ranging from Guam to the North Pole, and from California to the British Islands. They collected more than 500 samples, 167 of which registered above 1,000 cpm.[19]

Also on September 7th, AFOAT-1 began disseminating daily status reports. By the following day, the daily report began to reflect the initial onslaught of information flowing into the headquarters from the first big wave of sorties. At that time, the fresh debris samples indicated radioactive air was now present in the western Pacific near Japan and Kamchatka and in the eastern Pacific south of Alaska, and was beginning to move over northern Canada. The hottest sample from September 8th had registered a rate of 2,400 cpm. AFOAT-1 was concerned, though, that volcanic activity in northern Japan may have impacted many of the air samples.[20]

The September 9th status report indicated more sorties were now following the aforementioned air masses and that

60 hot filters had arrived registering as high as 4,650 cpm. This report was significant because it strengthened AFOAT-1's initial confidence that the event was most likely a fission detonation. With three samples fully analyzed at this point, AFOAT-1 estimated a detonation date between August 26th and September 1st. This report also ruled out any contamination concerns from volcanic activity.[21]

That same day, as the effluent air mass headed towards the United Kingdom, President Truman approved AFOAT-1's request to notify British authorities. Immediately after alerting their allies on Saturday morning, September 10th, senior members of AFOAT-1 conducted a lengthy teletype conference with their British counterparts in the American embassy in London. They conveyed information that the air mass was about to pass just north of Scotland, and encouraged the British to independently sample and assess its significance.

Just like the American AWS, the British weather service flights, equipped with similar filters, were already flying standard routes. Those flights had become routine only several months earlier. The sampling aircraft, a mixture of Mosquitos, Lincolns, and Halifaxes, were based at airfields in Scotland, Northern Ireland, and Gibraltar. These locations allowed the UK to surveil an area bounded by a longitude running from Greenland to the western European coast, and a latitude between the Arctic Circle and Gibraltar. The northern zone was codenamed BISMUTH, the southern zone NOCTURNAL. That day, the British launched a special flight to 70°N, and another on September 12th out of Gibraltar. Then other UK flights were vectored to conduct additional collections. Between September 10th and 16th, the British collected 21 significant samples. The initial British report, written on September 17th, placed the date of fission between the 26th and 30th of August. Their analysis coincided with the LASL and NRL results, which offered no "smoking gun" but were confident a fission detonation had taken place. Still, based on the data, they informed the British Prime Minister, Clement Attlee, on September 17th, that the Soviets probably had an atomic bomb.[22]

Recognizing the gravity of the information coming in from the ongoing activities, Vandenberg, now Chief of Staff of the Air Force, began assembling a panel of experts to review the data collected by the sampling sorties and AFOAT-1's report. He also requested that the British scientists attend the meeting to provide supplemental data to complement the American analysis. Giving them adequate time to travel, Vandenberg set the date of the meeting for September 19th at AFOAT-1's headquarters in Washington, D.C.

At 10 AM on September 19th, 29 key personnel met in the War Room at AFOAT-1 headquarters to evaluate the Soviet event in order to achieve consensus that, in fact, a nuclear detonation had occurred in the Soviet Union and that the analyses were accurate. The panel consisted of many of the world's top nuclear scientists and organizational leaders. In addition to Vandenberg, they included Dr. Vannevar Bush, former U.S. Director of Research and Development during the Second World War (serving as Chairman), and Dr. J. Robert Oppenheimer, former Director of the Manhattan Project. AFOAT-1 members included Major General Morris Nelson, who had taken command of AFOAT-1 only days earlier, as well as Northrup, Olmstead, Crocker, and Urry. Tracerlab sent technical director of Tracerlab Dr. Frederick G. Henriques, Dr. Lloyd R. Zumwalt, Dr. A. F. Stephens, and Nobel Laureate Dr. Harold Urey.[23]

From the AEC came commissioners Dr. Robert F. Bacher and Sumner Pike, as well as William Webster (who also served as chairman of the MLC). Other AEC members were Carroll L. Wilson, recently named AEC General Manager, and Spofford English, a former group leader in the Manhattan Project and now Chief of the chemistry branch in the AEC's Research Division. Dr. George Weil, assistant director of the AEC's Reactor Development Division, was also present.

Air Force Meteorology and LASL were represented by Colonel B. G. Holzman, recently chief meteorologist of JTF-7 at Operation SANDSTONE, and Dr. R. W. Spence, respectively.

Senior military officers present were Vandenberg's deputy, Air Force Vice Chief of Staff Lieutenant General Lauris Norstad, as well as Major Generals David M. Schlatter and K. D. Nichols. Schlatter was Assistant Deputy Chief of Staff, Operations, for Atomic Energy, and also served as the senior Air Force member on the MLC. Nichols, also a member of the MLC, was the Deputy Director for Atomic Energy Matters within the Plans and Operations Directorate on the Army General Staff. Included as well was Rear Admiral William Parsons, a member of the RDB's CAE. Parsons had previously served as a Deputy Commander for Leslie Grove's AFSWP, and Operations CROSSROADS and SANDSTONE.

Importantly, the British contingent included Sir William Penney. During the war, Penney was a key member of the British delegation to the Manhattan Project and had worked closely with Oppenheimer. Now he was the head of Britain's High Explosive Research Project—the beginning of the United Kingdom's nuclear weapons program.

Arguably, the most important British attendee was Commander Eric Welsh. Welsh headed the Atomic Energy Intelligence Unit (AEIU) within the Ministry of Supply and would become the key figure in fostering the close working relationship with AFOAT-1 going forward (see Annex 1: Allied Cooperation in the AEDS). Penney also brought Lord Arthur W. Tedder and Dr. Wilfrid B. Mann. As Vandenberg's counterpart, Tedder was Chief of the Air Staff, Royal Air Force, and had served as General Eisenhower's wartime deputy (i.e., Deputy Supreme Commander). Mann, working out of the British embassy in Washington, was Welsh's liaison to the Americans and would eventually be known as the greatest radionuclide meteorologist in the world.[24]

Bush called on Vandenberg to open the meeting. The Air Force Chief of Staff emphasized that their discussions and conclusion were of "national importance." At length, he acknowledged the tremendous brain power assembled and thanked everyone for their quick response in coming together. Bush then asked the AFOAT-1 commander to provide an opinion of the overall collection. Nelson, who had assumed command from Hegenberger only days earlier, was careful to point out that the fledgling Interim Net (i.e., the AEDS) could not yet yield a definitive answer. He told the panel AFOAT-1 was "still trying to develop that sort of system. Our techniques, our procedures and the scientific analyses and functions which to us is the critical phase of the processing [of] this physical data is also, in our opinion, still undeveloped to a point where we consider it reliable." Nelson noted he had told Vandenberg only one week earlier that the Soviets would most likely achieve a nuclear device no earlier than 1953. He now bowed to the scientists to verify the error of his assessment.[25]

Taken at face value, Nelson's remarks would seem to paint AFOAT-1's efforts in confirming the event as somewhat inadequate. In retrospect, however, he was making a point that the Interim Net was not satisfactorily mission capable. At this point in time, AFOAT-1's budget had been so severely slashed that it left almost no funding for geophysical R&D, something Nelson and Northrup believed was imperative in order to field an effective AEDS by mid-1951. Indeed, this very issue had led Johnson to resign from AFOAT-1 a year earlier.

Following Nelson's remarks, Bush, in the role of chairman, voiced his strong appreciation for the British involvement and provided a critical atmosphere of candor for the subsequent conversations. Very aware of the restrictions imposed by the McMahan Act, which prohibited or restricted the sharing of classified information, Bush wanted to reference its existence but did not wish to stifle any dialog between the American and British scientists. "We are collaborating with the British in this under Area 5, I believe, but naturally in our discussions we will confine ourselves to that. . . ."[26] Area 5 was a designation evolving out of a meeting between AFOAT-1 (Northrup) and the AEIU (Welsh) in September 1948. Those discussions between the two LRD units were driven by deliberations conducted at the Anglo-American-Canadian Combined Policy Committee meeting in December 1947, to explore areas of collaboration that did not violate the new law. Area 5 allowed for the "detection of a distant nuclear explosion, including: meteorological and geophysical data; instruments (e.g., seismographs, microbarographs); and air-sampling techniques and analysis."[27]

The McMahan Act was named for its sponsor, Senator Brien McMahan. Truman had signed the act into law in August 1946, and it had taken effect on January 1, 1947, although most of the provisions did not go into effect until September 18, 1947, the day after Forrestal became the first Secretary of Defense. Now called the Atomic Energy Act of 1946, it formed the basis for one of the most significant reorganizations of the federal government in American history. The act laid to rest the contentious debates over who would control nuclear weapons development and nuclear power by establishing the AEC—a civilian (rather than military) agency. It also took the air arm out of the Army to create the Air Force, and renamed the Department of War to the Department of the Army. It then placed all three services—Army, Air Force, and Navy—under the Secretary of Defense. Importantly, the act also created the National Security Council and the CIA. On this day, however, the act served as a barrier of sorts. The impetus for the act arose out of concerns the Soviets were stealing classified nuclear information through British Commonwealth spies (a valid concern soon to be proven correct). Recently, as Congress finalized the bill, a former Manhattan Project participant was identified as having passed information to the Russians.[28] Consequently, the act placed significant restrictions on sharing information with Canadian and British allies. It did so by creating a security classification designation called "Restricted Data" or RD. The originating authority for RD was the AEC (later the Department of Energy). RD encompassed a wide range of topics related to nuclear weapons tests and characteristics, thus making collaboration difficult, especially for the pioneers of LRD.

Northrup largely facilitated the remainder of the morning session. He began by recounting the events of September 3rd and used his graphics to explain the initial collection routes. At that point, Oppenheimer asked for additional information, such as the location of subsequent routes, samples collected and their counts per minute, altitudes flown, wind velocities, and the paths of the air masses. Northrup then asked Welsh to recount the British collection sorties. Welsh first described their normal flights, which they conducted every 48 hours, and the first indication of a suspicious sample on September 11th. He then recounted the subsequent sorties, which were able to collect at an altitude of 30,000 feet—almost twice the height of the American missions.[29]

Following the retelling of the British and American collections, Northrup distributed his report simply entitled "Joe" for everyone to read. "Joe," derived sardonically from the name of the Soviet leader, Joseph Stalin, was essentially Northrup's Technical Memo #37 dated that same day, September 19th. The three-page report summarized all of the technical data related to the age determination and radiochemical analyses, and referenced the results of the other participating labs. It concluded the event was a fissionable detonation (plutonium) "similar to the Alamogordo bomb." The test occurred somewhere on the Asiatic land mass between the east 35th meridian and the 170th meridian within the August 26th and 29th timeframe. Importantly, for anyone who might suggest an alternative scenario, Northrup wrote that "a variety of alternative explanations have been proposed. Upon examination, none of these turn out to be technically likely." At this point in the meeting, the assembled scientists began confirming that assessment. Tracerlab's Dr. Harold Urey began his remarks by addressing Northrup's last point about analytical uncertainty. "We are in the business of indicating an atomic event at a great distance. We cannot get away from indicating [i.e., considering] other events [as well]."[30]

While acknowledging the labs found inconsistencies in the half-life measurements, which they should all "look upon with suspicion, their absence offered "no hidebound conclusion." He added "the only hidebound conclusion, the only way we can arrive at a conclusion is to separate the fissionable products, identify them and measure them." Urey assured everyone Tracerlab had precisely accomplished those measurements and emphasized "we are presenting it in a very unquestionable form." Urey noted how much learning they had accomplished since Operation CROSSROADS. He then called on his co-worker, Dr. Stephens, to explain Tracerlab's radiochemistry work.[31]

Dr. Stephens reminded his audience that they had functioned as AFOAT-1's lab since SANDSTONE and had analyzed most of the previous 111 alerts. In detail, he then walked everyone through the radiochemical processes and their discoveries, speaking at length about the particle analysis of the filter paper samples. He described their observations of the fission product ratios, and his estimation of a detonation time and date of 5 PM, August 29th, "plus or minus one day." Stephens explained how the data pointed to an instantaneous event rather than a reactor accident (a view Truman expressed). Oppenheimer now appeared convinced. "It seems to indicate that if it was a bomb, they are not doing too badly."[32]

Northrup then called on the British, LASL, and the Office of Naval Research (ONL) to comment on their analyses relative to the results Tracerlab presented. Dr. Penney remarked that the British analysis agreed with Tracerlab's, but not to the degree and depth of analysis the American lab had achieved. He added that their estimation of a detonation date was closely the same between August 27th and 31st.

Penney acknowledged that, despite their many limitations, they would have eventually picked up the detection out of Gibraltar. "If you hadn't been doing this we would have found it for the first time on September 12 from our flights."[33]

From LASL, Dr. Spence recounted their analysis from the two filter papers they had received from Tracerlab. He simply confirmed the samples contained fission products. For the first time that day, he addressed the possibility of a reactor accident but quickly discounted it. "I have almost completely ruled out a pile. It may be discerned that if it wasn't an atomic bomb, it acted like one."[34]

Mr. James W. Smith, from the Office of Naval Research or ONR (the parent organization of the NRL), likewise presented similar findings. In addition to the filter papers NRL received on September 13th, he spoke to their analysis of the rain water collected in Alaska. He stated while their "very rapid chemical analysis on these [samples] are not as efficient as they ordinarily are," he agreed a fission event occurred. Northrup, noting that Tracerlab and the NRL had reported similar conclusions, asked Urey to comment. Urey remarked that U.S. test data was useful because the NRL and Tracerlab had observed that the results came close to matching the Yoke shot in Operation SANDSTONE.[35]

By this time, Bush, like Oppenheimer, seemed convinced and urged Northrup to conclude AFOAT-1's report. Northrup stated some of his final remarks were "in a very high level, clear out of Area 5." Bush replied that Northrup could speak at that level in the afternoon executive session "but you will confine yourself to Area 5 now, please." Just as it appeared the morning session was ending, surprisingly, Oppenheimer asked, "Would it be all right to hear any alternative suggestions?" What followed was a short debate, largely led by Urey that "it is just barely possible that this could have been an explosion of a pile which had just started its operation." Over the next few minutes, Bush, Oppenheimer, Urey, and Penney discussed and then generally dismissed the idea. Bush, apparently eager to conclude the meeting, ended the discussion. "I think we might have a little recess and stretch our minds a little bit."[36]

At 1:45 PM, the leading Americans convened the Executive Session. Bush, Oppenheimer, and Bacher immediately praised Northrup's presentation and validated the morning's discussions. They then worked to draft a formal document for Vandenberg, cautious not to overburden the report with too much detail or scientific jargon. After much wordsmithing, they listed six points in their conclusion:

1. Fissionable material in quantity comparable to that used at U.S. "A" bomb tests underwent fission at some time between the 26th and 29th of August, at some point between the East 35th Meridian and 170th Meridian, and at some point on the Asiatic land mass.

2. The fission products resulted wholly or largely from the fission of plutonium.

3. Natural uranium was present in close proximity to fissionable materials at the time of reaction.

4. The observed phenomena are consistent with the view that the origin of the fission products was the explosion of an atomic bomb whose nuclear composition was similar to the Alamagordo bomb.

5. A variety of alternative explanations have been proposed. Upon examination, none of these turns out to be technically likely. Those which although unlikely are either consistent with the data, all call for the use of enough plutonium to make an atomic bomb.

6. We, therefore, believe that an atomic bomb was detonated as stated.[37]

Given the criticality of the Soviet event, Russian secrecy over its first nuclear test, and the realization of a new bi-polar nuclear world, some historians have questioned why Truman waited four days before announcing the Russian test. Within hours of concluding the September 19th meeting, Vandenberg informed the president about the unanimous assessment. Apparently, however, as the meeting was underway, the British Foreign Minister Ernest Bevin had just informed Truman the British Government would devalue its currency that same day. Because the devaluation could have a major impact on several governments and the world market, Truman opted to separate the two serious events by time. Indeed, nine governments were compelled to follow suit as a reaction to the 31 percent devaluation of the British pound.[38] Others have speculated that Truman and his Secretary of Defense, Louis A. Johnson, were very skeptical and thought it was most likely a reactor accident, as the confirmation meeting participants briefly discussed. Interestingly, Truman asked AEC Chairman David E. Lilienthal and the scientists to sign a statement confirming their findings.[39]

The day following the historic meeting, with the report in hand, Bush officially informed Vandenberg that, after careful consideration of the facts, the committee unanimously agreed the Soviet Union had exploded an atomic bomb during the latter part of August 1949. Three days later, on September 23rd, President Truman announced to the world that the USSR had exploded its first nuclear device. As the U.S. would later learn, a team of Russian scientists headed by academician Igor Kurchatov successfully detonated the first Soviet nuclear device on the morning of August 29, 1949, at the Kazakhstan Test Site. The Russians gave this test the code name "First Lightning." Detonated from a tower, the nuclear device was a plutonium implosion based on the American design (the "Fat Man" bomb). Years later, it was confirmed as a 22kt detonation.[40]

The events of September 1949 marked a major turning point in the Cold War. Until then, the United States was fairly confident it would maintain its monopoly and technological lead in the development of nuclear weapons for some time. On the eve of Joe-1, a general consensus had seemed to settle on a mid-1953 date as the most likely time for the Soviet Union to test its first atomic device. Confidence in that date grew as evident by the severe budgetary cutbacks levied on Johnson and Northrup's R&D program. Joe-1 reversed those views virtually overnight.

On October 17, 1949, Dr. Vannevar Bush wrote to General Vandenberg. He commended the Air Force Chief of Staff on the foresight exhibited by the Air Force in recognizing the need for confirming the Joe-1 conclusions and for initiating, in the short time available, the "very valuable confirmatory studies by the Los Alamos Scientific Laboratory, the Naval Research Laboratory and by the scientists in the United Kingdom."[41]

The discovery of Joe-1 was indeed fortunate. AFOAT-1 had not sufficiently advanced the LRD Program with adequate instrumentation to detect the blast by geophysical means. Moreover, it came at a time when the prestige of AFOAT-1 and the whole LRD Program had slipped and faced possible further cutbacks in funding. Taking the original readiness date of mid-1950 as a gauge, it was evident the progress made in the R&D program between December 1947, when General Kepner organized AFMSW-1, and September 1949, was very limited. The seriousness of the lag in R&D is indicated in the following table, which illustrates the anticipated expenditures within the limits of the Loomis Report, and AFOAT-1's dollar estimate on work already accomplished:

Programs	Projected Expenditure	Actual Expenditure
Geophysical Instrument. & Techniques	2,290,000	354,000
Long Wave Range Propagation	74,700	26,000
Properties of Explosion Products	373,000	46,000
Transport of Explosion Products	169,000	28,000
Collection of Explosion Products	623,500	135,000
Analysis of Explosion Products	446,000	180,000
Total	3,976,200	*769000
* In 2018 dollars: projected expenditure = 41.7MM, and actual expenditure = 8MM.		

While the LRD program was badly off schedule, AFOAT-1 had made some progress in the development of geophysical instruments and techniques. A number of studies of certain natural phenomena significantly contributed to the improvement of instruments, the design and manufacture of some new instruments, and the feasibility study of underwater techniques for the detection of large explosions. Further analysis of SANDSTONE data indicated long-range acoustic propagation was based primarily on very low frequency signal components. In detection and collection, researchers devised methods for collecting rain water, precipitating the particulate matter and analyzing it for content of fission products. The radiochemical analysis of samples taken from drone planes in the Baker shot at CROSSROADS substantiated their hypothesis about the size of particles produced in an atomic explosion.[42]

President Truman's announcement of the Soviet atomic detonation had far reaching consequences for the future of LRD. Soon thereafter, the RDB took steps to reevaluate and reinterpret the LRD mission and its methods. On October 12, 1949, the RDB requested the JCS to reevaluate its guidance on LRD in light of current discussions around the feasibility of developing thermonuclear weapons (discussed in Chapter 5).[43] The JCS did so on January 20, 1950, and five days later, the Air Force Chief of Staff was charged with the responsibility of "establishing, operating, and maintaining a long range detection system utilizing scientific means to detect future atomic explosions." The directive was brief and concluded "the Soviet atomic energy program was of major importance and might be [assessed] by a scientific detection system." In short, the Chief of Staff, USAF, and therefore AFOAT-1, was again formally charged with the responsibility of detecting future atomic detonations, the type of weapon used, and the time and place of the tests.[44]

As the new decade began and the Cold War rivalry escalated to a new level, AFOAT-1 re-assessed its R&D program for fiscal years 1951 through 1953. Northrup finalized the adjusted R&D priorities in a report to General Nelson on July 24, 1950, which superseded the original Statement of Capabilities document provided to the MLC on November 26, 1949. He noted the revision was based on advances made over the last eight months and the re-evaluation of national requirements, which positioned the detection of a Soviet thermonuclear bomb as AFOAT-1's top priority. He was frank about the current limitations of the AEDS. For the detection of fission detonations, AFOAT-1 could "with certainty" confirm the occurrences by radiochemical means, but they could only assess the total energy of a detonation within a factor of 2. He also noted the AEDS could only gather some data to help determine whether the event was an airburst, underwater, surface, or underground explosion. Northrup stated the geophysical techniques were "borderline" at best for fission weapons (i.e., not accurate beyond 1,800 miles). For seismic instrumentation, the array designs appeared to solve many problems. Arrays—a dispersed pattern of multiple seismographs—reduced the background noises from natural disturbances and provided azimuths of incoming signals. AFOAT-1 anticipated improved seismographs and microbarographs to enter the AEDS by the summer of 1951, the target date for a fully mission-capable system.[45]

Northrup's reassessment concluded that airborne debris collection, though highly validated with the detection of Joe-1, required further research given that AFOAT-1 would now look for evidence of a hydrogen bomb. Although the design of a thermonuclear weapon was still in a theoretical stage at the time, AFOAT-1 scientists understood what elements to look for in the collected air samples. The presence of helium, deuterium, or tritium would be indicators, as well as "neutron induced activities in the constituents of the atmosphere" such as carbon-14 or argon-37.

What were needed most were improvements in filter media and filter design in order to help perfect radiochemical techniques. The revised capabilities statement contained several basic diagrams to explain the improvements. For the air filter unit on top of the RB-29—nicknamed the "bug catcher"—the interior was re-designed to increase air flow and to reduce the contamination problem that was critical in analyzing microchemical and physical accuracies. A "super sampler" would occupy the forward bomb-bay and would be capable of collecting much larger samples for greater accuracy in radiochemical analyses. The improved foil and the super sampler would also contain newly designed wrap-around counters, which were four times more sensitive than the counters used at the Eielson lab to measure the radioactivity of the Joe-1 samples. The new counters replaced the old glass design with an extremely sensitive stainless steel Geiger-Mueller tube. For the aircraft, R&D efforts made the radioactivity detector 20 times more sensitive than those used during Operation SANDSTONE and they were unaffected by weather conditions. This scintillation-type beta-gamma detector read out to an instrument panel indicator. Another similar instantaneous radioactivity detector (an ion detector) also added to the RB-29 was even better than the scintillation type but was susceptible to adverse weather conditions.[46]

Established techniques in surface particle sampling did not promise more than partial fulfillment of mission objectives. Northrup's revised capabilities statement showed the crude design of the ground filter unit (GFU), noting it functioned as a "backup" system of sorts to augment airborne collections. The GFU was highly susceptible to ground-level contamination from natural radioactivity in dust and radon emanations. Northrup advised Nelson that no solution for the contamination problem was in sight.[47]

To advance these multiple experimental programs, it was imperative for AFOAT-1 to once again utilize the U.S. nuclear testing program. After the JCS revalidation of the LRD program on January 20, 1950, AFOAT-1 had almost an entire year to plan out its schedule of experiments before the next test series began. Operation RANGER, a series of five test shots, was scheduled to commence on January 27, 1951. However, Nelson's immediate concern was how to plan for the renewal of U.S. tests while simultaneously building up and improving the rudimentary AEDS. To fulfill the broad requirement of detecting future atomic explosions, his operational plan for 1950 required strategically positioned ground-level sampling equipment in areas over which Soviet air masses travelled. He also had to establish an air patrol intercept line between Guam and the North Pole across which most of the USSR air masses passed, and the capability of vectoring air patrols out of strategically located areas to intercept suspicious air masses crossing the Guam to North Pole intercept line. In addition, his personnel needed to monitor pressure waves in the atmosphere and in the earth's crust to detect unusual perturbations that were distinguishable from natural events. This capability required a ring of acoustic and seismic stations around and as close to Soviet testing areas as possible.[48]

Operationally, the AFOAT-1 plan was necessarily complicated. It provided for special vectoring services out of the Dhahran-Tripoli area, Hawaii, and McClellan AFB in California to support the Guam-to-North Pole airborne interception line on a once-every-24-hour basis. The plan called for 13 ground-level filtering stations, nine of which already existed, located in various regions where Soviet air masses were expected to traverse. A field radiochemical laboratory was planned for the Dhahran-Tripoli area in addition to the existing four laboratories already in operation. The plan called for nine seismic stations, four of which were already operational at Eielson, Frankfurt, Kyoto and Misawa. The reporting elements of the plan closely resembled the process used in the detection of Joe-1. That is, after splitting the airborne samples for redundant analyses, any positive collections would result in a meeting of top level scientists for validation before notifying the nations' principal leaders.[49]

On March 20, 1950, the Boner Panel completed a reassessment of the LRD R&D program and made several recommendations. The panel believed R&D should immediately develop techniques to assess and determine the detonation of a thermonuclear weapon. Despite Nelson's emphasis on improving radiological, seismic, and acoustic methods for the detection and location of future explosions of fissionable materials, the panel stated R&D for those methods should continue but at a reduced rate, with primary emphasis on instruments and techniques promising early operational application. The panel also recommended the approval of the eight LRD projects planned for the 1951 series of tests, and the funding of $3.5 million for the current year R&D program as well as $3 million earmarked in the Air Force FY51 R&D budget for LRD. The RDB, at its meeting on April 5, 1950, approved the panel's recommendation and directed a request be made to the Secretary of Defense for additional monies to immediately permit full implementation of the LRD R&D program then underway.[50]

With the RDB approvals in hand, Nelson announced on June 1, 1950 that AFOAT-1 would complete its plans for a fully operational AEDS by June 1, 1951. At that time, he said, the AEDS would consist of 6 bases capable of conducting airborne sampling sorties, 35 stations capable of collecting particulates on the ground, 9 locations capable of geophysical surveillance by seismic and/or acoustic methods, 9 laboratories, and 5 field offices for operational supervision of data collection stations. On June 13, 1950, Nelson provided the following budget to build and expand the AEDS:

FY	R&D	Ops & Maint	Total
1948	2,700K	0	2,700K
1949	5,500K	0	5,500K
1950	3,450K	310K	3,760K
1951	3,000K	3,250K	6,250K
1952	1,370K	2,500K	3,870K
TOTALS	16,020K	6,060K	*22,080K
			*2018 = 231MM [51]

By mid-summer 1950, Nelson and AFOAT-1 were now off and running as Joe-1 opened the gates to increased R&D funding. Nelson's challenge was to shake off the lethargy created by the slow pace of mission and budgetary debates, which had impeded progress since SANDSTONE. With a greater degree of urgency, he was now faced with the dual tasks of making the new AEDS fully operational by mid-1951 while preparing for the next series of U.S. nuclear tests critical to his R&D program. That sense of urgency, however, was fueled at the same time by the North Korean invasion of South Korea on June 25, 1950.

The North Korean invasion caught the United States off guard. While the Cold War was becoming colder with each passing day elsewhere in the world, the Truman Administration had viewed Korea as one country where the Soviets and the Americans would not fight a proxy war. That had not been the case immediately after the end of the Second World War when Soviet and American forces occupied the country in an arrangement of joint administration. They remained there for several years primarily just to restrain each other rather than for reconstruction activities. In retrospect, Korea was just one more step in a procession of confrontational events escalating tensions between the United

States and the Soviet Union. In late 1948, Soviet forces departed Korea, followed by the departure of American forces in the summer of 1949.[52]

Other serious tensions between the two super powers had mounted in the three-and-a-half years separating Operation CROSSROADS and the Joe-1 event. Beginning with Churchill's "Iron Curtain" speech in the spring of 1946, the ideological divide between East and West had resulted in a number of decisions and activities that would place nuclear weapons (and the need for the AEDS) at the center of American foreign policy. In early March 1947, the Truman Doctrine was implemented as a response to stop the spread of communism in Greece and Turkey. One week later, Vandenberg wrote his aforementioned letter (see Chapter 2) to the Departments of the War and Navy, the AEC, and the JRDB to express his concern for the need of an LRD capability. In June 1947, the United States announced the Marshall Plan that subsequently made the division of Germany more entrenched. However, when the Soviet Union invaded Czechoslovakia in February 1948 and blockaded Berlin in June, the West responded by creating the North Atlantic Treaty Organization (NATO) and permitting the creation of the Federal Republic of Germany (West Germany) in April and May. Lastly, as the various labs were processing samples during the Joe-1 event in September 1949, the Air Force terminated the 18-month Berlin airlift operation and, two weeks later, East Germany became the German Democratic Republic, a Soviet communist satellite state. Diplomatically, the United States watched another communist state form as the civil war in China ended and Mao Zedong declared a new communist government—the Peoples' Republic of China (PRC) on October 1, 1949.[53]

Four months prior to the start of the Korean War, in February 1950, the Soviet Union and the new PRC signed a mutual defense pact. This agreement resulted in both communist nations supporting the North Korean invasion several months later. During the Russian occupation, Stalin had developed a close relationship with Kim Il-sung, a Korean communist who had fought against the Japanese alongside Chinese forces and was trained in Moscow. While Kim and his southern counterpart, Syngman Rhee, both wanted to reunite the Korean peninsula, they realized they needed external support to do so. Although the Americans cautioned Rhee, Stalin gave Kim the "green light" in early 1950. Between June 25th and August 4th, Kim's army (with help from Soviet air support) drove American and South Korean forces all the way to the southern port of Pusan, a 90-square-mile defensive perimeter at the tip of the peninsula. However, when General Douglas MacArthur enveloped the North Korean Army with an amphibious landing at Inchon on September 15, 1950, the invaders quickly fled north.

Routed to the Yalu River, the boundary between China and North Korea, Kim's army received massive support on October 25th from their Chinese allies, who massed 500,000 soldiers to begin driving United Nations forces back toward the 38th parallel, the original border between North and South Korea.[54]

The ebb and flow of combat up and down the Korean Peninsula indirectly impacted the development of the AEDS. Truman faced a dilemma. How could the American nuclear monopoly be leveraged to manage the conflict in Korea? The president's foreign policy of containment (i.e., meeting Soviet expansionism and aggression primarily though economic and political counter forces) was revised and solidified on April 7, 1950, with National Security Council policy document NSC-68. His approval of NSC-68 authorized a dramatic increase of the military budget in order to build up both conventional and nuclear arms. Importantly, it called for the enhancement of American technical superiority through "an accelerated exploitation of [its] scientific potential."[55] At the time of the invasion, the U.S. possessed nearly 300 nuclear weapons and had more than 260 aircraft to deliver them.[56]

This large capability, however, appeared to offer no deterrent in Korea.[57] As retreating U.S. and South Korean forces were falling into the Pusan Perimeter on July 28, 1950, both Vandenberg and General Omar Bradley, Army Chief of Staff, advocated for the use of nuclear weapons targeting North Korean cities. Truman denied the request. For the remainder of the war, Truman and his advisors periodically considered the use of nuclear weapons, especially when China entered the war in October of 1950, and later when peace talks lost momentum. However, inevitably, the president dismissed the suggestions.[58] Joe-1 had certainly opened the door to more funding for LRD. Now, however, with NSC-68's emphasis on technological superiority encouraging innovation and R&D, the path to an effective AEDS appeared more clearly. As one Cold War historian noted:

> Everything changed with Korea. American diplomacy, defense budgets and military reach exploded across the globe in the aftermath of the invasion, as U.S. taxpayers and Congress alike gave the unstinted political support the strategic planners had hitherto sought with only limited success.[59]

For AFOAT-1, Russian support in Korea made rapid advances in LRD and improvements for the AEDS an imperative. On October 12, 1950, as U.S. Army and UN forces crossed the 38th parallel in their advance into North Korea, Northrup updated the LRD program guidance for fiscal years 1951 through 1953. In short, in his introductory statements to the plan, he argued for an "increase in the scope and intensity of research on [LRD] techniques." He noted other R&D programs were increasing across DoD as a reaction to the increased threat. "The invasion of Korea has revealed the imminence of the possibility of open conflict with the USSR, and has advanced the previous JCS estimate of the earliest possible date when open hostilities might commence." Without mentioning the agency's name, Northrup criticized the CIA for its inability "to define the status of the Soviet atomic energy program" by their "conventional means," which he labeled "inadequate." In getting to the point that LRD offered the nation the best surveillance system, he stressed "the capability of the USSR to wage full scale atomic warfare will depend upon the status of USSR <u>weapon technology</u>." He concluded by noting the proof of LRD's primacy was the thorough detection and analysis of Joe-1.[60]

Throughout 1950, with the increased frequency of international developments and conflicts (especially with Joe-1 and the "loss of China" the previous fall), Truman placed an emphasis on science to ensure a strong national security posture. The big "unknown" was how much progress the Soviets were making in developing their nuclear arsenal. Largely driven by fear of that unknown (after all, Joe-1 had arrived years earlier than most expected), Truman authorized research into the development of a hydrogen weapon in January. Although the scientific community was divided over the morality of such a destructive weapon, those wanting a thermonuclear bomb would win the day.

Sir William Penney

Penney had been a prominent member of the British contingent in the Manhattan project. He played a key role in developing LRD for the United Kingdom and in establishing routine airborne sampling routes similar to those used in the AEDS. He led the British team in the verification meeting of Joe-1 that was instrumental in validating the initial analyses. Soon thereafter, Penney headed the United Kingdom's nuclear weapons program. *Photo courtesy of Los Alamos National Laboratory*

Dr. Vannevar Bush

First chairman of the RDB that provided oversight of LRD. He was a former president of the Carnegie Institute and founder of the Raytheon Company. During the war, he headed the U.S. Office of Scientific Research and Development, responsible for more than 6000 scientists. He chaired the September 19, 1949 meeting of experts that validated the analyses of Joe-1. *Photo courtesy of the Office of Emergency Management*

RB-29 with the original "bug catcher" on top of the fuselage
Photo source: AFTAC archives

Louis Johnson

Shown here being sworn in as Secretary of Defense following Forrestal's resignation in April 1950. He was at first skeptical of the Joe-1 analyses, believing (like Truman) that it was actually a nuclear pile accident in the USSR.

Photo courtesy of the Truman library

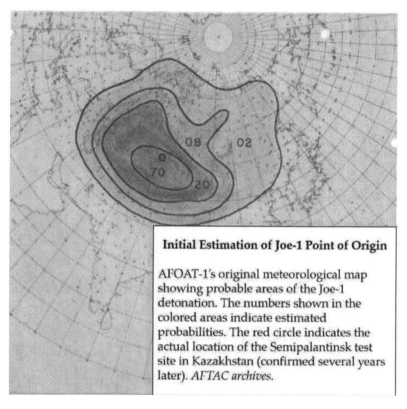

Initial Estimation of Joe-1 Point of Origin

AFOAT-1's original meteorological map showing probable areas of the Joe-1 detonation. The numbers shown in the colored areas indicate estimated probabilities. The red circle indicates the actual location of the Semipalantinsk test site in Kazakhstan (confirmed several years later). *AFTAC archives.*

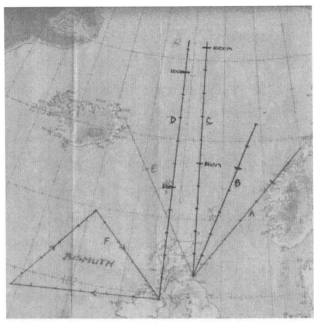

Original Maps Depicting UK Sampling Zones
Like the standard U.S Loon Charlie flights between Alaska and Japan, the UK surveilled an area bounded by a longitude running from Greenland to the western European coast, and a latitude between the Arctic Circle and Gibraltar. The northern zone was codenamed BISMUTH and the southern zone NOCTURNAL. *AFTAC archives.*

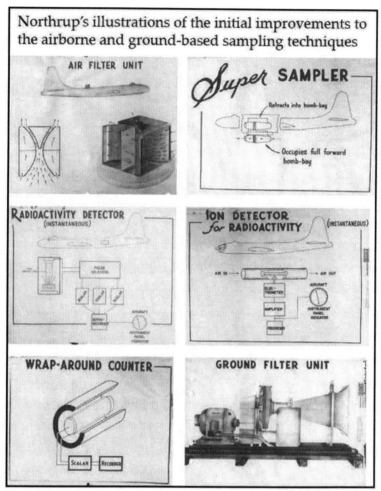

Northrup's illustrations of the initial improvements to the airborne and ground-based sampling techniques

Included in technical memo #43, Northrup to Nelson, Subject: "Status of Technical Capabilities of long Range Detection of Atomic Operations," July 24, 1950.

AFTAC archives

Notes to Chapter 4

[1] Unpublished manuscript, AFTAC archives.

[2] Ibid., 25.

[3] Doyle L. Northrup and Donald H. Rock, "The Detection of Joe-1,"unpublished manuscript [date unknown], AFTAC archives.

[4] Unpublished manuscript, AFTAC archives

[5] AFOAT-1 discovered later that the samples which had triggered the previous alerts came from the contaminated aircraft themselves or were due to background radiation caused by previous U.S. tests.

[6] "Scientific Report No. 2," September 8, 1949, AFTAC archives.

[7] "Scientific Report No. 3," September 9, 1949, AFTAC archives.

[8] "Obtained on Operation Joe - Presented to the Joint Chiefs of Staff December 1949," December 13, 1949, AFTAC archives

[9] George B. Olmstead, Memorandum for Record, "Telephone Conversation with Captain Brittain," September 7, 1949, AFTAC archives.

[10] George B. Olmstead, Memorandum for Record, "Telephone Conversation with Captain Brittain," September 8, 1949, AFTAC archives.

[11] The prestigious NRL was established in 1923, and produced critical technologies over the years to include RADAR and the first satellite. It became a subordinate organization of the Office of Naval Research (ONL) when President Truman created the ONR in August 1946.

[12] Technical memo #40.

[13] Ibid., Report, from T. A. Solberg, Chief of Naval Research to Nelson, Subject: "Analysis of Sample of Airborne Material," September 26, 1949, AFTAC archives; Report, from R. W. Spence, LASL, "Identification of Radioactivity in Special Samples," October 4, 1949, AFTAC archives.

[14] Herbert Friedman, Luther B. Lockhart, and Irving H. Blifford, "Detecting the Soviet Bomb: Joe-1 in a Rain Barrel," *Physics Today*, November 1996, 40-41; Also, NRL Memo from M. K. Fleming to Nelson, Subject: "Collection and Identification of Fission Products of Foreign Origin," September 22, 1949, AFTAC archives.

[15] U.S. Weather Bureau Report on Alert 112 of the Atomic Energy Detection System, Washington, D.C., September 29, 1949, AFTAC archives.

[16] Ibid.

[17] Zumwalt took 15 fellow scientists from ORL with him. Charles Ziegler, "Waiting for Joe-1: Decisions Leading to the Detection of Russia's First Atomic Bomb Test," *Social Studies of Science*, SAGE, London, Vol. 18 (1988), 197-229.

[18] Correspondence from Lloyd Zumwalt to Walt Singlevich, October 3, 1988, draft conference paper entitled "Analysis of Fission Products from Russia's First Atomic Bomb Test," Submitted for presentation at ANS-NBS Conference on Fifty Years with Nuclear Fission, AFTAC archives. Also, Report from TracerLab to AFOAT-1, September 10, 1949, AFTAC archives (also available in the Truman Library).

[19] *History of Long Range Detection: 1947-1953*, 223.

[20] AFOAT-1 Scientific Report No. 2, Dr. Donald H. Rock, Chief, Analysis Section, September 8, 1949, AFTAC archives.

[21] AFOAT-1 Scientific Report No. 3, Dr. Donald H. Rock, Chief, Analysis Section, September 9, 1949, AFTAC archives.

[22] Technical memo #37, from Northrup to Nelson, September 19, 1949, AFTAC archives. Note that this memo essentially formed the report entitled "Joe" that Northrup presented at Vandenberg's meeting on September 19th. AFTAC archives. See also Michael E. Goodman, *Spying on the Nuclear Bear: Anglo-American Intelligence and the Soviet Bomb.* (Stanford: Stanford University Press, 2007), 45-47.

[23] Harold Urey won the 1934 Nobel Prize in Chemistry for the discovery of deuterium. During the war, he led the group at Columbia University that developed isotope separation using gaseous diffusion. Some confusion with AFOAT-1's Urry and Tracerlab's Urey. Both directly communicated with each other quite frequently.

[24] Memo from Nelson to Chairman, Military Liaison Committee, Subject: "U .S. and UK Conference of 19 September 1949 on Operation Joe," January 24, 1950, AFTAC archives.

[25] Transcript, "Notes from Special Meeting Held on September 19, 1949," 1-3, AFTAC archives.

[26] Transcript, 3.

[27] Goodman, *Spying on the Nuclear Bear,* 44-45.

[28] There were several beginning to emerge, most recently English physicist Allan Nunn May. See Goodman, *Spying on the Nuclear Bear,* Chapter One.

[29] Transcript, 8-10; William Penney, "An Interim Report of British Work on Joe," September 22, 1949, AFTAC archives.

[30] Technical memo #37. Henceforth, this Soviet test and all others would be named Joe.

[31] Transcript, 11-12.

[32] Transcript, 12-17.

[33] Transcript, 19.

[34] Ibid., 20. "Pile" refers to a plutonium production facility.

[35] Ibid, 19-22.

[36] Transcript, 24-26.

[37] Transcript, Executive Session, 1-5.

[38] Goodman, *Spying on the Nuclear Bear,* 49.

[39] *Defense's Nuclear Agency: 1947- 1997* (Washington, D.C.: Defense Threat Reduction Agency, U.S. Department of Defense, 2002), 73.

[40] Gordon, *Red Cloud* a *Dawn,* Chapter Seven.

[41] Letter from Bush to Vandenberg, October 17, 1949, AFTAC archives.

[42] Ibid.

[43] *History of Long Range Detection: 1947-1953,* 37.

[44] Ibid., 107.

[45] Technical memo #43.

[46] Ibid.

[47] Ibid.

[48] *History of Long Range Detection: 1947-1953,* 225-226.

[49] Ibid.

[50] Ibid., 44-46.

[51] Ibid., 48.

[52] Gaddis, *We Now Know,* 70.

[53] Ibid., 70-72; Walker, *Cold War,* 66-78.

[54] Gaddis, *We Now Know,* 71

[55] See https://history.state.gov/milestones/1945-1952/NSC68

[56] 195 David Alan Rosenberg, "U.S. Nuclear Stockpile 1945 to 1950," Bulletin of the Atomic Scientists, Vol. 38, No. 5 (May 1982),

p. 26, as cited in Roger Dingman, "Atomic Diplomacy during the Korean War," *International Security*, Vol. 13, No. 3 (Winter, 1988-1989), pp. 50-91.

[57] See Gaddis, *We Now Know*, 99: "Washington officials never transformed their atomic monopoly into an effective instrument of peacetime coercion." During the Truman administration, the nuclear monopoly/superiority was always assumed to be a deterrent.

[58] Roger Dingman, "Atomic Diplomacy during the Korean War," pp. 50-91.

[59] Walker, *Cold war*, p. 77.

[60] Emphasis is Northrup's. Technical memo #49. , Northrup to Nelson, Subject: "Program Guidance in the Field of Atomic Energy," October 12, 1950, AFTAC archives.

CHAPTER 5

Operations RANGER and GREENHOUSE

For some time, scientists had speculated about the development of a hydrogen weapon—a super bomb or H-Bomb. Prestigious physicists such as Dr. J. Robert Oppenheimer and Albert Einstein were opposed to its development, believing such devastation was unnecessarily destructive and morally unconscionable. Theoretical physicist Dr. Edward Teller ("father" of the hydrogen bomb), however, did not agree with his colleagues and strongly advocated for the creation of a weapon based on thermonuclear fusion. In fact, he had speculated about the possibility since 1941, but had met strong resistance from his fellow physicists involved in the Manhattan Project who were totally focused on developing a fission weapon.[1] With third and fourth generations of fission weapons tested at CROSSROADS and SANDSTONE, few seemed interested in thermonuclear fusion until Joe-1 occurred. Given the advantageous growing stockpile of fission weapons in the U.S. and with Russia only demonstrating an initial fission test, why was there a sudden sense of urgency from the highest levels of government to proceed with an expensive but theoretical R&D thermonuclear program? What would the existence of thermonuclear weapons mean to the LRD mission and the capabilities of the AEDS?

In 1950, AEC chairman Lewis Strauss, Secretary of Defense Louis Johnson, and Chairman of the JCS General Omar Bradley, expressed their support for the development of a hydrogen weapon, citing concerns over the speed of Soviet technological advances. Despite his abhorrence of nuclear weapons derived from his authorization of their employment against Japan to end the Second World War, President Truman accepted his advisors' recommendation to develop the H-bomb. On January 31, 1951, Teller received welcomed news from Truman that the U.S. would proceed with the development of a fusion weapon. Two months later, Truman declared the program a high priority for U.S. national security.[2] Most likely, this was not a difficult decision for the president. For some time, Truman had lost hope in achieving any form of nuclear arms control. As early as July 1949, just weeks before Joe-1, he had remarked to congressional leaders, "since we can't obtain international control [of atomic energy] we must be strongest in atomic weapons."[3]

Certainly, the detonation of Joe-1 several years earlier than expected caused concern about how rapidly the Soviets were advancing their nuclear weapons program. Indeed, Joe-1 was viewed as an intelligence failure. Remarkably, the day *after* the Vandenberg Joe-1 confirmation meeting on September 19, 1949, the CIA completed its top secret assessment of the Soviets' nuclear weapons program, which projected mid-1953 as the most probable date of a Russian fission test.[4] Subsequently, the director of the newly created CIA, Roscoe Hillenkoetter, soon lost his job, in part over the perceived failure of the nation's espionage agency to ascertain Soviet advances in nuclear physics.[5] The problem was the Soviets had reaped an enormous amount of information about the American nuclear weapons program through a number of spies deeply embedded in various research projects. The result was an accelerated learning curve for the Soviet physicists.[6]

Conversely, the CIA had no human intelligence capability in the Soviet Union capable of penetrating the Soviet nuclear weapons program. Consequently, the Americans could only gauge Soviet progress by comparing Joe-1 and its developmental time line with the American experience during and after the Manhattan Project. This method, of course, would continue to produce estimates that miscalculated and underestimated Soviet progress. Even within AFOAT-1, it soon became common practice to assess the detected detonations by comparing the AEDS data against the data from the Nevada and Pacific shots. In reality, no one felt secure with a method based purely on the recent evolution of nuclear science. The loss of the American atomic monopoly propagated a sense of insecurity throughout the nuclear physics community and the federal government.[7]

The surprise of Joe-1 immediately jolted members of the Joint Committee on Atomic Energy (JCAE) to take action. On the same day as the president's announcement in the press of Joe-1 (September 23, 1949), the JCAE began holding a series of meetings proposing an enormous effort to surge the U.S. nuclear R&D and testing programs, especially in support of developing a thermonuclear weapon. They devoted their September 29, 1949 meeting to discussing the feasibility of "an all-out effort to build a hydrogen bomb." General Omar Bradley also appeared before the JCAE to strongly encourage development. AEC Commissioner Lewis Strauss, a strong proponent as well, pressed the president to approve the program. His general manager, Carroll Wilson, who had been at Vandenberg's Joe-1 verification meeting only four days earlier, proposed an all-out R&D effort to use the 1951 GREENHOUSE operation to test "an implosion bomb boosted with deuterium and tritium." Such a test would constitute an important step in developing an H-bomb.[8] At that time, Los Alamos was in the early planning stages for Operation GREENHOUSE, and Strauss and Wilson saw the test series as an important opportunity to advance the fusion weapons program.[9]

Not everyone agreed with the recommendations. As Oppenheimer continued to voice moral opposition, other important personalities in the scientific community urged caution and advocated for a delay in testing a thermonuclear device until well after the November 1952 presidential elections. At one point, Vannevar Bush approached Truman's secretary of state, Dean Acheson, to convince the president to postpone any such test until 1953. Bush argued that the new president (either Eisenhower or Adlai Stevenson, the Democratic candidate), would suddenly be placed in a dangerously changed world dynamic that could permanently terminate any chances for nuclear arms control. Oppenheimer, Conant, and Bethe all concurred with Bush. Truman, however, was not dissuaded. To underscore his decision, he did not renew Oppenheimer and Conant's appointments to the GAC and replaced them with proponents of fusion development.[10]

As time went on, Strauss, Bradley, and Teller gained enough support to convince the president to approve the program. In the months leading up to Truman's decision in January 1951, they recruited important allies such as Ernest Lawrence and Luis Alvarez of the Berkeley Radiation Laboratory.[11] In fact, Lawrence utilized his Berkeley lab to begin the initial development of a fusion device, as he perceived the AEC was moving too slowly to do so. In truth, the AEC was itself divided on the issue, especially when Strauss resigned (on the very day that Truman approved the thermonuclear program) and was replaced by David Lilienthal, who opposed the development of an H-bomb. Concurrently, the Air Force agreed with Lawrence's efforts and Teller's view that the AEC plans were vague and "threatened to open its own laboratory if AEC refused."[12]

As various Cold War events unfolded throughout 1950 and 1951, apprehension over the "unknowns" surrounding the Soviet program turned to fear. Lawrence and Alvarez actually believed the Russians already had the lead on the development of the H-bomb. "They declared that for the first time in their experience they are actually fearful of America's losing a war, unless immediate steps are taken."[13]

Adding to the sudden high-level support of thermonuclear R&D in early 1951 was speculation that the Russians, possibly facing a shortage of uranium, were moving rapidly toward developing such a device. By now, it appeared fission weapons were limited to a maximum yield of approximately 250kt due to the risk of an uncontrolled, premature nuclear reaction. An H-Bomb, however, promised to produce yields in large *megaton* ranges, thus making the existing U.S. stockpile of almost 200 fission bombs inadequate. An unconfirmed Soviet test in late 1950 added further impetus for its development. Allegedly, an AWS sampling flight had picked up a positive sample that left analysts feeling "chilly and strange." While the detonation was never confirmed (the second Soviet test would not occur until September 24, 1951), the scientists had found evidence of fusion and had only notified Truman, Acheson, and the MLC. The information was kept from Oppenheimer, Teller, and the AEC.[14]

Throughout 1950, as momentum grew for the support of an H-bomb and anxieties increased over the unknown state of the Soviet nuclear weapons program, the LRD researchers accelerated their preparations for two upcoming nuclear test series, codenamed GREENHOUSE and WINDSTORM. The former had been planned since July 1949 and was scheduled to begin in April 1951, and the latter was scheduled for autumn 1951. WINDSTORM was conceived by AFSWP and the AEC in early 1950 as a test of two 20kt devices detonated at the surface and subsurface levels. Whereas Operation GREENHOUSE focused on weapons design, WINDSTORM would be dedicated to weapons effects, an emphasis DoD strongly pushed. The operation would be conducted at a new site—Amchitka Island in the Aleutian chain of Alaska.[15]

The prospects of an underground test greatly excited Northrup. Long frustrated by the slow progress of geophysical R&D, the underground test, the first one ever conducted, would provide an unprecedented environment for seismic and acoustic research. For the first time, AFSWP was driving the test objectives and had requested project proposals from all stakeholders in DoD. In response, on June 19, 1950, Northrup presented Nelson with a detailed plan for the LRD experiments. He identified five projects encompassing 14 technical objectives:

Project 51.1: Atomic Cloud Diffusion

Project 51.2: Atmospheric Pressure Waves from A-bombs

Project 51.3: Seismic Pressure Waves from A-bombs

Project 51.4: Collection and Analysis of A-bomb Debris

Project 51.5: Aerial Survey of Terrain Consideration[16]

However, LRD budget restrictions were still largely in place. Expecting serious limitations to the AFSWP operational budget, Northrup proposed using 50 percent of AFOAT-1's contingency funds. He estimated a total cost of $565,000 (2021 = $6.4 million).[17] Nelson awaited word from AFSWP before acknowledging Northrup's budget recommendation. In mid-November, Nelson learned AFSWP had approved projects 51.1, 51.4, and 51.5. To Northrup's disappointment, they disapproved his two most important projects—acoustic and seismic. Northrup and Nelson received the news at a meeting of a CAE working group that had met to review AFOAT-1's supplemental budget request for FY51. The good news was the CAE would recommend AFOAT-1's 400K supplemental request to the RDB. However, based on remarks from D. Z. Beckler, executive secretary of the CAE, much skepticism still existed about the utility of the geophysical LRD techniques. Implying that the upcoming GREENHOUSE tests, less than six months away, would determine the future of the geophysical techniques, Beckler stated "GREENHOUSE should be the end to the litigation on the feasibility of geophysical means of the LRD program." It was clear that, for WINDSTORM, the AFSWP was not willing to absorb funding for geophysical techniques they believed held little promise. Fellow CAE member Colonel Holzman, ever the defender of LRD, countered "it would be fallacious to scrap the geophysical program simply because the required results were not forthcoming from GREENHOUSE." Recognizing the political precariousness of their position, Nelson and Northrup appeared to take the news in stride. A bit disingenuously, Northrup stated he was "quite anxious to establish a definite viewpoint either in the affirmative or in the negative."[18]

Unfortunately, additional site surveys of the proposed test area in Alaska determined the terrain was unsuitable for an underground test in terms of providing quality data. With that determination, AEC decided to conduct the WINDSTORM tests in Nevada. Renamed Operation JANGLE, the operation was merged with Operation BUSTER. Operation BUSTER-JANGLE was scheduled to begin on October 22, 1951, with the two new JANGLE shots targeted for late November. Regrettably for the geophysical researchers, an underground shot was no longer planned. Instead, JANGLE would consist of two surface detonations, with one occurring in a shallow crater. Northrup would have to wait five more years for an underground test.[19]

On January 4, 1951, Brigadier General Raymond C. Maude assumed command of the 1009th SWS/AFOAT-1 from Nelson.[20] A 1926 West Point graduate (a classmate of Nelson's), Maude was initially commissioned as a Signal Corps officer. During the Second World War, he served as the communications officer of the Ninth Bomber Command (the medium bomber component of Ninth Air Force), and then as the director of communications for the 29th Tactical Air Command as Allied forces advanced eastward across Germany to end the war. Staying on in Europe after the war, Maude served as the communications officer for U.S. Air Forces, Europe, until he returned to the U.S. to become chief of staff of the Airways and Air Communications Service in Washington, D.C. After transferring to the new U.S. Air Force, Maude served as the communications officer of Continental Air Command at Mitchell AFB, New York, and then in several staff assignments on the Air Staff before assuming command of the 1009th SWS/AFOAT- 1.[21]

Within AFOAT-1, as planning was underway for WINDSTORM, Northrup faced the most serious problem in his tenure thus far as technical director. In a sharply worded six-page memo to General Maude on March 30, 1951, he informed his new boss that a personnel crisis at Tracerlab threatened the operations of the AEDS and important LRD research. Sometime that month, Tracerlab's Technical Director F. C. Henriques and four senior scientists left the Eastern Division of the lab under apparently unfavorable circumstances, and the future employment of an additional 50 workers was in question. The root of the personnel problem was a serious financial deficit the lab faced in its contract with the Air Force, projected to reach $312,000 (2021 = $3.25 million) by February 1952, the termination date of the contract. Northrup viewed this development as a serious crisis because, at this point in time, Tracerlab had become the most important operational arm of the AEDS. In his memo, Northrup outlined a number of capabilities critical to the radiochemical analyses of gas and particles that would now be lost or seriously imperiled. Not only would the AEDS lose an invaluable capability, but that loss would jeopardize AFOAT-1's participation in GREENHOUSE only one week away and, more importantly, leave the nation "blind" in the event of a second Soviet nuclear test.[22]

Northrup was desperate to salvage GREENHOUSE. He told Maude that Tracerlab could default on their contract at any time, but he had reassured the lab on March 13th that AFOAT-1 would not hold them to any default as Tracerlab dealt with its problems. Northrup also expressed frustration that Tracerlab rejected his proposal to "reinstall all five individuals in Tracerlab in the status of Air Force consultants with responsibility of making their brains and good-will as far as Tracerlab is concerned available to carry out [GREENHOUSE]." For Northrup, the Tracerlab crisis demonstrated a serious concern he had expressed for some time—the inability of the Air Force to recruit and pay for the scientific expertise needed to develop LRD. He viewed this situation as a grave problem, writing that they were faced with the inability "to conduct expeditiously certain nuclear researches required by the Air Force to accomplish its mission in the approaching era of atomic warfare." Three months earlier, on January 3rd (Nelson's last day in command), Northrup had pleaded for a dedicated Air Force nuclear research laboratory. Realizing civil service pay could not match salaries in the private sector, he advocated for an Air Force lab that could hire or contract for periods longer than one year (a current limitation) for the necessary skills required to fulfill AFOAT-1's LRD mission. In his proposal, Northrup pointed to the success of the Joe-1 analysis, the risk associated with total reliance on Tracerlab, and the significant increase in U.S. tests as important justifications. Emphasizing that no one knew the speed of progress in the Soviet nuclear program, Northrup used the term "present national emergency" several times in his proposal. He "considered it essential that this laboratory be completed by October 1951," and envisioned a long-term integration with Tracerlab, whereby many contract scientists would work in the Air Force labs.[23]

Northrup proposed lab locations close to Tracerlab's eastern and western divisions. He estimated $500,000 (2021 = $5.2 million) for lab construction at Cambridge, Massachusetts, and $250,000 for Berkeley, California. For long-term planning, he calculated a $2 million annual operating budget. Obviously frustrated with government bureaucracy, Northrup concluded his proposal by stating:

> It cannot be too strongly emphasized that creative work on the frontiers of research with which the personnel of this laboratory will be dealing cannot be achieved in an atmosphere of regimentation. No significant scientific achievement has ever been accomplished in such an atmosphere. It is believed that in the present critical situation, there is no time to await the slow moving trend in present Civil Service laboratories toward less regimentation.[24]

With WINDSTORM planning in flux and subordinate to the primacy of developing a thermonuclear weapon, GREENHOUSE was the main event for 1951. While Northrup dealt with the Tracerlab crisis, his LRD researchers continued to focus their efforts on developing the required instrumentation to detect and locate future fission and (now) thermonuclear bomb detonations. For AFOAT-1 and the AEDS, the problem of detecting a hydrogen blast by geophysical methods seemed somewhat easier than detecting a fission detonation because a hydrogen bomb was expected to release upwards of 1,000 times the energy of a fission bomb. However, actual identification of a hydrogen bomb, it was assumed, would still depend upon the capture and analysis of gases in the atmosphere.

As they had done with CROSSROADS and SANDSTONE, LRD researchers planned to deeply embed themselves into the series of GREENHOUSE tests at the Pacific Proving Grounds in the Enewetak Atoll. Indeed, they began planning for GREENHOUSE even prior to the completion of the SANDSTONE data analyses. However, as weapons research designers planned for GREENHOUSE, LASL scientists advocated for a small, preliminary test series that would produce important data for use in determining design criteria for the nuclear devices earmarked for the GREENHOUSE shots in the Pacific. They had yet to determine how to initiate a thermonuclear reaction with a fission "trigger." In requesting a short-notice test, they argued that variations in the compression of the critical material could affect the yields of the GREENHOUSE devices. Only five weeks after LASL first expressed concern, President Truman approved Operation RANGER on January 11, 1951.[25]

Although AFOAT-1 personnel were consumed with GREENHOUSE preparations, AFOAT-1 suddenly found itself hastily engaged in RANGER when LASL (AEC) announced it could not conduct three important experiments due to a lack of time and resources as they shifted priorities to work H-bomb issues. To salvage their test plans, LASL personnel approached AWS and AFOAT-1 on December 18th to request their sponsorship of the planned experiments.

Seizing the opportunity to advance LRD, Northrup quickly drafted requirement plans and recommended AFOAT-1 participation to General Nelson on December 28th.[26]

LASL and AFOAT-1 goals aligned. The first and most important of those experiments concerned the meteorological tests designed to analyze radioactive samples taken from clouds resulting from the detonations (i.e., "fractionation of cloud particles by sheering winds"). LASL requested flight tracking out to 600 miles. LASL also informed the Air Force it could not provide adequate aerial survey work because the short time frame precluded the requisition of drone aircraft.

Northrup was especially interested in this unexpected opportunity to advance radiochemical analysis and to exercise the AEDS. The RANGER shots would consist of sub-critical amounts of fission material producing low yields. Fully aware the political winds were shifting toward thermonuclear weapons, Northrup informed Nelson "these tests, in all probability, will be the last U.S. shots of pure Pu-239 and U-235." With recently improved instruments, such as the super sampler, the collection of "large samples of these will permit determination of yield curves and further calibration data," which would accelerate the R&D program by four or five months. The shots would also allow AFOAT-1 to advance the meteorological research program, and the longitude and latitudes of the drifting effluents would "provide criteria for distinguishing Russian sub-critical bomb tests." AWS and AFOAT-1 authorities accepted responsibility for those tasks and rapidly produced a detailed plan of operations. According to post-RANGER reports, the Air Force support "provided more complete and more expert coverage than was at first thought possible."[27]

The five shots of Operation RANGER were conducted in rapid succession from January 25 to February 6, 1951, marking the first time nuclear devices were detonated within the continental United States since the Trinity test in July 1945. B-50 bombers dropped all five of the nuclear devices over

Frenchman Flats of the Nevada Proving Grounds from altitudes ranging between 1,100 and 1,450 feet. The AEC invited many members of Congress to view the detonations to demonstrate the AEC's ability to conduct safe nuclear testing within the continental United States. Compared to future tests, all of the shots were small and occurred without mishap. Two produced yields of 1kt, and two others yields of 8kt. The final test, Shot Fox on February 6th, was the largest of the series with a yield of 22kt.[28]

Despite extensive efforts directed toward planning for the April GREENHOUSE series in the Marshall Islands, LRD researchers utilized Operation RANGER to significantly advance aerial sampling. The sampling missions provided the Air Force the first opportunity to use manned sampling aircraft on a routine basis. Based on Major Fackler's demonstration during SANDSTONE and his subsequent advocacy for manned close-in aerial sampling, the Air Force now concluded such sampling was safe and feasible. The close-in sampling capability was important because analysts could develop decay curves for those immediate collections and then compare them against collections from the same cloud made at extended distances many days later. The 3425th Special Weapons Group and the 374th Reconnaissance Squadron (Very Long Range Weather) conducted the close-in and long distance sampling missions. The 374th RS, operating out of Nellis AFB, collected the samples utilizing two RB-29s, each with 10 crewmen. Each aircraft was equipped with two impact filter paper collectors, approximately 8-by-10 inches. These collectors were mounted in boxes, one on top of the fuselage behind the wing, and the other on the bottom of the fuselage forward of the tail skid. After landing at Nellis, the samples were split, with one-half going to TracerLab and the other to LASL.

Long distance sampling was successful for shots 2, 4, and 5 as the aircraft detected debris as far as the Persian Gulf. AEDS routine filter flights between Japan and Alaska—the

LOON CHARLIE missions—also detected debris from those shots. The two 1kt shots (1 and 3), however, were not detected beyond the borders of the United States. The Air Force continued to fly test sorties throughout February and March, during which time they collected more than 3,000 samples. However, due to the ongoing crisis with TracerLab, only 100 samples received radiochemical analysis. The analyses of those samples revealed many more extremely fine particles than previously observed.[29]

RANGER also allowed AFOAT-1 to exercise the AEDS. During Operation RANGER, five 1009th SWS officers manned the AEDS control center at Nellis AFB to coordinate operations with their field units. These units included ground filter stations at Wright-Patterson AFB, Ohio; Tinker AFB, Oklahoma; Rapid City AFB, South Dakota; and Offutt AFB, Nebraska. The AWS provided five RB-29 sampler aircraft for special vector flights from Barksdale AFB, Louisiana, and Robins AFB, Georgia. Their mission was to analyze the path of the nuclear contamination in areas outside the continental United States. The AEDS also employed additional ground filter stations and cloud sampler aircraft in Alaska, Japan, Guam, and Saudi Arabia.[30]

Northrup reported that the Nevada shots "created serious problems for the Atomic Energy Detection System" because the debris air stream from the Soviet tests (crossing the Japan to Alaska LOON CHARLIE track) overlapped the NTS debris trajectories across North America at the same altitude levels. Northrup also noted wintertime meteorological conditions tended to hold the clouds intact for a longer period of time. This concerned AFOAT-1 because the Soviets could conceivably "hide" behind the American atomic clouds to detonate bombs of their own, especially if the tests were underground. In other words, small Soviet tests could intermix with the RANGER debris, making identification impossible. AFOAT-1 specifically looked for evidence that such an event had actually occurred. Decay counting produced dates of origin where, "in every case, the time between origin and collection of the debris was much greater than calculated travel time from suspect areas in the USSR to the point of collection." Therefore, they concluded no Soviet test had occurred during the RANGER shots. Still, Northrup remained concerned that the increased use of the Nevada Test Site "will result in some compromise of effectiveness of the Atomic Energy Detection System operations." Future tests, he said, would create "periods of contamination." However, he believed modifications of existing counting equipment could simultaneously enable decay curve calculations on a large number of filters. He also emphasized the importance of collecting close-in samples and further development of particle analysis. His final recommendation was that the AFSWP and the AEC plan all future tests "in a manner to produce the least interference with Atomic Energy Detection System operations."[31]

Only two months separated the RANGER and GREENHOUSE operations. The four tower shots planned for GREENHOUSE began on April 8, 1951, with Shot Dog and were followed by shots Easy (April 21st), George (May 9th), and Item (May 25th). From a weapons-design viewpoint, the hastily-planned RANGER series paid off in dividends during the GREENHOUSE operation. RANGER had tested important triggers needed to test Teller's work on the H-bomb. Then, for shots George and Item, the next steps in determining the feasibility of achieving fusion was to use a fission detonation to increase yield (shot Item), and to heat deuterium to a point where it would burn to initiate a thermonuclear reaction (shot George). Those tests proved successful. George produced the United States' first thermonuclear reaction with an unprecedented yield of 225kt, proving an H-bomb was certainly possible.[32]

While the overall objective of GREENHOUSE was to test the efficiencies of weapons designs, especially in the development of an H-bomb, the extensive LRD experiments

138

were generally structured as they had been for SANDSTONE. LRD researchers found themselves embedded within a large joint task force where their experiments constituted only one of eight experimental programs. LRD was again officially designated as an independent program within JTF-3. The JTF-3 fell under the command of Air Force Lieutenant General Elwood R. (Pete) Quesada.[33]

The framework for the different LRD experiments at GREENHOUSE became the model and template for most of AFOAT-1's experiments in future tests. In GREENHOUSE, the LRD Program included eight experiments:

1. Radiochemical, chemical, and physical studies of atomic bomb debris

2. Infrasonic wave propagation studies

3. Location of an A-Bomb cloud by observation of air currents

4. Collection of bomb debris by airborne filters

5. Instantaneous detection of A-Bomb debris by balloon borne detectors

6. Collection of bomb debris at ground level and analysis of samples

7. Seismic wave propagation studies

8. Detection of A-Bomb debris by atmospheric conductivity[34]

The researchers in the LRD test program closely followed the procedures developed in SANDSTONE. Again, the experiments primarily focused on collecting bomb debris. Despite positive results from the detection of Joe-1, the radiochemical methods employed for the analyses of the long-range samples of SANDSTONE were not satisfactory. First, from the vast majority of the collections, only ultra-micro quantities of effluents were available, which made accurate quantitative analysis difficult. Also, the dissolution of the filter paper containing the samples presented many difficulties. The scientists required improvements in the ultra-micro methods for the detection of fissionable material to permit a lower level of observation.[35]

During GREENHOUSE, B-17 and T-33 drones were again used to penetrate the clouds, although they were part of Program 8 that experimented with the blast effects on aircraft. For LRD, WB-29s from the 55th and 57th Strategic Reconnaissance Squadrons from McClellan AFB, California, and Hickam AFB, Hawaii, respectively, conducted aerial sampling and cloud tracking for 36 hours following the detonations. All of the WB-29s carried the new C-1 box filter system, now referred to as the "Shoebox," which allowed better air flow and changeable filter papers. Crews changed one of the system's two filter frames every 10 minutes when the aircraft flew above 2,000 feet. Protective clothing was minimal as the person changing the filters only wore gloves, a respirator or oxygen mask, a dosimeter, and a film badge. The WB-29s which flew out of Kwajalein first stopped at Enewetak to remove the samples and to conduct decontamination "if needed."[36]

Like the shortcomings with radiochemical methods, acoustic measurements during SANDSTONE revealed the inadequacy of existing instruments and techniques. Microbarographs lacked sufficient sensitivity, low frequency response, stability, and adequate recording and analyzing systems. In addition, the test revealed the need for a systematic study of the nature of noise background and for the development of noise-reducing techniques. AFOAT-1 researchers sought to address these problems in GREENHOUSE.[36]

For the seismic and hydroacoustic experiments of the LRD program, AFOAT-1 planned to use the young AEDS. Unlike SANDSTONE, researchers already had approximately 25 monitoring stations in position to observe the detonations. Still, AFOAT-1 resources were stretched thin as the initial stations of the AEDS were new or in the process of becoming fully mission capable. Consequently, AFOAT-1 established an additional 50 temporary stations world-wide specifically for Operation GREENHOUSE. This was an important expansion in terms of preparation because the primary function of the seismic component of the AEDS was to serve as an early warning device to alert other components in the AEDS, and to formulate meteorological trajectories—methods that had failed to detect Joe-1.[37]

Overall, the LRD program of experiments in GREENHOUSE yielded a much greater volume of test data than had SANDSTONE. This was largely due to the dramatic increase in equipment utilized for collection, both as part of the AEDS and from temporary locations. GREENHOUSE, however, did not significantly expand beyond the triad of seismic, hydroacoustic, and radiochemical research. At the time of the operation, efforts remained focused on R&D, and the tests allowed AFOAT-1 to employ new or improved instruments. While funding had remained minimal, R&D personnel had made significant progress since SANDSTONE in operationalizing advances in all three areas. For example, they had developed geophysical instruments for detecting a disturbance of atomic magnitudes by measuring pressure waves. The most significant development was the new array configurations for acoustic and seismic data collection techniques that proved successful during GREENHOUSE. The arrays succeeded in reducing background noise levels and provided the azimuths of incoming signals. Researchers eagerly awaited new seismograph and microbarograph components that were expected soon. At the time of GREENHOUSE, during Shot George, new seismographs (Benioff instruments) replaced the ineffective DTMB and Stanley systems and were installed at some AEDS stations as well as many of the temporary sites. Some worked so well that they were considered fully operational and finally efficient enough to provide full-time monitoring of the Soviet Union.[38]

In advancing airborne sampling, all of the new equipment and instruments identified in Northrup's July 1950 capabilities statement were successfully tested at GREENHOUSE. The instantaneous radiation detectors—the scintillation type beta gamma detector and the atmospheric conductivity type (the ion detector)—all performed well in positioning the aircraft more rapidly and accurately on the targeted effluents. The samples collected with the new C-1 filter boxes improved the contamination problem and resulted in greater degrees of accuracy of the micro chemical and physical analyses. The new "super-sampler" collected larger samples that also contributed to an increase in the accuracy of radiochemical analyses.[39]

With more aircraft now flying routine sorties on the periphery of the Soviet Union, AFOAT-1 required a much larger laboratory capability. While Tracerlab of New Jersey had established a lab at McClellan AFB for GREENHOUSE (by modifying building S-34 and procuring equipment from local vendors), AFOAT-1 planners were busy bringing the field radiochemistry labs in Alaska, Guam, and Japan (established shortly before RANGER) up to an efficient level of operation. Several more were planned for the near future.[40]

Overseeing the expansion now was AFOAT-1's new commander, Major General Donald J. Keirn, who assumed command from Maude on September 16, 1951.[41] Originally a field artillery officer in the Army, Keirn completed pilot training in 1930. However, with an advanced degree in aeronautical engineering, he spent much of his long career serving in engineer or nuclear R&D assignments. In April 1946, General Keirn became liaison officer for the Manhattan Engineering District and, while serving in that capacity, participated in the atomic bomb tests at Operation CROSSROADS. In May, he was appointed Special Assistant to the Director of the Division of Military Application at the AEC. While in that position, Keirn served on the initial LRD committee that met on May 14, 1947 to organize the nation's LRD effort. Just prior to assuming command of the 1009th SWS/AFOAT-1, Keirn was appointed Deputy Chief of Staff for Research of the newly organized Air Research and Development Command (ARDC). Keirn's selection to lead AFOAT-1 was critical at this point in time because of ARDC's recently gained authority over AFOAT-1 's R&D budget that carved away some of Northrup's budgetary controls (discussed in Chapter 6).[42]

By early fall of 1951, as LRD researchers finalized planning for the next series of tests scheduled for late October (codenamed BUSTER-JANGLE), airborne sampling remained the most prominent technique in the AEDS. The AWS flight patterns around the periphery of the Soviet Union were now operating at peak efficiency as Northrup and his team anxiously awaited a second Soviet nuclear detonation. Keirn and Northrup were now quite aware that the high anxiety expressed by upper-level decision-makers made the AEDS critically important. Two years had passed since Joe-1, and the Air Force realized it was only a matter of time until a second test occurred. AFOAT-1 AEDS operators and AWS aircrews did not have much longer to wait.

Major General
Raymond C. Maude

Commander, 1009th SWS/AFOAT-1 from January 4 to September 16, 1951. Maude was initially commissioned in the Signal Corps. During WWII and after, Maude served mostly in communications assignments. Only in command for eight months, Maude led AFOAT-1 in the GREENHOUSE experiments and significantly advanced radiochemical analysis capabilities. *Photo source: AFTAC archives*

Major General Donald J. Keirn

Commander 1009th SWS/AFOAT-1 from September 16, 1951 to August 16, 1952. An aeronautical engineer, Keirn largely served in engineer or nuclear R&D assignments. He participated in Operation CROSSROADS and served on the initial LRD committee that met on May 14, 1947 to organize the nation's LRD effort. Keirn's selection to lead AFOAT-1 was critical in securing adequate funding for R&D. *Photo source: AFTAC archives*

Dr. Ernest Lawrence

Was well-known for his work on uranium-isotope separation during the Manhattan Project. He won the Nobel Prize in 1939 for his invention of the cyclotron. In the 1950s, he allied with Teller and Strauss to promote the development of the H-bomb. With support from the Air Force, the three scientists resisted Eisenhower's attempt to negotiate an arms control agreement with the USSR.
Photo courtesy of the Nobel Foundation

Dr. Edward Teller

A close associate of Lawrence, Teller was an early member of the Manhattan Project. After the war, he became Director of the Lawrence Livermore National laboratory. Known as the "father" of the hydrogen bomb, teller was an ardent anti-Russian and advocated for a large U.S. nuclear arsenal. With Lawrence, he opposed Eisenhower's quest for a nuclear arms control treaty. *Photo courtesy of the DoE Office of Scientific and Technical Information*

AFOAT-1's first laboratory

Authorization to convert building S-34 at McClellan Air Force Base was received on August 18, 1950. *Photo source: AFTAC archives*

Initial filter mounted under fuselage of the RB-29s

Photo source: AFTAC archives

C-1 foil mounted on top of fuselage (rear view)

Photo source: AFTAC archives

145

Team 407 of the 1009th SWS/AFOAT-1
Taking a break during Operation GREENHOUSE, May 1951.
Photo source: AFTAC archives

Notes to Chapter 5

[1] Because thermonuclear weapons were thousands of times more destructive than fission bombs and would produce casualties in horrific numbers, the scientific and military communities became embroiled in huge debates over the moral and ethical possession of such weapons. These debates seriously divided the scientific community causing many scientists to walk away from government work. See especially Richard Rhodes, *The Making of the Hydrogen Bomb* (New York Simon and Schuster, 1995); and Herbert York, *The Advisors: Oppenheimer, Teller, and the Superbomb* (Stanford, CA: Stanford University Press, 1976). At the center of all of this was the Oppenheimer-Teller feud. See especially, Herken, *Brotherhood of the bomb*.

[2] As he made his decision, Truman's only question to the experts was "can the Russians do it?" Bowie and Immerman, *Waging Peace*, 16. Also, Divine, *Blowing on the Wind*, 14-18; and Gaddis, *We Now Know*, 231.

[3] As cited in *Defense's Nuclear Agency: 1947-1997*, 73. Coincidentally, that was the same time estimate that Nelson had given Vandenberg just prior to Joe-1.

[4] Ibid.

[5] Truman removed Hillenkoetter from the CIA in mid-August 1951 due to the surprises of Joe-1 and the North Korean invasion. After Joe-1, Congress lambasted Hillenkoetter, and the rank and file within the CIA lost confidence in their Director.

[6] Richard Rhodes, *Dark Sun*, 363. Note that the best detailed accounts of the Soviet spying efforts are found in this book. The most serious breach was caused by Klaus Fuchs, a member of the British contingent at Los Alamos for the Manhattan Project. See *Klaus Fuchs, Atom Spy* (Cambridge: Harvard University Press, 1987).

[7] Decades later, after the Cold War, Russian test data from the early atmospheric Joe shots revealed that the AEDS data frequently overestimated the yields.

[8] Rhodes, *Dark Sun*, 379.

[9] Herken, *Brotherhood of the Bomb*, 224.

[10] Ibid., 248-251.

[11] Rhodes, *Dark Sun,* 496-497. Lawrence would soon move his lab to a new site at Livermore, California; a former Second World War base soon to be called the Lawrence Livermore Laboratory.

[12] Ibid. Also Herken, *Brotherhood of the Bomb,* 248.

[13] William L. Borden notes, JCAE Classified Document No. 66, National archives, as cited in Rhodes, *Dark Sun,* 385.

[14] Miller, *Under the Cloud,* 110-113.

[15] DTRA, *Defense's Nuclear Agency: 1947-1997,* 84.

[16] Technical memo #44, Northrup to Nelson, Subject: "U.S. Air Force Program in Connection with Proposed Underground Test," June 19, 1950, AFTAC archives.

[17] Ibid.

[18] Memorandum for Record, Subject: "Report on Meeting before the Hosford Group of the CAE on the Fiscal Year 1951 Third Supplemental Request of AFOAT-1," John Lewkovich, CAE Administrative Officer, November 22, 1950, AFTAC archives.

[19] Technical memo #49, Northrup to Nelson, Subject: "Program Guidance in the Field of Atomic Energy," October 12, 1950, AFTAC archives.

[20] General Orders Number 1, January 4, 1951, AFTAC archives.

[21] Biography, Major General Raymond Coleman Maude, http://www.af.mil, last accessed June 4, 2018.

[22] Technical memo #60, Northrup to Maude, Subject: "Modifications in Work Load-AFOAT-1 R&D Program," AFTAC archives.

[23] Technical memo #57, Northrup to Nelson, Subject: "Requirement for an Air Force Nuclear Research Laboratory," January 3, 1951, AFTAC archives.

[24] Ibid.

[25] Miller, *Under the Cloud,* 84.

[26] Technical memo #54, Northrup to Nelson, Subject: "Proposed Atomic Tests for Early 1951," December 28, 1950, AFTAC archives; Technical memo #55, Northrup to Nelson, Subject: "Proposed Atomic Tests for Early 1951 (AFOAT-1 Requirements)," December 28, 1950, AFTAC archives.

[27] Ibid., Also Thomas L. Shipman, M.D., "Report of Rad-Safe Group," April 17, 1952, Operation RANGER, Los Alamos Scientific Laboratory, Los Alamos, New Mexico, 55. Attending the December 18th meeting was Major Gerard Leies, who would later serve as the

Technical Director of the Air Force Technical Applications Center from 1973 to 1982.

[28] Miller, *Under the Cloud,* 83-106.

[29] Technical memo #62, Northrup to Maude, Subject: "Effect of U.S. Atomic Tests on Reliability of the Atomic Energy Detection System," May 9, 1951, AFTAC archives.

[30] Ibid.

[31] Technical memo #62.

[32] Rhodes, *Dark Sun,* 457; Miller, *Under the Cloud,* 113.

[33] Quesada had made his mark as a tactical airpower pioneer credited with developing precision close air support and the employment of forward air controllers (FACs). Quesada earned the Distinguished Flying Cross, alongside Carl Spaatz, in the 1929 successful demonstration of air-to-air refueling. In 1946, Quesada became the first commander of the new Tactical Air Command (TAC) but quickly became disillusioned as General Curt LeMay's Strategic Air Command (SAC) dominated Air Force planning and resources. GREENHOUSE would be Quesada's last command. Several months after the operation, Vandenberg approved his request for early retirement.

[34] L.H. Berkhouse, S.E. Davis, F.R. Gladeck, J.H. Hallowell, C.B. Jones, E.J. Martin, F.W. McMullan, and M.J. Osborn, "Operation GREENHOUSE: 1951 ," Technical Report. (Washington: Defense Nuclear Agency, June 15, 1983), 176-178.

[35] Ibid.

[36] *History of Long Range Detection: 1947-1953,* 56-57. Goals were directly derived from Tech memo #43 which was revised after Joe-1.

[37] Ibid.

[38] Romney, *Recollections,* 111-116.

[39] *History of Long Range Detection: 1947-1953,* 52.

[40] Ibid.,156.

[41] General Orders Number 13, September 16, 1951. AFTAC archives.

[42] Biography, United States Air Force, Major General Donald J. Keirn, http://www.af.mil last accessed September 11, 2018.

CHAPTER 6

Refining the Experiments

There was little doubt RANGER and GREENHOUSE had successfully tested operational improvements in instrumentation to make the AEDS more effective. Until Operation GREENHOUSE, AFOAT-1 had placed primary emphasis on designing an AEDS rather than operating the system as a world-wide surveillance network. To date, AFOAT-1 had served principally as a planning center for R&D, especially with the number of outside organizations contributing to the LRD experiments. Following GREENHOUSE, it became clear LRD demanded a steep learning curve and was not as easy to develop and operate an AEDS as policymakers believed.

Operationalizing new science was extremely challenging and, as the Cold War quickly escalated, AFOAT-1 came under intense pressure to simultaneously expand the AEDS both in operational size and with new or improved techniques. Operationally, AFOAT-1 was on track to build capabilities to accomplish the mission that the RDB had articulated in April 1950: "to establish, operate and maintain an atomic energy detection system utilizing specific means for the detection of future atomic explosions." AFOAT-1 was still adhering to the 12-month "Interim Plan for Atomic Energy Detection Operations," published on June 1, 1950, on the eve of the Korean War. That plan was designed to meet the adjusted target date of mid-1951 for a fully operational AEDS. In retrospect, it was an impressive scope of expansion considering the workload generated by the RANGER and GREENHOUSE data collections, the planning for Operation BUSTER-JANGLE, high-level pressures generated by Joe-1, fears of Russian advances, and the decision to produce a thermonuclear weapon.

AFOAT-1 defined the requirements of a mission-capable AEDS under six operational parameters:

1. The collection of airborne particles and gas from six air bases deployed around the world;

2. Sampling of surface level particles with approximately thirty-five stations;

3. Surface level gas sampling at approximately nineteen stations;

4. Geophysical surveillance by seismic and acoustic methods, or both, at nine locations;

5. Maintenance of laboratories at nine locations for the analysis of data produced by various collection techniques; and

6. Maintenance of five field offices for supervision of data collection stations.

The field offices were especially critical in supervising the growing AEDS and in moving the LRD program forward on a permanent basis. In the fall of 1950, AFOAT-1 created a command and control facility at McClellan Air Force Base, called the Western Field Office (WFO), which acted as an extension of the Washington staff to administer the increasing field responsibilities. The budget histories (overall and for LRD) also reflected this shift in emphasis:[1]

OVERALL LRD BUDGETS			
FY	R&D	O&M	Totals[2]
1948	2.7M	0	2.7M
1949	5.5M	0	5.6M
1950	3.45M	310K	3.76M
1951	3M*	3.25M	6.25M
1952	1.37M	2.5M	3.87M
*$3 million in 1951 = $32 million in 2021			

[2]

LRD RESEARCH AND DEVELOPMENT PROGRAMS					
Program	FY50	FY51	FY52	GREENHOUSE	Totals
Nuclear	355K	355K	50K	1,191K	1,191K
Acoustic	200K	0	100K	335K	635K
Seismic	119K	140K	50K	365.6K	674.8K
Meteorological	215K	145K	0	154.5K	514.5K

Emphasis on transitioning to a stronger operational posture was underscored on September 24, 1951, when the Soviet Union tested its second nuclear device. Conducted from a tower at the Kazakhstan site, this second test produced a yield of 38kt. Three weeks later, on October 18th, just four days before the first shot of Operation BUSTER-JANGLE, the Soviets detonated their third nuclear device. This was their first aerial deliverable bomb, at a yield of 42kt.

Considering its early stage of development, the AEDS performed well in detecting the two detonations. For the September Soviet test, the AEDS acoustic stations located in Japan, Alaska, and Turkey immediately noticed a large disturbance within the Soviet Union. Using the acoustic results from Operation GREENHOUSE for comparison, AFOAT-1 determined the explosion occurred at 1018Z hours on September 24th in the vicinity of Lake Balkhash, Kazakhstan. AFOAT-1 meteorologists forecast the air masses to cross the AEDS flight line on September 28th. AWS flights vectored from Yokota, Japan, successfully intercepted the cloud and collected two filter samples registering approximately 30,000 cpm. Subsequent long-range sorties of the cloud at 25,000 feet over Japan, the Pacific, and the United States yielded additional samples. In total, the AWS flew 44 sorties. For the October test, the U.S. Army Signal Corps' Central Evaluation Unit reported a series of acoustic signals indicating a nuclear detonation in the USSR at 0720Z on October 18th in the same area as the previous test. On the basis of these signals, AFOAT-1 vectored a WB-29 from Eielson AFB to Yokota AB on October 20th. This aircraft used both short- and long-duration papers to collect samples of 15,000 cpm and 38,000 cpm respectively. From October 20 to 30, daily sorties produced 211 samples with cpm rates above 5000, and six samples measuring above 1,000,000 cpm. In total, AWS flew 55 sorties.[3]

Many LRD personnel were surprised that the acoustic technique detected both detonations so well. Until now, planners had expressed little faith in the fidelity of the acoustic system simply because it had not reported Joe-1. Although a subsequent review of the data from the September 1949 event revealed evidence of a detonation, it had gone unnoticed by the Army operators due to the immaturity of the system at that time, and because of the heavy emphasis placed on aerial sampling. Consequently, no funds were allocated to acoustic R&D for FY51 (see budget chart above), but that would soon change.[4]

The back-to-back Soviet detonations in September and October 1951 seemed to cap a year of international anxiety, spurring LRD officials to mature the AEDS as quickly as possible. Within the context of a rapidly escalating Cold War, AFOAT-1 sought to expand the AEDS using two approaches: to expeditiously expand the AEDS and to work closely with Air Material Command (AMC) to procure AEDS equipment. The latter posed a challenge for AMC due to the fact LRD instrumentation and other logistical requirements were very unique. However, AMC now understood that the AFOAT-1 mission was one of the nation's top priorities. They expeditiously met the extensive number of requests, largely through commercial contracts.

One of the most successful of these contracts involved the initial production, procurement, and fielding of a ground-based whole-air sampler. Taking its name from Johnson's 1949 list of 38 short- and long-term LRD projects, "B/20", the whole-air sampler functioned simply by capturing air in cylinders that would then be shipped to a laboratory for analysis. Work to develop the sampler had begun in March 1950 but had progressed slowly in the face of other priorities. At that time, the Air Force had allocated $185,000 (2021 = $2.1 million) for the modification of seven Type A-1 oxygen-generating units, already in the Air Force inventory, to fabricate the B/20. The fabrication involved substituting a suitable diesel engine for the existing column and attaching a special unit furnished by the New York-based Air Reduction

Sales Company (Airco).[5] While AFOAT-1 had deployed eight B/20 units by the time of the second and third Soviet tests, the stations experienced difficulties in keeping the equipment operating at all times. The units were prone to maintenance problems, largely because of the diesel drives, and required the local purchase of spare parts. Indeed, replacement generators and spare parts for the eight stations consumed an enormous sum of money—$1.26 million in 1951 (2021 = $13.1 million).[6]

After the Soviets detonated their third device, AFOAT-1 made the expansion of the B/20 program a high priority. Plans for the further expansion of B/20 operations during FY52 required procurement of additional oxygen generators and compressors. Consequently, AMC was directed to procure four additional skid-mounted electric units, three compressor kits, and seven additional attachments. This project totaled $415,000 (2021 = $4.3 million), of which $66,000 was for additional spare parts. The urgency was such that AFOAT-1 made every effort to place the contract without delay with delivery to occur as soon as possible. AFOAT-1 authorized the highest priority for material and the payment of overtime to expedite the procurement. In response, AMC designated a high-ranking individual to serve as the B/20 Program Manager (PM). The PM had full authority to take any action necessary to meet AFOAT-1 requirements for systems and spare parts.[7]

Similar efforts were underway for the ground-based particulate sampler—the Ground Filter Unit (GFU). Although 10 crude GFUs were fielded in late 1949, they did not fully meet mission requirements (i.e., they were far less efficient than airborne particulate samplers), and were prone to maintenance problems as well. In early 1950, AFOAT-1 researchers designed an improved GFU. They also identified an available turbocompressor manufactured by the Spencer-Turbine Company that met all operational requirements. Unlike the B/20 air sampler, the GFU was less expensive and more rudimentarily designed. Key to the new design was the requirement to utilize AFOAT-1's existing Type-5 sample paper. The Spencer-Turbine Company succeeded in matching the proper turbo-compressor to the system at a cost of $1,600 per unit (2021 = $167,000). However, while the 29 units were being manufactured, research resulting from SANDSTONE led the LRD researchers to develop a new type of filter paper. The new requirement caused several months of delay as the modifications were developed. Then, as GREENHOUSE was underway, AFOAT-1 requested another change, wanting AMC to design and procure a prototype gasoline or diesel engine to drive the new GFUs. Both AMC and Spencer-Turbine accomplished the modifications within a few months. Just prior to the second Soviet detonation, AFOAT-1 positioned the first two new GFUs in Greenland and Alaska.[8]

The second strategy employed to rapidly expand the AEDS was to solicit Canadian and British assistance (see Annex 2: Allied Cooperation in the AEDS).

The elevation in LRD priorities for both research and operations throughout 1951 cannot be understated. These dual priorities consumed limited resources as the U.S. raced to keep ahead of the Soviets in weapons design programs and in improving the AEDS to monitor any Soviet detonations. For LRD researchers, expectations and political pressures were high as they began their experiments in Operation BUSTER- JANGLE.

BUSTER-JANGLE, conducted from October 22 to November 29, 1951, consisted of seven nuclear detonations. Four of the shots were airdrops, while the other three consisted of one tower, one surface, and one crater detonation. The surface and crater tests were the first of either type at the Nevada Proving Grounds (NPG). The series tested several new nuclear devices and was utilized to improve military tactics, equipment, and training. In regard to the latter, approximately 6,500 DoD personnel took part in the Army-sponsored exercises Desert Rock I, II, and III.[9]

The LRD Program consisted of six experiments. The primary R&D emphasis remained on the radiochemical analysis of nuclear debris. At BUSTER-JANGLE, LRD researchers continued to rapidly build their knowledge base, making significant progress. At TRINITY, they had learned that characteristic properties of atomic explosions could be determined by measuring the efficiency of the utilization of the fissionable material. At SANDSTONE, the scientists had employed new techniques to do so. Although the SANDSTONE measurements were limited, they were sufficiently positive to warrant investments in specialized techniques to obtain sound and reliable quantitative radiochemical assays of extremely minute amounts of atomic bomb debris. At GREENHOUSE, the researchers were at a point in the maturity of analysis that they began to look for departures from previously measured quantities as an indication of any special features in an explosion. BUSTER-JANGLE was an extension of that effort. Here, researchers advanced their work on determining the condition of an explosion (i.e., underground, tower, or air burst).[10]

BUSTER-JANGLE also allowed significant advancements with other detection techniques. The second experiment observed seismic wave propagation of bombs detonated over a land mass. Researchers measured amplitude, wave type, frequency, propagation velocity, and the attenuation of elastic waves generated by the detonation. Project personnel obtained extensive data from six onsite stations as well as a number of offsite stations in the AEDS. The new data allowed comparisons with existing seismic data and observations. It also allowed inferences about the capability of establishing scaling laws and of determining the energy in the blast by seismic measurements taken at long range. The goal was to produce criteria for distinguishing between natural and artificial seismic events. The second experiment also explored the advantages to long-range detection of the seismic and acoustic techniques working in concert.[11]

The third project tested the range and reliability of acoustic detection equipment for continental nuclear explosions of various yields. The scientists took measurements of the airborne low-frequency sounds from the detonations out to distances exceeding 1,500 nautical miles. The results confirmed the feasibility of using the acoustic technique to detect and locate distant atomic explosions of various calibers detonated in the air, on the ground, or in shallow underground depths.[12]

In terms of the latter environment, the last shot of BUSTER-JANGLE, Shot Uncle, was only a crater shot. Detonated at a shallow depth of 17 feet, it did little to advance geophysical R&D. Uncle produced a yield of 1.2kt and created a 53-foot crater in which unprecedented, extraordinarily high radiation levels were measured.

Concerned about intense radiation drifting toward populated areas, the Air Force launched a sampling sortie to follow the cloud over the Continental Divide. While sampling over South Dakota, the radiation readings exceeded 40,000 cpm. From this moment on, concerns over the dangers posed by fallout would grow exponentially among the public as more tests were conducted in the years ahead. The urgency of predicting the cloud path also served to underscore the difficulties of meteorology and the need to improve meteorological modeling—a capability which would prove critical after all nuclear testing in the world went underground in 1980. The meteorologists involved in Shot Uncle somberly noted "even over a region with relatively dense, reliable, and promptly available upper-air meteorological data, it is not always possible to accurately forecast the path of the debris."[13]

With little time to digest all of the data obtained in BUSTER-JANGLE and with Operation TUMBLER-SNAPPER only days away, Northrup received unwelcomed news. On January 23, 1950, the Air Force had established the Air Research and Development Command (ARDC) to centralize all R&D. Having resolved the inadequacies of AMC's previous management of AFOAT-1's research programs by granting AFOAT-1 full acquisition authority, Northrup strongly objected to the Air Staff's plans, which would now place his R&D programs under ARDC. He argued it was almost impossible to carve out AFOAT-1's R&D budget from the operational budget of the AEDS. Such an attempt would severely handicap AFOAT-1 because there was no clear dividing line between the two. The AEDS was essentially an R&D surveillance system in constant flux as newly-developed instruments and LRD techniques were added to the network or as existing instruments received innovative upgrades from other government agencies or private industry. Unfortunately, Northrup lost the argument and received notification on March 4, 1952, that ARDC would manage some of the AFOAT-1 R&D budget. Consequently, despite the heavy workload, much manpower was diverted to plan out how that split would be determined. The end result was that ARDC took control of approximately 10 percent of AFOAT-1's 1952 and 1953 R&D budget. The impact was not so much a financial one but rather an issue of effectiveness. Every project had pieces carved away. Improving the AEDS would now take longer due to bureaucratic accountability.[14]

As Northrup dealt with the crisis over R&D funding, LRD researchers launched another robust series of experiments with the start of Operation TUMBLER-SNAPPER. This operation consisted of eight shots and ran from April 1 to June 5, 1952. The first four tests were airdropped bombs testing weapons effects and the effect of height of burst on overpressure. The latter four tests were weapons design tests. TUMBLER-SNAPPER was a large DoD operation with 10,600 participants. Approximately 70 percent of that total took part in the Army-sponsored exercise Desert Rock IV.[15]

Four of the five LRD projects for TUMBLER-SNAPPER were essentially a continuation of those conducted in BUSTER-JANGLE and extended the primary experiments from earlier tests. Project 2, "Detection of Airborne Low-frequency Sound from Atomic Explosions" (which was Project 3 in BUSTER-JANGLE), sought to determine the accuracy of long-range acoustic detection methods. The Army's Signal Corps Engineering Laboratories conducted this LRD test with operating stations in Alaska, Hawaii, Kentucky, New Jersey, Texas, and Washington. Results from the project reinforced conclusions drawn from previous test series. The detection range of acoustical equipment depended upon yield of the detonation, atmospheric conditions, existing noise levels at each recording station, and the sensitivity of the sound-receiving equipment. These test results led LRD researchers to recommend this project as a priority experiment in future test series.[16]

Project 3, "Radiochemical and Physical Analysis of Atomic Bomb Debris" (Project 1 in BUSTER-JANGLE), was conducted by AFOAT-1 personnel in conjunction with Program 13, "Radiochemistry Cloud Sampling," performed by the 4925th Test Group (Atomic) of Kirtland AFB. The Air Force flew B-29, T-33, and F-84 aircraft to collect the air and particles. The sampling aircraft penetrated the rising clouds at different altitudes ranging from 11,000 to 40,000 feet. In the initial shots, the pilots penetrated the clouds only three or four times. However, during Shot Dog, the samplers entered the cloud 34 times. Following the detonations, the sampling aircraft tracked the drifting clouds for approximately five hours.[17]

AFOAT-1 and the Coast and Geodetic Survey conducted Project 4, "Seismic Waves from A-Bombs Detonated over a Desert Valley." The objective was to determine the seismic properties of the geological structure of the test area. Four hours after the announcement of recovery hour, AFOAT-1 personnel retrieved the seismic records from stations located in nearby Yucca and Frenchman Flats, and at remote sites ranging out to a distance of 440 miles. The project confirmed results obtained at Operation BUSTER-JANGLE, that less than five percent of the energy entering the ground as seismic waves is transmitted to remote locations.[18]

The one new project in TUMBLER-SNAPPER was Project 1, with two subcomponents. Project 1a, "Electromagnetic Effects from Atomic Explosions," was conducted by the National Bureau of Standards, Air Force Cambridge Research Center, Air Weather Service, and the Geophysical Laboratory of the University of California at Los Angeles. They designed the project to study the electromagnetic pulses produced by a nuclear detonation. Data were evaluated as a means of determining the location of distant nuclear explosions. Onsite stations were located at Frenchman and Yucca Flats, and offsite stations were in Colorado, Florida, Georgia, Massachusetts, New Mexico, Virginia, Bermuda, Germany, and Puerto Rico. Project 1b, "Long-range Light Measurements," was conducted entirely offsite by AFOAT-1 personnel and contractors from Edgerton, Germeshausen, and Grier, Inc. (EG&G). AFOAT-1 and EG&G established light-detecting stations in Arizona, Idaho, Texas, and Washington. AFOAT-1 manned each station with 10 personnel and two EG&G contractors, who took measurements for six hours before and one hour after each detonation.[19]

TUMBLER-SNAPPER produced a large volume of data. However, there was little time to analyze and report much of what the scientists collected. LRD planners only had four months before the start of the next test series codenamed Operation IVY. For AFOAT-1, the operational tempo had never been higher.

During this interlude between operations TUMBLER-SNAPPER and IVY, the tempo in American politics had also risen to a high level as the presidential elections approached. During the summer and early fall of 1952, the election campaign highlighted an ugly and destructive climate of anti-communist sentiment across the nation. The two contenders for the Republican nomination, Dwight Eisenhower and

Howard Taft, took on Truman's policy of containment as ineffective and too costly, and directly challenged the administration's efforts to confront Soviet expansionism. Their assertions resonated with the American people. A Gallup poll recorded on February 14, 1952, showed that Truman had an approval rating of only 22 percent. The "too soft on communism" perspective was exacerbated to an extreme by Wisconsin Senator Joseph McCarthy, who had created turmoil over the last two years by claiming the State Department and other government organizations had hundreds of communists working among their ranks. The assertions led to more than 100 congressional investigations, most notably by the House Committee on Un-American Activities. Those investigations resulted in the conviction of hundreds of people and more than 10,000 people losing their jobs. Although Eisenhower stayed aloof from McCarthyism, especially after he won the party nomination on July 11th, the majority of Americans supported the investigations. As David Halberstam noted: "McCarthy's carnival-like four-year spree of accusations, charges, and threats touched something deep in the American body politic. . . . McCarthyism crystallized and politicized the anxieties of a nation living in a dangerous new era."[20]

International events fueled such anxieties and contributed to this "red scare." In Europe—the front lines of the Cold War—the division of Germany appeared more permanent with each passing day. On March 10, 1952, Stalin offered German unification and free elections if the new Germany remained a neutral state. Not trusting the Soviets after witnessing communist takeovers in Eastern Europe, the United States signed the European Defense Community (EDC) treaty in May, which permitted a re-armed German military to be integrated into a multinational armed forces. In July, the West rejected Stalin's proposal, which in turn precipitated Russia's approval for the East German Communist Party to implement programs of socialization.

Several weeks later, in June, SAC began exercising alert deployments of B-36 and B-47 long-range nuclear bombers to overseas bases where they could strike Moscow without refueling.[21]

By the summer of 1952, most Americans had grown tired of the stalemate in Korea. Eisenhower promised to end the war if elected, and to go there personally to assess how to do so. During his campaign, he embraced John Foster Dulles's views on ending the policy of containment and adopting a new policy of retaliation and liberation in confronting Soviet expansionism. Dulles and Eisenhower's policy of massive retaliation, which would become known as the "New Look" after his election, would also allow the new president to pursue his austerity program by permitting extensive cuts in conventional forces.[22] The new policy would place far greater reliance on nuclear weapons and give prominence to the Air Force. Within this context, the AEDS took on greater importance. It was now clear the U.S. had to stay ahead of the Soviet Union in weaponizing nuclear energy. Knowing where the Russians were positioned in their nuclear program now became an even greater national imperative. In 1952, LRD was the best method to accomplish that task.

For the 1009th SWS/AFOAT-1, Brigadier General William M. Canterbury would lead that effort going forward in the waning days of the Truman Administration and well into the Eisenhower era.[23] On August 16, 1952, he assumed command from General Keirn, who returned to ARDC to become chief of the Aircraft Reactor Branch within the Reactor Development Division. Canterbury was a 1934 graduate of West Point and soon thereafter became a pilot assigned to the Third Pursuit Group at Clark Field in the Philippine Islands. During the war, he became an expert on radar systems and formed the first Army Air Corps radar school. In September 1944, Canterbury commanded the 346th Bomb Group at Dalhart, Texas, and then joined Second Air Force at Colorado Springs, Colorado, to lead radar bombardier-navigator training. After the war, he became the assistant deputy chief of staff for R&D and participated in Operation CROSSROADS. His follow-on assignments prior to taking command of the 1009th SWS/AFOAT-1 were director of R&D of AFSWP at Sandia Base, Albuquerque, New Mexico; and ARDC's first director of development. In retrospect, given Northrup's struggles with his loss of R&D budgetary control to ARDC at that time, his new boss was fully aware of AFOAT-1's concerns and issues.[24] Canterbury only had a few weeks to acquaint himself with AFOAT-1's testing plans before he found himself once again in the Marshall Islands for a two shot operation codenamed IVY.

For Operation IVY, the LRD Program of experiments largely mirrored the format of the previous two tests, with several projects essentially constituting the continuation of previous LRD experiments. In conducting a project entitled "Electromagnetic Effects from Nuclear Explosions" (the first project in TUMBLER-SNAPPER), AFOAT-1 personnel positioned reception equipment in Hawaii, Guam, Alaska, Colorado, and Virginia to detect the electromagnetic pulses created in shots Mike and King. By now, the existence of electromagnetic phenomena in connection with atomic explosions was definitively established. AFOAT-1 researchers also believed the electromagnetic pulse was closely associated with the phenomenon of primary gamma radiation. Preliminary experiments with TNT explosions by LASL, the University of Texas, and the Naval Electronics Laboratory indicated that within a few feet of a TNT explosion, some sort of electromagnetic phenomenon was detectable, though not as pronounced as that associated with atomic explosions. In detonations with energy release in the order of several hundred kilotons, as in the Nevada shots, electromagnetic radiation was detected at distances of several thousand miles. AFOAT-1 hoped electromagnetic radiation might eventually be used to determine the energy release of the detonation. Even lacking success in this attempt, the electromagnetic phenomena were important in establishing the atomic nature of a disturbance detected either by seismic or acoustic evidence without the measurements of radioactivity.[25]

The second project, "Airborne Low-Frequency Sound from Atomic Explosions," involved personnel from AFOAT-1 and the Army Signal Corps. The airmen and soldiers established acoustic detection stations in Japan, Hawaii, Alaska, Washington, Arizona, and Washington, D.C., to attempt the recording of low-frequency sound waves generated by both Mike and King. The primary objective was to record the airborne acoustic waves with equipment at a sufficiently wide broadband so that observers could establish the true character of the signal at a number of remote locations, thus covering a variety of distances and directions from the explosion site. Researchers expected their analysis to reveal signal characteristics to distinguish detonations of various sizes. Another important objective was to determine the capabilities and limitations of standard detection equipment in the detection, location, and identification of very large nuclear shots. Finally, they wanted to interpret the acoustic data in terms of the structure of the atmosphere.[26]

Shots Mike and King occurred on October 31 and November 15, 1952.[27] Shot Mike was unique and marked a major milestone in the evolution of U.S. nuclear testing as the culmination of the work of Edward Teller and Stanislaw Ulam on the development of a hydrogen bomb. Detonated only days before Eisenhower was elected president, Mike was the United States' first experimental thermonuclear device. Constructed on Elugelab Island, Mike was a 22-foot long, 5-foot diameter cylinder weighing 65 tons. When it detonated with a yield of 10.4 megatons, it obliterated Elugelab and "ripped a hole in the atoll big enough to fit several buildings the size of the Pentagon and deeper than the height of the Empire State Building." The detonation also revealed two new elements, einsteinium (99) and fermium (100).[28] Shot King, dropped from a B-36 bomber, was a fission device designed to produce a yield larger than previously believed possible. At 500kt, Shot King became the largest fission device ever detonated.[29]

The huge yields produced by Mike and King provided LRD researchers with unique and important findings. AFOAT-1's 12 locations used both standard and special low-frequency sound recording equipment to collect data. This equipment detected signals at distances as far as 32,000 miles for Mike and 7,500 miles for King.[30] Project participants observed an important discrepancy between the two shots. Unlike King, the Mike signals produced an initial dispersive wave train of very low frequency. They also demonstrated larger amplitudes and longer periods.[31]

From both detonations, LRD researchers advanced their knowledge of nuclear debris analysis. While radiochemical analysis was not new as a primary project in every nuclear test series, the refinement of collection and analysis characterized projects 3 and 4. Project 3, "Calibration Analysis of Close-In Debris," partnered 16 recently modified F-84G aircraft with radiochemists to collect greater concentrations of particles and air, and marked the first time manned aircraft intentionally penetrated clouds in the Pacific testing area. Major modifications included four avionics systems and dual cloud sampling systems. A preliminary inspection of an F-84G aircraft had revealed that the cockpit did not have sufficient space for installation of filters at the terminals of the pressurization system. As a solution, a filter was placed in the defroster line, which was an independent system located under the windshield of the cockpit. One of the two types of sampling systems installed on the F-84Gs was called a "snap-bag." This consisted of a polyethylene bag mounted on the gun deck of the aircraft nose. A trigger switch on the control stick enabled collection of the gaseous samples through a sampling probe on the aircraft nose for 10 to 20 seconds. The second sampling system involved modifications to the wingtip fuel tanks to collect particulate matter from the nuclear clouds.

While the use of F-84s offered improvements in sampling operations, sample removal procedures had advanced only minimally. Upon landing, four-man teams (now dressed in better protective clothing) removed the particulate samples from the wing pods with long-handled tongs and a nine-foot pole. They used tongs to cut the retaining wire and the pole to hook the sample paper, which they then rolled to facilitate shipping. Once rolled, the team placed the rolled sample in a shielded enclosure called "the pig." The Air Force then flew the pig to the laboratory.[32]

To remove the air samples, two five-man teams used degassing equipment. This degassing equipment included a cart with a vacuum pump for each team, metal cylinders for bottling the gas samples, and hoses with nozzles inserted into the aircraft gas probe to form an airtight fit. The teams transferred the gas samples from the aircraft snapbag to the sample bottles by evacuating air from the bottles with the vacuum pump. They then used an external power unit to open a solenoid valve between the probe and the snap-bag mounted in the nose of the aircraft. The vacuum in the sample bottles drew the gas sample from the snap-bag. The teams used new sets of degassing equipment for each sample taken. They also opened the gun deck lid on the aircraft and removed the snap-bags.[33]

The fifth project, "Transportation of Airborne Debris," was basically designed to establish a database of scattered debris across a wide geographical environment. From two months before Mike to six months after King, LRD researchers collected water samples each week at 10 locations to detect the presence of any debris. For two weeks after the detonations, they took samples daily at Guam and Hawaii, project sites nearest to the proving ground. Data from the cloud tracking activities also fed into this project.[34]

The sixth project, "Detection of Fireball Light at a Distance," was essentially a continuation of the same experiment in TUMBLER-SNAPPER, with AFOAT-1 placing optical detection instruments on Johnston Island, Kwajalein, and in a C-47 flying over Kwajalein. The goal was to detect light from the Mike and King fireballs at very long distances. However, after carefully evaluating the optical measurements of the explosion, AFOAT-1 analysts concluded that further exploitation of the optical technique was not justified for the immediate future because the equipment could not detect the light beyond 500 or 600 miles. At greater distances, there was a sudden drop off in transmission, which appeared to be a serious limitation to the range of detection by surface equipment. Measurements made in the serious absorption of light from bomb explosions occurred in the atmosphere only below altitudes of 12,000 to 15,000 feet. In theory, detection stations located in mountains above 15,000 feet could conceivably detect light at greater ranges. However, for the AEDS, equipment for optical surveillance could not be located within 1,000 miles of the USSR proving grounds. A second factor contributing to the decision to terminate the optical program was a breakthrough in radiochemistry, allowing the determination of burst yield within an error of plus or minus 25 percent.[35]

Operation IVY proved to be a significant step forward for the LRD program and the improvement of the AEDS, especially for seismic and acoustic detection techniques. For the fourth project, "Propagation of Seismic Waves," AFOAT-1 leveraged AEDS seismic stations to detect seismic waves from both Mike and King. Prior to IVY, LRD seismologists had observed that both seismic and acoustic equipment could generally detect earthquakes, volcanic explosions, and meteoric explosions. They viewed seismic equipment as the most effective way for detecting earthquakes, while acoustic equipment was best suited for meteoric explosions. Both seismic and acoustic equipment were effective in detecting volcanic explosions.

Interestingly, examples of the usefulness of the seismic and acoustic elements of the AEDS occurred between the Mike and King shots. On November 4, 1952, only four days after Mike, a very large earthquake occurred near the Kamchatka Peninsula. At the time, LRD observers questioned whether the phenomena could have been a Soviet thermonuclear test. The geophysical data indicated a very large disturbance of the earth's crust in magnitude ranging from 8.2 to 8.5. IVY, however, offered the scientists a unique opportunity to compare and contrast the test data with the earthquake. They concluded that the magnitude of the primary phase, which was 10-27 ergs, plus the large number of aftershocks (many of which had an energy of approximately 10-22 ergs), made a man-made disturbance inconceivable.[36] Only earthquakes were known to produce such large and repeated phenomena. In addition, remote acoustic stations recorded signals indicating the waves had been transmitted at seismic velocities through the earth rather than at acoustic velocities through the atmosphere. They observed that the small vertical movements of a large area of the earth's crust in the vicinity of an acoustic station apparently acted as a piston, producing atmospheric pulsations which the microbarographs recorded. Experiences gained in IVY and the Kamchatka earthquake highlighted the difficulties in distinguishing between nuclear tests and earthquakes, difficulties LRD seismologists would face for decades to come.[37]

Soon thereafter, another example of the interplay between acoustic and seismic equipment took place. On November 13, 1952, just two days before Shot King, six AEDS acoustic stations reported large explosive waves whose azimuths intersected in French Equatorial Africa. Observers noted that the natural period of the record was similar to that of Shot Item in Operation GREENHOUSE. In fact, in all other respects, the signal characteristics were similar to those observed from large nuclear detonations in the air or on the surface of the earth. However, other AEDS systems failed to detect the event. One very sensitive seismic station, located close to the indicated source, recorded no data at all. In addition, ground-based samplers indicated no radioactivity was associated with the explosion. LRD analysts also ruled out a volcanic explosion since no energy was observed in the earth's crust. After a thorough assessment of all AEDS data, they concluded that a large natural explosion, probably meteoritic in origin, had occurred at some distance above the earth, especially given the signals detected as far as 6,000 miles from the intersection.[38]

The huge yields produced by shots Mike and King expanded the knowledge of LRD seismology by triggering detections throughout the entire AEDS.[39] The quality and quantity of data was also attributed to the new seismic equipment recently installed in the AEDS. For example, signatures recorded in AFOAT-1's Wyoming detachments were "off the scale." Acoustic results from Mike and King were equally gratifying. Indeed, the AFOAT-1 facility in Washington, D. C., detected the direct wave from Enewetak as the wave passed through the antipode and again as the direct wave rounded the Earth a second time.[40]

On March 17, 1953, four months after Shot King in Operation IVY, the 11-shot series codenamed Operation UPSHOT-KNOTHOLE commenced at the Nevada Test Site. Three shots were airdropped, seven were tower shots, and the Army fired one warhead from an experimental atomic cannon. Yields ranged from 200 tons to 61kt. For the fifth iteration of the Army's DESERT ROCK exercises, 21,000 military personnel participated in the operation. Operation UPSHOT-KNOTHOLE was designed to test new nuclear devices; to improve military tactics, equipment, and training; and to determine civil defense requirements. For this operation, which would produce lower detonation yields,

LRD researchers wanted to conduct the same or similar experiments from the previous, larger shots for comparison purposes.[41]

The five experiments of the LRD program mirrored the previous LRD projects in Operation IVY. In essence, the same experiments were repeated in order to gather more data. As with the cloud sampling project in IVY, AFOAT-1 researchers were especially interested in the close-in data gathered by the F-84G airborne samplers.

What was new in UPSHOT-KNOTHOLE, however, was the risk assessments and precautions for the F-84 aircraft and pilots. F-84 aircraft penetrated debris clouds in which peak intensity readings were as high as 80 Roentgens per hour (R/hr).[42] Consequently, the Air Force employed new safety measures to help reduce exposure rates. For example, monitors discovered F-84 sampler pilots received half of their doses during return flights to base. To reduce these rates, maintainers polished the aircraft skin so radioactive particles would not adhere to the aircraft. This procedure reduced the after-mission exposure by about 35 percent. They also lined the cockpits with 0.08 centimeter lead sheets. Because of a delay in obtaining the lining for the cockpits, not all F-84 samplers were lead-lined until the fifth shot (Ray). Finally, they outfitted the pilots with lead-glass vests. These measures substantially decreased the amounts of exposure received on the return flight and enabled the pilots to spend more time in the cloud.[43]

For long-range sampling, the WB-29 fleet had also received several improvements. Now standardized across all aircraft were the permanently mounted C-1 air foil boxes. Located on top of the aircraft, the boxes contained two filter papers that were exposed simultaneously. The second foil, the D-1, was a trapeze device mounted in the front bomb bay, which could be lowered and retracted in flight. The lower end of the assembly held several filter papers which were lowered and exposed during the sortie. B-31 units were also installed in the front bomb bay. These units consisted of five large metal cylinders mounted on a platform and connected to a compressor. When filled, they contained 275 pounds per square inch of compressed air. Finally, the WB-29s also housed the B-199 system. Installed in the aft pressurized compartment, this instrument detected the intensities of radiation outside the aircraft. The B-199 operator used the system to direct the pilots into the areas of heaviest concentrations.[44]

By June 1953, the AEDS was well-established to support the LRD experiments within the now routine framework of the U.S. nuclear test series. While AFOAT-1 would continue to construct temporary sites based on the parameters of an experiment, the AEDS could now significantly contribute to the various test projects while simultaneously standing watch over all Russian tests. Operationally, AFOAT-1 had a clear understanding of the fidelities and thresholds of the AEDS detection techniques, especially with seismic and acoustic capabilities. Northrup documented those capabilities in a technical memo to General Canterbury on February 13, 1953. The AEDS could now:

1. Detect and locate deep underground or underwater explosions of 20 KT or more anywhere in the Soviet Union.

2. Detect and locate surface and shallow underground bursts greater than 100 KT anywhere in the Soviet Union. It was also believed that shots less than 50 KT could be detected perhaps twenty-five per cent of the time anywhere in the Soviet Union.

3. Establish an estimate of energy in the earth at the source.

4. Establish the true nature of large earthquakes or meteoric explosions as opposed to man-made atomic explosions.

5. Give reasonable assurance of detecting shots greater than 20 KT over most of the Soviet Union.

6. Establish whether the explosion is deep underground or high above the surface.[45]

Compared to future AEDS capability assessments, this February 1953 report would prove to be an overestimation of the AEDS capabilities. Still, this particular assessment gave AFOAT-1 a new level of confidence at the time. The seismic technique, while still challenged to distinguish between detonations and earthquakes, was considered the best source for determining time and geographical location of burst. The acoustic method was now established to provide location information of certain events not detectable by seismic units, and remained the prime source for yield measurements. The radiochemical analysis of particulate filter samples was capable of quite accurate time measurements plus other analytical information not otherwise obtainable. The data derived from whole air samples could also provide corroborating information. In short, by mid-1953, the AEDS was now an integrated "system of systems" capable of providing reliable information under a variety of conditions.[46]

By late summer 1953, R&D in the electromagnetic field had advanced beyond experimentation, enabling AFOAT-1 to produce prototype equipment—the Electromagnetic Pulse Technique (designated "Q"). This new detection system recorded electromagnetic pulses radiating from nuclear explosions in the atmosphere or at high altitudes. For some time, the Defense Research Laboratory at Austin, Texas, had been under contract to AFOAT-1 to develop equipment suitable for translating the electromagnetic phenomena occurring in nuclear explosions into data that would augment other AEDS produced analyses. Within the AEDS, AFOAT-1 initially placed the Q systems at Larson Air Force Base, Washington; Williamson-Johnson Airport in Duluth, Minnesota; and Dow Air Force Base, Bangor, Maine. This net became operational on September 1, 1953.[47]

Command and control of the AEDS also expanded. In addition to the WFO, AFOAT-1, now with an authorized strength of 800 personnel, established the Delta Field Office at Brockley AFB in Alabama; the Central Field Office (CFO) at Wheelus AFB in Libya; and the Eastern Field Office (EFO) in Tokyo, Japan. These command centers oversaw an AEDS now consisting of 10 acoustic stations, 17 seismic stations, 5 air bases from which airborne sampling aircraft operated, 6 ground-based sampler stations, and 5 radiochemical laboratories. These stations and air bases were located in Turkey, Libya, Eritrea, Liberia, Japan, the Philippine Islands, the Marshall Islands, Alaska, Germany, Greenland, Pakistan, Iran, Scotland, England, Iceland, the Continental United States, Korea, Guam, Bermuda, Brazil, Chile, and the Panama Canal Zone.[48]

While the maturation of the AEDS looked promising at this point, LRD planners remained concerned about coverage of the Soviet Union. The USSR encompassed an enormous geographic area; and with airborne sampling comprising the heart of the AEDS, AFOAT-1 recognized that, depending on where it occurred, a Soviet test could be undetectable. It was a physical impossibility to devise an aerial filtering system capable of monitoring all of the air emanating from the USSR. Air masses under the tropopause in the northern region just east of the Ural Mountains flowed northward, circulating only three to five percent of the time over the Scandinavian countries. AFOAT-1 accepted the risk of leaving this area of the troposphere uncovered. If the Soviets tested southwest of their known test site at Semipalatinsk, AFOAT-1 faced a greater risk by not monitoring the area south of the Caspian Sea and Aral Sea, because the frequency of air flow through that corridor occurred only about a quarter of the time in the summer. If alerted, the 53rd Weather Reconnaissance Squadron (WRS) could provide limited coverage from Dharan, Saudi Arabia, within 72 hours of notification. Even with the routine daily flights out of the Japan-Alaska region, there was still a possibility of missing debris clouds from a small explosion. AFOAT-1 briefly considered flying that route twice daily but dismissed the proposal based on expense and minimal risk. Based on these risk assessments, AFOAT-1 decided to discontinue sampling sorties from Guam, Hickam AFB, and McClellan AFB, and to conduct filtering operations as close as possible to the Asiatic mainland between Formosa and 85° north latitude.[49]

Finally, LRD planners were most concerned about any future detonations of hydrogen weapons. They believed the explosions of multi-megaton devices would result in all of the debris ascending into the stratosphere, thus leaving an inadequate amount of debris below the tropopause for collection and analysis. The constraint was the altitude limitations of the B-29 and B-50 aircraft. These two aircraft could fly to maximum operational ceilings of approximately 25,000 and 30,000 feet respectively. To address this limitation, AFOAT-1 contracted with General Mills in 1952 to develop a high-altitude balloon capable of collecting samples. Unfortunately, the first prototype did not become available until the fall of 1953. Codenamed Project GRABBAG, the balloons would be tested in the next U.S. test series— Operation CASTLE.[50]

To date, Shot Mike was an anomaly. It would take time and more experiments to understand the characteristics of fusion weapons. Following the completion of Operation UPSHOT-KNOTHOLE in the summer of 1953, everyone recognized it was only a matter of time before the Soviet Union developed a thermonuclear weapon. The Soviets would not wait long—1954 would be the year of the hydrogen bomb.

B-31 Unit

Also installed in the front Bombay of the RB-29. These units consisted of five large metal cylinders mounted on a platform and connected to a compressor. When filled, they contained 275 pounds per square inch of compressed air.

Photo source: AFTAC archives

Major General
William M. Canterbury

Commander, 1009th SWS/AFOAT-1, from August 1952 to June 1954. During WWII, he was an expert on radar systems and formed the first Army Air Corps radar school. After the war, he served as the Assistant Deputy Chief of Staff for R&D, participating in Operation CROSSROADS. Prior to assuming command, he was the director of R&D for AFSWP, and served as ARDC's first Director of Development. His extensive experience in R&D proved critical at this stage of building the AEDS.

Photo source: AFTAC archives

Shot Mike in Operation IVY (November 1, 1952)

Photo courtesy of the United States Army

Placing a hot filter sample into the "pig."

Photo source: AFTAC archives

The B-199 system

It detected the intensities of radiation outside of the aircraft and was installed in the aft pressurized compartment. The B-199 operator used the system to direct the pilots into the areas of heaviest concentrations of nuclear debris.

Photo Source: AFTAC archives

Air sampling equipment ("the squeegee") mounted in the F-84

Photo source: AFTAC archives

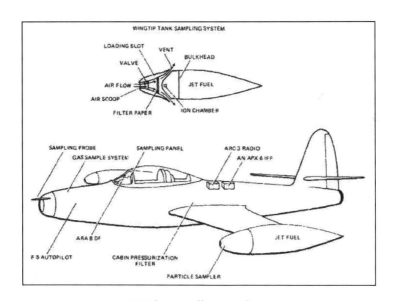

F-84G sampling equipment

Diagram courtesy of the Defense Nuclear Agency

The EMP "Q" System
(Added to the AEDS in September 1953)

Photo source: AFTAC archives

Notes to Chapter 6

[1] *History of Long Range Detection: 1947-1953,* 47-48. As the AEDS expanded, other FOs were to follow. First envisioned in the initial war plans (see Annex 3), the WFO exercised regional command and control. Yet to come were the Central, Southern, and Eastern WFOs.

[2] Technical memo #41.

[3] H. E. Menker, "Analytical Results Obtained by Air Force Field Laboratories during Operation DOGFACE," November 15, 1951 (Berkeley: TracerLab Report), AFTAC archives. Note that beginning with the September 1951 detonation, AFOAT-1 codenamed their surge operations (i.e., flying more sorties) against Russian tests. That test was Operation DOGFACE and the October test was Operation FLATFOOT. See also L.J. Beaufait, A. DeHaan, L. Leventhal, and L.R. Zumwalt, "Interim Report of DOGFACE and FLATFOOT Radiochemical Analysis Results," November 15, 1951 (Berkeley: TracerLab Report).

[4] Memorandum for the Technical Director, Olmstead and Crocker to Northrup, Subject: "Capabilities of Seismic and Acoustic Techniques of the AEDS, January 5, 1953, AFTAC archives.

[5] Technical memo #34, from Northrup to Hegenberger, Subject: "Research Required to Reach the Objectives of the Long Range Detection Program," June 3, 1949, AFTAC archives. In February 1949, Doyle Northrup, Assistant Technical Director at that time, submitted a three phased, long-range R&D plan to Hegenberger. The plan envisioned 30 projects for the nuclear, acoustic, and seismic components of the AEDS.

[6] Letter, AFOAT-1 to AMC, Subject: "AFOAT-1 Project Authorization No. B/200/AMC," June 13, 1951, AFTAC archives.

[7] AFOAT-1 Project Authorization No. B/58/P/AMC, September 22, 1950, and AFOAT-1 Project Authorization No. B/102/C/AMC, April 30, 1951, AFTAC archives.

[8] AFOAT-1 Project Authorization No. B/100/0p/WB, March 28, 1951, AFTAC archives.

[9] Jean Ponton, Stephen Roher, Carl Maag, and Jean Massie, "Shots Able to Easy: The First Five Tests of the BUSTER-JANGLE Series, 22 October - 5 November 1951," Technical report (Washington D.C.: Defense Nuclear Agency, 1982). Also Miller, *Under the Cloud,* 119-134.

[10] Walter Singlevich, and E. M. Douthett, "OPERATION BUSTER-JANGLE Radiochemical, Chemical, and Physical Analysis of Atomic Bomb Debris (Buster), (Jangle)" (Washington, D.C.: Armed Forces Special Weapons Project, May 28, 1952). DTIC ADA995009.

[11] Operation BUSTER-JANGLE: Seismic Waves From A-Bombs Detonated Over a Land Mass, DTIC ADA374648.

[12] Jean Ponton, et al, "Shots Able to Easy."

[13] Miller, *Under the Cloud,* 134. See also "Operation BUSTER-JANGLE 1951," DTIC ADA123441.

[14] Technical memo #68, Northrup to Major General John A Samford, Subject: "Classification of Work Presently Labeled R&D in the AFOAT-1 Program," February 27, 1952, AFTAC archives; also Technical memo #69, Northrup to Samford, Subject: "Classification of USAF (AFOAT-1) Projects," March 25, 1952, AFTAC archives.

[15] Miller, *Under the Cloud,* 137-156.

[16] Jean Ponton and Craig Maas, "Shots Able, Baker, Charlie, and Dog: The First Tests of the TUMBLER-SNAPPER Series, 1 April-1 May 1952, Technical Report, Defense Nuclear Agency, Washington D.C., June 15, 1982.

[17] Walter Singlevich, "Radiochemical and Physical Analysis of Atomic Bomb Debris," Report, Operation SNAPPER, Headquarters U.S. Air Force, Office of Atomic Energy, DCS/O, AFOAT-1, March 18, 1953, AFTAC archives.

[18] Ponton and Maas, "Shots Able, Baker, Charlie, and Dog."

[19] Jean Ponton, Carl Maag, Mary Francis Barrett, and Robert Shepanek, "Operation TUMBLER-SNAPPER, 1952," Defense Nuclear Agency, Washington D.C., June 14, 1982, DTIC ADA122242.

[20] Halberstam, *The Fifties,* 52.

[21] Walker, *Cold War,* Chapter 4, and Larson, *Anatomy of Mistrust,* Chapter 2.

[22] Divine, *Eisenhower and the Cold War,* 15-19.

[23] General Orders Number 13, August 16, 1952. AFTAC archives.

[24] Biography, Major General William M. Canterbury, www.af.mil, last accessed June 12, 2018.

[25] Gladeck, et. Al. , "Operation IVY: 1952."

[26] George B. Olmstead, "Detection of Airborne Low-Frequency Sound from Nuclear Explosions," Report to the Director, AFOAT-1, HQ, USAF, Washington DC, September 1953, DTIC ADA363576.

[27] F. R. Gladeck, J. H. Hallowell, E. J. Merton, F. W. McMullan, R. H. Miller, R. Pozega, W. E. Rogers, R. H. Roland, C. F. Shelton, and L. Berkhouse, "Operation IVY: 1952," Technical Report, Defense Nuclear Agency, December 1, 1982.

[28] Miller, *Under the Cloud,* 115-118; Rhodes, *Dark Sun,* 487.

[29] Gladeck, et. Al., "Operation IVY: 1952."

[30] For Mike, the signals travelled around the Earth (i.e., the second arrival of the direct wave).

[31] George B. Olmstead, "Detection of Airborne Low-Frequency Sound from Nuclear Explosions," Report to the Director, AFOAT-1, HQ, USAF, Washington, DC, September 1953, 21. DTIC ADA363576.

[32] Ibid., 96-99.

[33] Ibid.

[34] Ibid.

[35] Ibid. Proven only weeks before in monitoring the October 1952 British test (described in Annex 1).

[36] An erg is a unit of energy in the centimeter-gram-second system. It is the amount of energy expended in moving a body one centimeter against a force of one dyne (981 dynes exert the same force as one gram mass in the earth's gravitational field). Definition from http://www.google.com, last accessed September 11, 2018.

[37] Technical memo #79, Northrup to Canterbury, February 13, 1953.

[38] The Mike shot was detected at every seismic station in the AEDS with the exception of Camp King, Germany, which was so situated in the "seismic core shadow zone."

[39] The Mike shot was detected at every seismic station in the AEDS with the exception of Camp King, Germany, which was also situated in the "seismic core shadow zone."

[40] Technical memo #79. "Antipode" means that part of the Earth diametrically opposite.

[41] Operation UPSHOT-KNOTHOLE 1953, p-1. DTIC ADA121624. Also Walter Singlevich and Charles K. Reed, "Calibration Analysis of Close-In A-Bomb Debris ," Operation UPHOTKNOTHOLE, Report of the Test Director, Headquarters

U.S. Air Force, Office of Atomic Energy, DCS/O, AFOAT-1, September 1955, AFTAC archives.

[42] Meaning Roentgen per hour. A Roentgen was a unit of measurement for the exposure of X-rays and gamma rays.

[43] Crew protection would remain a concern for decades to come. Operation UPSHOT- KNOTHOLE 1953, p. 137-141. DTIC ADA121624.

[44] Historical Supplement to the 56th Strategic Reconnaissance Squadron from July 1953 through December 1953, author unknown, AFTAC archives.

[45] Technical memo #79.

[46] At the time, AFOAT-1 was under extreme pressure to build a war plan that would support the USAF and SAC war plans (see Annex 3). Like the politicians, higher level planners believed that AEDS capabilities were more extensive and efficient than they actually were. Northrup's February 1953 capability assessment undergirded those war plans.

[47] Letter, 1009th SWS to OIC, Team 101, Subj: "Project 8/180-A," April 20, 1953. AFTAC archives.

[48] See various project authorization documents in AFTAC archives, folder 12, "Historical Data, 1949-1955."

[49] *History of the Air Force Office for Atomic Energy-One (AFOAT-1): 1 January- 31 December 1954,* author unidentified (Washington, D.C.: The Air. Force Office of Atomic Energy-One, Headquarters, United States Air Force, 1954), 10-11.

[50] Ibid., 39-40.

CHAPTER 7

The Game Changer

The busy test schedule between mid-1951 and mid-1953 reflected the sense of urgency and the high premium the nation's most senior leaders placed on implementing a robust, effective AEDS in the shortest time possible, especially in anticipation of more Soviet tests. Indicative of the priority given to R&D for LRD was the fact that, despite the ultra-high level of security placed on U.S. weapons development (the primary focus of the test series), LRD research was shrouded with even higher levels of classification. The LRD experimental program was so compartmentalized that most members of the Joint Task Forces (JTFs) were unaware of the LRD experiments. During those two years, the Air Force effectively shielded the LRD program from any outside observation due primarily to AFOAT-1's direct subordination to the Air Force Deputy Chief of Staff for Operations who allowed extensive freedom of action.

During the final years of the Truman Administration, escalating Cold War events elevated the importance of the AEDS and made LRD capabilities an important component of American nuclear diplomacy. Severe distrust of the Soviet Union appeared to have stressed Truman's policy of containment. As the AEDS slowly gained some levels of effectiveness in the summer of 1951, the heaviest fighting in Korea seemed to have subsided as United Nations and Communist forces agreed to negotiate an end to the war. However, with both the Soviet Union and China directly participating in the fighting, U.S. policymakers remained concerned over the possibility of the conflict spreading beyond Korean borders as negotiations to end the war failed on several occasions, resulting in renewed combat. While Truman had briefly contemplated the use of nuclear weapons when U.S. forces were pushed back into the Pusan perimeter, he never truly considered the atomic bomb to be a viable component of his containment policy. It was assumed that, with a significant stockpile of fission weapons at hand, the American nuclear monopoly was deterrent enough, even after Joe-1. In any case, the AEDS's ability to monitor Russian nuclear tests and assess their effectiveness served to inform and reassure policymakers.

The expensive plans of NSC-68 forced Truman to abandon his austerity program. The requirements for both the war in Korea and a large conventional capability to meet Soviet expansionism anywhere in the world demanded large expenditures. To help offset the large expense, the U.S. sought assistance from its allies.

In September 1951, the U.S. formulated a number of treaties to bolster containment. Australia, New Zealand, and the U.S. entered into a mutual defense pact with the signing of the ANZUS Treaty on September 1st, a development that coincidentally would soon help with the expansion of the AEDS into the territories of the new partner nations. One week later, American and Japanese officials signed the Japan-United States Security Treaty that allowed for the stationing of U.S. military forces in Japan once the occupation ended; making Japan another key strategic location that would greatly enhance the AEDS for decades to come. In Europe, NATO admitted Greece and Turkey as members on September 20th, with the latter becoming a key geographical location for seismic stations in the AEDS.

In 1952, both the Army and the Air Force expanded. Fueled by a $50 billion defense budget (2021 = 513 billion), the Army doubled in size to three million personnel, while air groups doubled to a total of 95. The Air Force deployed many of those groups to new air fields in the United Kingdom, Morocco, Libya, and Saudi Arabia. On April 15th, the Air Force flew its first B-52 bomber, an airframe that would now allow long-distance nuclear strikes into the Soviet Union. On June 14th, the U.S. Navy began construction of the world's first nuclear submarine, the *U.S.S. Nautilus*.

On the domestic political front, the vast majority of Americans had lost confidence in their president and believed he was weak on communism. Faced with an approval rating of only 26 percent, Truman announced on March 29, 1952, that he would not seek reelection. He then threw his support to Governor Adlai Stevenson, II, for the Democratic nomination. However, with the Red Scare at its highest peak and the perception that the Democrats were soft on communism, Truman's endorsement carried little weight. On November 4, 1952, Dwight Eisenhower won the presidential election with a landslide victory of 442 electoral votes compared to Stevenson's 89. With 55 percent of the voters selecting Eisenhower, the American people clearly favored a tougher foreign policy to alleviate their growing fears of the Soviet threat.

Eisenhower assumed the presidency determined to end the stalemate in Korea and to significantly reduce defense spending. While he seriously disliked leveraging American nuclear capabilities, Eisenhower was compelled to do so to end the Korean War. In May 1953, he implied he was prepared to use tactical nuclear weapons to end the war if North Korea continued to stall the peace negotiations. Consequently, the war came to an end only two months later when the armistice was signed on July 27, 1953. The conclusion of the war would allow some defense cuts, and Eisenhower sought even more reductions. However, only 16 days after the peace agreement, Joe-4 appeared to propel American national security policy in a direction which threatened his planned austerity measures.[1]

When Eisenhower took office, he immediately replaced all of the service chiefs of staff and appointed Admiral Arthur Radford as Chairman of the JCS. Concurrently, he tasked the NSC to reassess American foreign policy, paying particular attention to the U.S. defense posture. Meanwhile, John Foster Dulles, Eisenhower's secretary of state and a strong hawk when it came to U.S.-Soviet relations, advised the president to rely on nuclear retaliation to deter communist expansionism in the world. While Eisenhower is remembered for his numerous attempts to de-escalate the Cold War, in the early days of his first term in office he had few options in matters of defense, especially after the fourth Soviet test. Joe-4, with a detonated yield of 400kt, was a game changer. When that test occurred on August 12, 1953, the Cold War became much colder.[2]

On that day, between 1430Z and 1910Z, the AEDS detected a series of significant acoustic and seismic signals revealing a Russian nuclear explosion had occurred at 0131Z. AEDS station personnel immediately recognized it as a large detonation. The AFOAT-1 Seismic Analysis Center estimated the size of the explosion to be one-tenth of IVY-Mike. AFOAT-1 notified the Eastern Operations Center in Japan and alerted three AWS reconnaissance squadrons for collection missions. Other AEDS stations were directed to begin daily rainwater, humidity, and ground filter unit operations as well.[3]

The AWS sampling sorties made initial contact with the cloud on August 15th between 22,000 and 28,000 feet. The hottest long-duration filter paper, a three-hour exposure, measured 800,000 cpm, and the hottest short-duration filter paper, a 65-minute exposure, 500,000 cpm. Between August 14th and 23rd, the sampling missions collected 293 filter papers, 67 of which measured in excess of 1,000,000 cpm. The radio-chemical analysis indicated the Russian device contained a thermonuclear component.[4]

Like Joe-1, Joe-4 was another wake-up call. Had the Russians really advanced their nuclear weapons program so rapidly? The Los Alamos Lab and AFOAT-1 were most concerned that the initial radiochemical analysis had discovered the presence of fusion products. This revelation perplexed nuclear physicists Hans Bethe, Enrico Fermi, and Carson Mark, all of whom questioned the initial conclusion. Their analysis concluded the bomb was detonated by a high explosive and augmented with a contributing fusion reaction.

Therefore, while the yield was indeed large, it was not truly a hydrogen bomb like the U.S. Shot Mike.[5]

Two months after Joe-4, the NSC advised the president to continue Truman's containment policy but place greater reliance on nuclear retaliation. Joe-4 had shaken the JCS. In their view, the Soviet Union was rapidly catching up to the U.S. in its nuclear weapons program. In response, they endorsed the NSC recommendation and, conservatively, concurred with Dulles's hawkish views. Within NATO, the JCS had struggled with the formulation of European defense policy. At the time of Joe-4, the Soviet Union had a conventional force of 6.2 million personnel, of which 60 percent were ground forces, with the majority focused on securing Eastern Europe and defending against NATO.[6] The U.S. nuclear stockpile totaled 1,350, while estimates of the Soviet stockpile ranged between 20 and 30 bombs.[7]

Eisenhower, elected on his promise to reduce the U.S. budget deficit of $43 billion (2021 = $435 billion), faced an unsustainable expense in maintaining a large conventional force. The president inherited from Truman an armed force of more than three million personnel. Faced with an economic versus security dilemma, Eisenhower accepted the NSC recommendation, formalized as NSC 162/2, which emphasized the use of nuclear weapons as offensive striking power. Concurrently, however, he became "furious" with the joint chiefs upon receiving their recommendation of *increasing* the defense budget and bringing the armed forces to a total strength of 3.5 million personnel. The president immediately directed his secretary of defense to reduce the size of the Army, Navy, and Marine Corps, with the goal of achieving an end-strength of 2.8 million by FY57. The downsizing of the three services, combined with NSC 162/2, resulted in a new American foreign policy known as the "New Look." In essence, it recognized the U.S. could not match Soviet conventional strength; therefore, greater reliance on nuclear weapons would allow defense reductions while maintaining adequate security. To effect this strategy, the Air Force, as America's predominant nuclear force, would increase its size by 30,000 personnel and expand its capabilities to 137 wings. In time, this strategy would become known as "massive retaliation."[8] In essence, Eisenhower had opted for a policy of deterrence to replace containment—the threat of force as a way to avoid limited wars and unlimited defense expenditures.[9]

The greatest fear Joe-4 generated was the realization that the Soviets were moving ahead of the U.S. in developing thermonuclear weapons. If true, only a handful of hydrogen bombs would negate the United States' relatively large stockpile of fission bombs. The JCS feared that, in comparison with the huge IVY-Mike device, Joe-4 was much smaller and therefore possibly weaponized. They concluded as much from AFOAT-1's analysis, which discovered the presence of lithium in the debris. Unlike the Mike device, lithium required no refrigeration. Suddenly, there was a real possibility the Soviets could place small hydrogen weapons on missiles. This initial fear would lead to the well-known "missile gap" anxieties, which would persist through the end of the decade.[10] Joe-4 gave the United States an increased impetus to advance the development of thermonuclear weapons. Consequently, only one test series was scheduled for 1954— almost all fusion devices.

By the beginning of 1954, LRD researchers were positioned to conduct comparative analyses of newly collected data with known values of previous data. Those comparative analyses allowed them to gauge the accuracy and effectiveness of LRD techniques employed in the AEDS as well as those still undergoing R&D. In other words, they were making significant advances that would allow analysts to more accurately determine the time, place, height of burst, and yield of nuclear detonations. At the time, the LRD scientists referred to the comparative analyses as a process of "calibration." They were especially enthusiastic about the 1954 test series. Those tests would focus on calibration, which would then help to refine detection equipment and techniques.

The six-shot series of Operation CASTLE, (five at Bikini and one at Enewetak) were all detonations related to thermonuclear weapons designs. All but one of the CASTLE shots produced enormous yields in the megaton ranges. The day of the thermonuclear weapon had permanently arrived.

On February 28, 1954, the U.S. initiated Operation CASTLE. AFOAT-1 was once again the sponsor of the LRD program, with Colonel Paul R. Wignall as the program director, and Walter Singlevich as his technical deputy. Singlevich had joined AFOAT-1 in 1952 to lead Technical Directorate Four (TD-4), Radiometries. Prior to joining AFOAT-1, Singlevich worked for DuPont but was assigned to the Manhattan Project at Oak Ridge, Tennessee, where he conducted pilot-plant studies prior to full-scale construction of the Hanford plutonium production facility at Richland, Washington. While at Oak Ridge, he received graduate-level training under the auspices of the University of Chicago, with his mentors being Nobel laureate Dr. Enrico Fermi and Dr. Arthur Compton. In 1945, Singlevich transferred to the Hanford Engineer Works where DuPont was responsible for the design, construction, and operation of the first full-scale plant to produce weapons-grade plutonium. As chief of the Radiological Sciences Department, he remained at Hanford until 1951, when he returned to DuPont headquarters at Wilmington, Delaware. During his last year at DuPont, he participated in the design and construction of the Savannah River Atomic Energy plant near Augusta, Georgia.[11]

Within Operation CASTLE, the LRD program of experiments originally consisted of four projects. However, seismic operations, designated Project 3, were cancelled because of a lack of appropriate detection sites in the short and intermediate ranges.[12]

AFOAT-1's J. Allan Crocker served as project director over Project 1, "Electromagnetic Radiation Calibration." This project was the culmination of the studies that began with Operation TUMBLER-SNAPPER. By 1954, AFOAT-1 was close to operationalizing this EMP capability. The primary objectives for the electromagnetic program were to obtain additional information to aid in the production of self-contained units of equipment for independent operation in remote locations. More specifically, Crocker and his team wanted to investigate electromagnetic propagation to determine attenuation over water, over land, along the earth's magnetic lines, and perpendicular to the earth's magnetic lines. In addition, they planned to conduct close-in measurements to provide diagnostic data and basic norms of undistorted pulses for comparative analysis with existing data. By understanding the electromagnetic emanations from atomic detonations of known composition and characteristics, they hoped to become more capable in interpreting the pulses from unknown devices. Assisting Crocker were the National Bureau of Standards (NBS), and the Defense Research Laboratory of the University of Texas.[13]

Crocker's colleague from AFOAT-1, George Olmsted, supervised Project 2, "Acoustics." Like Project 1, Olmsted's project was an extension of the former acoustic projects in previous test series. Olmsted and his team directed their efforts to the gathering of calibration (i. e., comparative) data to help establish variations in acoustic signals from atomic explosions as a function of seasonal location and yield of the device. Other participants included the U.S. Army Signal Corps Engineering Laboratories, the National Bureau of Standards, and the U.S. Navy Electronics Laboratory.[14]

In addition to his deputy director duties, Singlevich also oversaw the third project "Calibration Analysis of Close-in Atomic Device Debris." This project, now a routine LRD experiment in all U.S. test series, was simply the collection of debris for radiochemical analyses. Despite its title, the project also included long-range sampling from aircraft operating out of McClellan AFB, Hickam AFB, and Guam. Other participants included personnel from SAC, AWS, the 4926th Test Squadron (Atomic), and numerous 1009th SWS field elements.[15]

On the morning of February 28, 1954, the U.S. detonated the largest nuclear device it would ever detonate. Similar to IVY-Mike, the Bravo shot's device was located inside a small protective building that "resembled a small shack at the end of a dock." It weighed approximately 12 tons. The control point for Bravo was a three-foot-thick concrete bunker under 10 feet of earth located 20 miles away. When the shot detonated at precisely 7 a.m., the crew of a small Japanese fishing trawler called the *Fukuryu Maru* (Fortunate Dragon), located 87 miles from ground zero, "was startled to see the darkness vanish . . . and the orange-red sun rise in the *west.*" Minutes after the detonation, the men in the bunker began to receive dangerously high levels of radiation from the fallout and had to be emergency evacuated. Aboard the *Fukuryu Maru,* all experienced radiation sickness and one crewmember later died. CASTLE Bravo was big. Indeed, it produced a yield of 15MT—three times larger than expected.[16]

Overall, OPERATION CASTLE provided AFOAT-1 researchers with new data from which they could make certain improvements in detection techniques and equipment. Crocker's Project 1 (electromagnetic) involved a total of 17 stations located at various distances. Those stations were able to analyze 74 sets of data from the 102 sets collected during the operation. Importantly, the data confirmed the previous theoretical conclusion that for close-in wave forms, the predominant frequency became lower as the yield increased. Analysis of the pulses received at greater distances showed that the higher frequencies were attenuated relatively more; and at several thousand miles, the close-in differences disappeared. CASTLE proved that an approximation of the detonation yield could be obtained at distances by measurement of low-frequency field strength. The test equipment functioned well; in most cases, the stations recorded detonations within two milliseconds of the actual detonation times. The wave-form information led to modifications of the operational equipment that first entered the AEDS in late 1953. These refinements strengthened confidence in the use of the new "Q" technique for detection and diagnostic work.[17]

The CASTLE shots also boosted advancements in acoustics from Olmstead's Project 2. The acoustic net for CASTLE consisted of 15 stations at remote locations, covering a variety of distances and directions from the Pacific Proving Ground. Project 2 measured airborne low-frequency sound from the CASTLE detonations to understand the relationships between the signal characteristics and the energy released over yields ranging from one to 15 megatons. The stations employed standard and very low frequency sound-recording equipment responsive to small atmospheric pressure variations in targeted frequency ranges. The stations detected signals at ranges exceeding 2,800 miles for explosions larger than five megatons (shots Bravo, Romeo, Union, and Yankee); 18,750 miles for the 1.7MT Shot Nectar; and 6,250 miles for the 110kt Shot Koon. All megaton shots produced the initial dispersive wave train of very low frequency previously noted for IVY-Mike. Analysis of acoustics signal periods gave a fourth-power relationship between signal period and yield.

Yield estimation was one of the most important products of the acoustic technique. Like the advancements made in EMP ("Q") technique, the CASTLE results led to similar advancements in acoustic ("I") technique equipment. Although the Army had operated acoustic stations since 1948, the stations did not officially enter the AEDS as a dedicated mini-network until the fall of 1953. Interestingly, the AFOAT-1 commander, General Canterbury, was so confident in the "I" technique that he invited various government dignitaries to witness the actual arrival of acoustic signals into the National Bureau of Standards building where a temporary station was monitoring the CASTLE signals.[18]

Singlevich's Project 4, "Calibration Analysis of Close-in Atomic Device Debris," involved F-84, WB-29, and B-36 aircraft. Both LASL and AFOAT-1 personnel controlled the technical directions of the project, with the former conducting particulate sampling and the latter the gas collections. Several laboratories analyzed the samples, to include the University of California Radiation Laboratory, AFOAT-1's McClellan Central Laboratory (MCL), Tracerlab, and the Argonne National Laboratory. CASTLE gave AFOAT-1 an excellent opportunity to compare and evaluate two different types of gas sampling devices in order to determine the best quality and optimum size sample. Singlevich was especially interested in reducing cross-contamination. The end result was that radio-chemical procedures in the laboratories to detect certain elements were revised and refined.[19]

Although Project 3 (seismic) was cancelled, CASTLE proved to be a good test of the operational fidelity of the growing AEDS. Every seismic station in the AEDS detected waves from the CASTLE shots, including the experimental station at Fort Sill, Oklahoma. Shots larger than Nectar (1.7 MT) produced waves large enough to be detectable after traveling through the core of the earth. The smallest detonation, Shot Koon (110kt), was detected at five seismic sites. Importantly, measurements of seismograms established the fact that the peak amplitude of the longitudinal seismic waves varied at the first power of the yield for surface shots.[20]

The CASTLE shots conducted from February to May 1954, proved to be game changers for several reasons. First, the AEDS incorporated significant advancements in technology for its primary techniques as a result of the LRD experiments, which confirmed equipment viabilities. Indeed, by late 1954, a considerable inventory of specialized equipment was in use throughout the AEDS. It was an exciting year for the AFOAT-1 staff as they constantly strove to improve current equipment or found new equipment to collect more complete and accurate data. For example, CASTLE provided the first opportunity to operationally test the new P-84 gas sampling system. For some time, LRD researchers had been frustrated by the altitude limitations of the RB-29 aircraft (25,000 feet). They had long sought to capture air samples at much higher altitudes, where detonation debris revealed unique characteristics. Although they had been experimenting with balloons, that technique (designated "Db" in the AEDS) carried significant limitations. During CASTLE, the F-84s used the new gas samplers for close-in collections at 36,000 and 42,500 feet. The new system consisted of two electrically driven high-pressure compressors taking air from an intermediate-stage compressor in the F-84's engine. The dual system was capable of capturing 60 standard cubic feet of air compared to only three standard cubic feet of air of the polyethylene bag technique on the RB-29s. The superiority of the new system made the time and money for research and development seem well spent.[21]

During CASTLE, airborne sampling also improved with the new P-36 compressor system used on the B-36 bombers, which were converted for close-in collection missions. This system consisted primarily of six electrically driven high pressure compressors mounted on a platform suspended in the bomb bay. The compressors were operated in parallel and exhausted individually into a 900-cubic-inch 3000 psi sphere. Using this system, the B-36s captured 100 standard cubic feet per bottle at altitudes up to 52,000 feet.[22]

In late 1954, LRD researchers and engineers improved the long-range sampling equipment used on the RB-29s by adapting the P-36 compressor system. This development occurred simultaneously with plans to replace the RB-29 with the RB-50 aircraft as the primary aircraft for airborne sampling. When that replacement effort began in early 1955, the modified P-36 system was essentially renamed the P-50 system.[23]

The new RB-50s also received a new particulate foil. Designated as the E-1 foil, this was a boxlike foil mounted on the top of the fuselage easily accessible from the inside. Using a single-fold detection filter assembly, mounted in a V-shape, it was capable of providing radioactive particulate samples of nearly twice the size of its predecessor, the C-1 foil. The E-1 foil also collected samples of much higher concentrations. Nicknamed the "shoebox," it was used in conjunction with a Geiger counter. In areas of heavy concentration of radioactivity, an aircrew member could replace papers during the sortie. The prototype of the E-1 foil tested during CASTLE proved so efficient that immediate procurement was ordered for installation on all new WB-50 aircraft.[24]

As LRD engineers improved the gaseous and particulate sampling equipment, a similar advancement occurred within the laboratories. Better samples resulted in increased accuracy in the area of physical measurements. After a thorough study, AFOAT-1 procured a prototype of a methane end window (MEW) flow counter from Tracerlab most suitable for reproducible measurements of radioactive nuclides. The MEW counter improved the internal consistency of radiochemical data and provided accuracy in the cross calibration between field laboratories, MCL, and Tracerlab. On May 4, 1954, AFOAT-1 directed AMC to procure 18 CE-2 MEW flow counters with associated equipment, spare parts, and manuals at an estimated cost of $37,000 (2021 = $370,000).[25]

Although Crocker's EMP experiments yielded very positive results confirming the viability of the "Q" technique, the initial prototype systems entering the AEDS in late 1953 had yet to prove themselves. While Crocker's team analyzed the CASTLE data throughout the summer of 1954, real confidence in the "Q" technique developed in early fall. Between September 14th and October 30th, the Soviets conducted 10 nuclear tests. Those detonations marked the first time the Q systems recorded quality measurements. The results were so good that AFOAT-1 worked with the Computer Division of NBS to design and engineer a final electromagnetic system for standard use in the AEDS.[26]

At this time, AFOAT-1 also made improvements to the seismic ("B") technique equipment in the AEDS. Technicians installed a system for remote calibration of seismic instruments in the seismic vaults. They also established a central source of electrical power, thus eliminating the use of batteries at each instrument outpost. To address the serious lightning problem that frequently brought down seismic sites by blowing out galvanometers, AFOAT-1 procured newly designed lightning arresters, which resolved the problem. All of these improvements greatly increased operational reliability and simplified maintenance procedures.[27]

One of the most important advances made in 1954 was the development of the seismic "Zippogram." Developed by Beers and Heroy Geophysical Corporation of Dallas, Texas, the Zippogram provided real-time transmission of seismic activity being recorded. Although still under development throughout 1954, a successful demonstration led AFOAT-1 to place a high priority on adopting the system as soon as possible.

For the acoustic stations in the AEDS, NBS provided new detection and recording equipment, which was installed throughout the acoustic net. NBS also produced new magnetic tape recorders for AFOAT-1 detachments in Turkey and the Philippine Islands. This equipment provided a means for recording received data in a form suitable for analysis on correlating equipment. A central power system similar in function to that installed in the seismic net was established throughout the acoustic system.[28]

As the AEDS expanded and became more efficient at collecting data and samples, MCL required expansion as well, both in terms of personnel and bench space. Based on the increased work load following CASTLE, MCL commander Colonel Robert E. Belville requested and received $87,000

(2021 = $870,000)—$35,000 for expansion and $52,000 for new equipment. Construction began in mid-September and was completed in December.[29]

The expansion of the AEDS and its increased capability to gauge Russian advances in nuclear technology gained importance in 1954 when the American public reacted with great anxiety to widespread reports on the dangers of nuclear weapons. The CASTLE Bravo shot (larger and more impactful than anyone anticipated) and the revelations about the thermonuclear Joe-4 test six months earlier, generated the worst fears in the minds of many Americans. In April, when the public viewed the freshly released film footage of the 10MT IVY-Mike shot from October 1952, domestic and "world-wide furor over the H-bomb reached a peak." Fears of the vast destructiveness of hydrogen weapons, combined with the health hazards of their extensive fallout, compelled Eisenhower to directly assuage Americans' concerns. Throughout the spring of 1954, Eisenhower conducted several press conferences to reassure all Americans that the U.S. would not proceed with developing large-megaton weapons. He argued the existing nuclear stockpile of fission bombs was more than adequate to undergird his policy of massive retaliation, and large megaton weapons were unnecessary in meeting military requirements and contingency plans. Still, many leaders across the world and their constituents called for an immediate ban on all testing.[30]

Thus began Eisenhower's six-year quest to achieve nuclear disarmament. In May, as the CASTLE shots were still underway, Eisenhower initiated an interdepartmental study to consider implementing a unilateral test ban. However, in June he dropped the idea on the recommendations of his advisors, who considered the Soviet threat to be too great. The United States had to stay ahead of the Soviet Union in developing nuclear technologies. Dulles and the JCS were strongly against any test ban, fearful the Soviets were already taking the lead on placing thermonuclear warheads on missiles. In fact, at that time, the top priority within the Department of Defense was the development of the Atlas missile system. New missile technology would require extensive testing.[31]

By the fall of 1954, a number of international events exacerbated fears of a thermonuclear war and radiation fallout. Communist expansionism in Southeast Asia prompted the president to warn the world that the French loss of Vietnam would lead to the fall of neighboring countries like a "row of dominoes." In referencing his policy of massive retaliation, he warned China not to intervene in the region, as the U.S. was prepared to consider nuclear options. In the months ahead, he once again threatened the use of the nuclear option as mainland communist Chinese forces engaged in hostilities with the exiled nationalist Chinese on the island of Formosa.[32]

By late 1954, genuine concern permeated the DoD that war with the Soviet Union was a real possibility. Communist expansionism, combined with all of the unknowns surrounding the state of the Soviet nuclear weapons program, seriously challenged political and military planners. Politically, the United States sought to contain the expansion of communism in Southeast Asia by creating a NATO-like alliance of eight nations. The result was the Southeast Asia Treaty Organization (SEATO) signed in September. In Europe, the U.S. now recommended the rearmament of West Germany (until now unthinkable) and its admittance to NATO. With the signing of the Paris Agreements in October, the Allies ended their occupation of Germany and approved NATO membership for their former enemy. Finally, the president and Congress were unsure of the new leader of the Soviet Union, Georgy Malenkov, and how he would deal with the West. Malenkov had assumed premiership upon Stalin's death on March 5, 1953. He was trusted by Stalin and had assisted the brutal Russian leader in the extensive purges of the late 1930s, resulting in more than a million executions.

During the war, Malenkov was solely responsible for the Soviet missile program. Prior to becoming premier, he had chaired the Council of Ministers Special Committee on rocket technology. Malenkov knew missiles—at a time when the Russians were quickly attempting to place fusion warheads on long-range missiles.

Simultaneously, the Russians were also accelerating their testing program. On September 14, 1954, Russia initiated a 10-shot test series, which began with a major military exercise similar to the U.S. DESERT ROCK exercises. The 40kt detonation occurred at the Totskoye military maneuver area in the Southern Ural Military District and involved 45,000 soldiers.

Within this context, the U.S. Air Force was in the driver's seat on employing the technologies and assets that helped make Eisenhower's "New Look" (i.e., massive retaliation) a viable policy. What everyone feared the most was a surprise attack. Even within AFOAT-1, Northrup had expressed such fears in pushing hard for a viable AEDS. Especially in the 1951-1953 time period, it appeared to him to be the only capability to detect Soviet preparations for the use of nuclear weapons. The AEDS's performance in detecting the Soviet tests in September and October underscored Northrup's view. Indeed, the five acoustic sites and eight seismic sites experienced no problems in detecting the 40kt explosion at the new Totskoye site located 1,300 miles northwest of the Khazakstan site. The first positive EMP measurements of a Soviet detonation were made against that shot. Additionally, it was the first time AFOAT-1 vectored a B-29 from Dhahran, Saudi Arabia, into the target area, thus collecting samples much earlier and enabling more quality analyses in the lab. AFOAT-1's new Eastern Operations Center in Japan had recently acquired the capability to vector flights more rapidly as well.[33]

Despite these successes, the fall Soviet tests also revealed some limitations. For yields greater than 20kt, the various components of the AEDS performed well. Several of the detonations, however, were very small at less than 3kt. Those shots were almost missed and only detected by radiochemical analysis, which discovered different dates in the half-life measurements. Several detonations occurring in quick succession also seemed to confirm AFOAT-1's concerns voiced during Operation RANGER that the Soviets could "hide" behind other tests—U.S. or their own. More ominously, the small yields led analysts to believe the Russians were experimenting with tactical uses of nuclear devices.[34]

With the nuclear arms race in high gear and fears of a "missile gap," General Curt Lemay's Strategic Air Command (SAC) was now poised to assure American strategic deterrence. In June, SAC began flying "reflex alerts." The goal was to prevent a Soviet surprise attack by keeping one-third of the entire command on standby, ready to fly at a moment's notice with a full complement of wartime munitions. The crews exercised frequently, as their *Convair* B-36's and B-47 *Stratojets* were now capable of striking the Soviet Union from overseas bases. SAC planners developed extensive studies on how to prevent surprise attacks against U.S. bomber bases and produced detailed war plans. In doing so, SAC had its sights aimed at using the AEDS as an integral component of those wartime contingencies (see Annex 2: The AEDS as a Wartime System).[35]

As 1954 came to an end, many Americans believed the nuclear arms race was beginning to spiral out of control. Not only was a nuclear war with the Soviet Union actually conceivable (DoD believed it was inevitable before the end of the decade), fears of radioactive fallout from the tests were becoming more pervasive. Fallout fears even invaded popular culture. For example, Warner Brothers' highest grossing film in 1954 was a movie entitled *Them!* starring James Arness and James Whitmore. This science fiction film depicted giant atomic-irradiated ants invading a small town in New Mexico. The final three sentences aptly captured that angst when

James Arness, the FBI agent, asks, "If the first atomic bomb caused all of this, what about all of the other tests since then?" The scientist involved replies, "The atomic bomb opened up a new world. What we eventually find in that new world, nobody can predict." *Them! w*as one of the first of more than one hundred science fiction films produced over the next two decades to depict the potential horrors of radioactive fallout.[36]

Eisenhower was sympathetic to Americans' concerns. Not long after the detonation of Joe-4 and the Soviet's apparent possession of a thermonuclear weapon, the president felt compelled to inform the public about the facts of the arms race. He believed his constituents had a right to know why and how their government would be spending more money on defense. To do so, he initiated Operation CANDOR, a series of six-minute radio and TV information talks by various administration officials. This plan, however, quickly evolved into a greater effort encompassing more than just U.S. citizens. On December 8, 1953, the president delivered his famous "Atoms for Peace" proposal before the United Nations General Assembly. In short, Eisenhower proposed that the three nuclear powers contribute some of their fissionable material to a UN-sponsored "International Atomic Energy Agency." The nuclear powers would provide that material to other nations in order to address energy challenges throughout the world. The response from the General Assembly was overwhelmingly positive and without precedent. As Eisenhower's biographer Stephen Ambrose noted, "he had replaced fear with hope."[37]

Although Eisenhower really believed the Russians would welcome the proposal, they did not respond for almost two years. By that time,

> a great opportunity had been lost. Eisenhower's proposal of atoms for peace was the most generous and the most serious offer on controlling the arms race ever made by an American President. All previous offers, and all that followed, contained clauses about on-site

inspections that the Americans knew in advance were unacceptable to the Russians.[38]

Once again, the irreconcilable issue of inspections for nuclear arms control had raised its ugly head.

Walter Singlevich

Joined AFOAT-1 in 1952 to lead Technical Directorate Four (TD-4), radio metrics. Prior to joining AFOAT-1, Singlevich worked for DuPont in support of the Manhattan project during the war. He also pioneered air sampling techniques and was instrumental in developing the balloon technique ("Db"). Throughout the 1950s, Singlevich led all LRD radiochemical experiments at the U.S. nuclear tests.

Photo source AFTAC archives

The WB-50D

Throughout 1956, AWS expedited its transition from the WB-29 to the WB-50 as the primary airborne collection aircraft. Though an improvement, the altitude limitation of 30,000 feet fell far short of the desired altitude for gas (air) collection samples.

Photo source: AFTAC archives

The E1-Foil

The E1-Foil was capable of collecting radioactive particulate samples of nearly twice the size as the C-1 Foil. The E1-Foil also collected samples of much higher concentrations. It was used in conjunction with a Geiger counter. Easily accessible from the inside, an aircrew member could replace filter papers during the sortie; especially useful in areas of heavy concentration of radioactivity.

Photo source: AFTAC archives

The F-50 Foil

Photo source: AFTAC archives

The D-1 Foil mounted on the B-36

Photo source: AFTAC archives

The D-1 Foil
(Also referred to as "the trapeze")

Photo source: AFTAC archives

Notes to Chapter 7

[1] Divine, *Eisenhower and the Cold War*, 29-31. See also Walker, *Cold War*, 95

[2] Ibid, 33-39.

[3] *History of Long Range Detection: 1947-1953*, 249.

[4] Ibid

[5] Rhodes, *Dark Sun*, 524.

[6] Abraham S. Becker and Edmund D. Brunner, 'The Evolution of Soviet Military Forces and Budgets, 1945-1953, RAND Corporation, September 1975.

[7] Saki Dockrill, *Eisenhower's New-Look National Security Policy, 1953-1961* (New York: St. Martin's Press, 1996), 52.

[8] Ibid, 54.

[9] Divine, *Eisenhower and the Cold War*, 39.

[10] Walker, *Cold War*, 96.

[11] Bio, Walter Singlevich, AFTAC archives. Overtime, Walt Singlevich would become the most beloved member of AFOAT-1 and AFTAC. In 2014, AFTAC named its new headquarters building after Singlevich, who died while working at AFTAC on August 7, 1992.

[12] *History of Long Range Detection: 1954*, author unidentified (Washington, D.C.: The Air Force Office of Atomic Energy-One, Headquarters, United States Air Force, 1954), 14-16. AFTAC archives.

[13] Ibid.

[14] Ibid.

[15] Ibid.

[16] Miller, *Under the Cloud*, 188-194; Also, Rhodes, *Dark Sun*, 541. Shot Bravo made the Bikini chain of islands permanently uninhabitable.

[17] *History of Long Range Detection: 1954*, AFTAC archives, 16-17.

[18] "Operation CASTLE-Detection of Airborne Low-Frequency Sound from Nuclear Explosions May 1954"; also *History of Long Range Detection: 1954*.

[19] "Op CASTLE-Nuclear Calibration Analysis of Atomic Device Debris;" also *History of Long Range Detection: 1954*.

[20] *History of Long Range Detection: 1954*, 19, AFTAC archives.

[21] Ibid., 52.

[22] Ibid., 52-53.

[23] Ibid., 53

[24] Ibid., 53-54

[25] Ibid., 54.

[26] Ibid., 19.

[27] Ibid., 56.

[28] Ibid., 57.

[29] Ibid.

[30] Divine, *Blowing on the Wind*, 11-21.

[31] Ibid., 24.

[32] Divine, *Eisenhower and the Cold War*, 48-61. On September 3, 1954, the communist Chinese began shelling the island of Quemoy (part of Formosa). Two Americans died in the artillery barrages.

[33] *History of Long Range Detection: 1954*, 24-25.

[34] Ibid.

[35] "Strategic Air Command and the Alert Program: A Brief History," Office of the Historian, Headquarters, Strategic Air Command, Offutt Air Force Base, Nebraska, April 1, 1988.

[36] James Arness was best known for the long running television series *Gunsmoke*. Prolific actor James Whitmore was an Emmy Award winner and two-time Academy Award nominee.

[37] Ambrose, *Eisenhower*, 149.

[38] Ibid.

CHAPTER 8

Shifting Winds and
Political Ramifications

The CASTLE Bravo shot on February 28, 1954, which produced a yield of 15 megatons, exacerbated growing concerns throughout the world about the dangers of radioactive fallout. Until then, such concerns had been negligible. The Bravo device used a fission "trigger" to fire a fusion device, and the combined yield was approximately half fission and half fusion. Most of the deadly fallout was produced by the fission contribution. The Eisenhower Administration worked hard throughout the year to reassure the American public that the U.S. would limit the risk of fallout in future testing and had little interest in developing large thermonuclear weapons. Indeed, the next series of tests, scheduled for February and May 1955 and codenamed Operations TEAPOT and WIGWAM, were structured to test much smaller nuclear devices. However, throughout the rest of the world, demands for disarmament or the termination of all testing continued to increase. International voices grew louder in the fall of 1954 after the first crew member of the Japanese fishing vessel *Fortunate Dragon* died on September 23rd from the fallout of Bravo he had been exposed to seven months earlier. The other twenty-two crew members would remain hospitalized for more than a year. By early 1955, leaders from all over the world, including the Pope and America's principal ally, Great Britain, began calling for nuclear disarmament between the United States and the Soviet Union.[1]

Eisenhower realized fears of the destructiveness and health hazards of thermonuclear weapons would never fully dissipate. Since his first few months in office, he had genuinely sought ways to reduce tensions with the Soviet Union. Two years earlier, as Malenkov assumed control after Stalin's death, the president had formally proposed disarmament and told the Soviets the U.S. stood ready to enter into "solemn agreements."[2] While insisting on United Nations inspections, he advocated international control of atomic energy. Eisenhower believed the enormous expenditures on weapons programs could then be channeled into raising the standard of living around the world. Unfortunately, the strong anti-communist hysteria generated by McCarthyism and the hard anti-Russian views of Dulles, the AEC, and the JCS, had kept Eisenhower's aims in check. However, by early 1955, he was prepared to empower someone to carry that mission forward.

In March 1955, as Operation TEAPOT was underway, Eisenhower appointed former Minnesota governor Harold Stassen as Assistant to the President for Disarmament. This was a new position and one carrying cabinet level rank. Stassen had served three terms as governor of Minnesota but resigned during his third term to serve in the Navy as Assistant Chief of Staff to Admiral William Halsey during the war. He had been a close advisor to Eisenhower during his 1952 presidential campaign, and had helped draft many of the candidate's speeches and position papers. Stassen's appointment was important because it signaled Eisenhower's strong desire to formulate a nuclear arms control treaty. The president had grown tired of the propaganda wars between the U.S. and the USSR over the feasibility of disarmament. In short, the Soviets had continuously called for the elimination of all nuclear weapons but with no mechanisms of verification and treaty oversight. In contrast, the U.S. had always responded in favor of a disarmament treaty that could be verified by the AEDS, as well as in-country, on-site inspection systems. Routinely, though, the Soviets balked at any inspection system they perceived would violate their geographical sovereignty.

Stassen's job was to find a way to break through the obstacle of inspections. His unprecedented position carried significant authority to coordinate with a number of agencies to bring detail and substance to the prospects of a treaty. However, his work would carry him into the realm of foreign policy and, thus, he was frequently at odds with Secretary of State Dulles.

Just days before he appointed Stassen, Eisenhower dismissed the idea of a test ban treaty because he viewed it as a "piecemeal approach" to his real goal of total disarmament. Stassen accepted the decision but suggested to the president that mutual aerial inspections might be an important first step in breaking the disarmament deadlock.[3] Eisenhower was optimistic at this point because Premier Malenkov had recently expressed some willingness to unilaterally ease tensions with Western Europe and the U.S. Not long after assuming the premiership, Malenkov warned the Supreme Soviet that "cold war policies" could lead to a "new world war [and] destruction of world civilization." This was the first time any Soviet leader had admitted that a nuclear war would not solely destroy capitalists.[4] Unfortunately, just days prior to Stassen's appointment, Malenkov lost an internal power struggle and was replaced by Nikolai Bulganin (as Premier) and Nikita Khrushchev (as First Secretary of the Communist Party). While these two leaders, in theory, governed the Soviet Union as partners, it was Khrushchev who soon rose to the top of the political power structure.

Several weeks later, on May 10, 1955, the Western countries invited the Soviet Union to a summit to explore ways to ease international tensions. The Russians accepted the invitation on the same day and surprised the West with a new disarmament proposal that appeared to largely agree with previous U.S. stipulations, most notably the requirement for inspections. Held in Geneva on July 18th with the leaders and foreign ministers of France, Great Britain, the United States, and the Soviet Union, the talks centered on trade policies and international relations. Eisenhower, not yet responding to the Soviet announcement, surprised the attendees with his "open skies" proposal.[5] During his speech, he twice spoke to the primary reason behind every arms control proposal to date, either Soviet or American; that is, the fear of a surprise attack by either nation. For Eisenhower and Stassen, "open skies"— an agreement to permit reconnaissance flights over each other's territories—was a means by which to reduce or eliminate such fears.[6]

It remains unclear as to why Eisenhower did not explore the new Russian initiative and instead replaced it with the "open skies" proposal. At that time, unlike his generals, he was not concerned about maintaining military superiority. In fact, he "rejected the Air Force's doctrine of nuclear superiority and settled on a policy of 'adequacy,'" which he unveiled in early 1956.[7] Just three weeks before the Geneva Conference, in an NSC meeting with the JCS, the president pushed back at the chiefs' hawkish concerns over his disarmament ideas. He "heatedly" replied that the arms race was a "mounting spiral towards war."[8] In retrospect, Eisenhower and Dulles did not trust the Russians and therefore could simply not accept the risk. In September, Eisenhower told Stassen to place a "reservation" on all previous U.S. arms control proposals. In November, the Soviets responded by withdrawing their willingness to accept inspections in any future proposal. The May 1955 proposal to accept inspections was the only time during the Cold War when the Soviet Union offered to negotiate on-site inspections.[9]

One explanation for Eisenhower's inattention to the Soviet proposal and his counter offer of an "open skies" program was his recent interest in the Air Force's ability to conduct effective reconnaissance. Eight months earlier, he had approved the construction of a single-jet-engine, ultra-high-altitude aircraft, designated the U-2 (nicknamed the "Dragon Lady"). In the initial days of the Cold War, U.S. reconnaissance of the Soviet Union was crude and ineffective, limited to the use of uncontrollable balloons and quick "dashes" of aircraft into Soviet air space. These attempts at spying were feckless components of a fledgling intelligence program. Eisenhower, however, saw the U-2 as

> . . . a far-sighted attempt to shift reconnaissance out of the realm of espionage altogether. Arms control could not rely upon ground inspections alone, the President told the French and British delegates at the July 1955 Geneva summit, because with thermonuclear weapons it was possible to conceal enough explosive material to defeat a nation in a relatively small space.'[10]

Once it became fully mission capable, the U-2 joined the AEDS as the Air Force's principal means of conducting strategic reconnaissance using technological innovation.

When Stassen received his appointment in March, as what the press jokingly referred to as "Secretary of Peace," Stassen quickly formed "eight special task forces each chaired by a distinguished private citizen, to review all aspects of the arms race."[11] Dr. Ernest Lawrence, the director of the Radiation Laboratory at Berkeley and influential voice behind Teller's efforts to develop fusion weapons, headed one of Stassen's task forces. Lawrence's group was interested in exploring novel ways to enhance Eisenhower's "open skies" program. He turned to Northrup and the new AFOAT-1 commander for help.

Northrup's new boss, Major General Daniel E. Hooks, had assumed command of the 1009th SWS/AFOAT-1 from Canterbury in June 1954. Hooks was a pilot as well as a scientist, having earned a master of science degree in physics from Harvard University in 1932. During the Second World War, he served on the island of Guam from December 1944 through the end of hostilities until March 1946. After post-graduate work in physics at the Ohio State University, Hooks became chief of the Research and Development Field Office for Atomic Energy at Kirtland AFB, New Mexico. Prior to his arrival at AFOAT-1, he served as vice commander of the Air Force Weapons Center at Kirtland AFB.[12] Hooks would prove to be the most significant commander of all AFOAT-1 commanders because of his four year tenure in command and his extensive experience with scientific R&D.

Stassen and Lawrence turned to AFOAT-1 to thoroughly understand LRD seismic and airborne capabilities. On September 16, 1955, Stassen visited AFOAT-1 headquarters and received a briefing on those LRD techniques. In trying to understand the fidelity of the AEDS, Stassen asked if there were "any conditions under which the Soviets could explode a multi-megaton bomb which would make detection by normal AFOAT-1 methods either difficult or impossible." Northrup replied it was unlikely but conceivable if the detonation occurred at a high altitude. He told Stassen that at 75km height, a 10 to 20 megaton "bomb would appear to have the equivalent energy of a 70kt shot at ground level." Northrup added it was critical to national security to explore the problem of high-altitude testing (a problem which would take another four years to address).[13]

Two weeks later, Stassen posed five questions to Northrup and his team. In exploring how well the AEDS could function in detecting various evasive Soviet tests, he first wanted to know what types of earth densities could substantially dampen seismic signals generated by a detonation. Northrup replied coral and unconsolidated earth could do so. He noted that two U.S. tests had demonstrated this effect by detonating 1kt bombs at 17 and 67 feet in such material. The seismic signals did not extend beyond 125 miles. Northrup compared the significance of that diminishment with a 15kt tower burst over the same material, which carried detectable signals out to 1,800 miles.[14]

Stassen then wanted to know if tests could occur in locations associated with earth tremors. If so, could the tremors or earthquake activities "hide" or obscure the detonations. Northrup replied that it was virtually impossible to hide a detonation alongside a natural earthquake. Similarly, Stassen asked if false seismic tremors would be induced by a series of judiciously chosen bomb detonations.

Again, Northrup answered that it was conceivable one seismic station could be fooled, but not multiple stations looking at the same event.[15]

Northrup pointed to a thorough study of the geography in the USSR that illustrated some areas "where it would be extremely difficult if not impossible to differentiate between the event and a natural earthquake." He cited current LRD methods that were not yet definitive and the fact that natural earthquakes were frequent in some regions. He stated, of the hundreds of earthquakes detected each year in the region, 40 or so were ambiguous on average. Northrup conceded in some of the difficult regions a 100kt detonation might be hidden.

Finally, Stassen wanted to understand the connection between seismic records from earth tremors and those from nuclear detonations. Northrup remarked the AEDS could record signals from depths "unattainable by man," and that earthquakes frequently exhibited large shear waves, "which had never been observed from a nuclear blast."[16]

As part of his larger fact-finding efforts, Stassen also engaged with AFOAT-1 through Lawrence's task force. AFOAT-1's role within the task force was to look at new methodologies that could be instrumented and placed aboard the existing airborne collection aircraft, the RB-29 and WB-50. AFOAT-1 explored three new methods, all using U.S. facilities as targets. The first was an attempt to "sniff out" radioactivity from a DC-3 aircraft equipped with gamma and neutron detection devices, and a single-channel gamma spectrometer. A similar attempt had been made by General Groves in 1944, using Ninth Air Force A-26 medium bombers equipped with classified equipment to fly low-level flights over Nazi Germany. Groves's search for atomic energy facilities, however, proved fruitless.[17]

The second experimental method tested the feasibility of detecting the magnetic fields of powerlines at an operational altitude of 20,000 feet. Using Pasadena, California, as a test area, the experiments proved successful in measuring the electrostatic field up to maximum altitude. However, no LRD technique was ever developed from the successful test of this hypothesis.[18]

The third method comprised an extensive test of experimental infrared (IR) equipment to determine the feasibility of detecting, identifying, and estimating the operational levels of industrial and atomic energy facilities. Again, using U.S. facilities as targets, the B-29s and B-50s flew 15 missions between February 12 and April 23, 1956. AFOAT-1 first used the existing IR equipment, the Mark-12 (AN/AAR-12 IR receiving set), as they awaited the prototype Mark-9 (AN/AAR-9 IR mapping set) then being developed at the Aerial Reconnaissance Laboratory at the Wright Development Center in Dayton, Ohio. Despite its initial perceived promise, the IR Mark-9 only detected a steel mill. Consequently, no LRD technique was explored or developed from these flights, leading to the realization that the "open skies" program would largely depend on photography as the primary surveillance tool.[19]

In late May 1956, Stassen submitted his first report to Eisenhower in which he basically concluded that a total elimination of all nuclear weapons was simply not realistic or feasible. Instead, he advocated other forms of arms control. Importantly, the Stassen report strongly undergirded the president's emphasis on an inspection system—an emphasis that would have great impact on the AEDS in the near and intermediate future. Stassen's report served as a catalyst for serious consideration about achieving a test ban in lieu of the elimination of all nuclear weapons. Still, progress toward such a goal would be slow given Eisenhower's belief that only a comprehensive disarmament treaty would ensure security for all.[20]

During this same time, support for a test ban grew within the scientific community as well. While well-known physicists such as Robert Oppenheimer and Albert Einstein were already on record voicing strong moral objections to the development and proliferation of nuclear weapons, others began to emerge as proponents of a test ban. The first to do so was David R. Inglis, a senior scientist from Argonne National Laboratory and a former Manhattan Project participant. Only months earlier, Inglis and his lab had worked with AFOAT-1's Walt Singlevich in the CASTLE shots. In a series of magazine articles, Inglis drilled down on what was realistically possible. He concluded the Soviets would never accept inspections. Therefore, a test ban was a huge first step in de-escalating dangerous tensions in the Cold War, and inspections could be circumnavigated by technology. On-site inspections were not necessary, he argued, referring to the AEDS, because "instruments located outside a nation's borders could detect nuclear explosions at long range."[21] Having worked with AFOAT-1 and possessing some knowledge of the LRD techniques, Inglis probably believed the AEDS could provide flawed but adequate oversight. He thought if both nations possessed long-range detection systems, a test ban treaty would be self-enforcing. He was most likely aware of Northrup's December 1953 assessment of the Soviets' development of their own LRD system.

On December 1, 1953, Northrup provided Canterbury an assessment on Russian capabilities to detect U.S. nuclear detonations based on the data obtained from the GREENHOUSE and UPSHOT-KNOTHOLE shots. AFOAT-1 believed the Soviets could collect debris out to a distance of 2,500 miles, and electromagnetic and optical signals out to 200 miles. Likewise, they most likely possessed long-range seismic and acoustic techniques similar to the capabilities of the AEDS. While aerial debris collection was largely confined to Soviet-controlled territories, submarines could police up debris near the test sites in the Marshall Islands. Submarines could also operate the electromagnetic and optical detection equipment. Northrup also reported the Soviet embassy in Washington was probably collecting rainwater. In short, the

Russians could detect 20kt shots from the Pacific Proving Grounds, but had not yet acquired the ability to do so from the Nevada test site.[22]

The convergence of Inglis's public advocacy for a test ban, the proliferation of thermonuclear devices, the worldwide fear of radioactive fallout, the establishment of Russian LRD, and the growing awareness of LRD in the U.S., all placed the AEDS at the heart of American nuclear diplomacy. From this point on, the AEDS took on greater and greater importance as a means of reliable oversight of Soviet weapons testing. The key word here is "reliable." Could the AEDS, in its state of development in 1955, reliably detect all nuclear detonations within the vast borders of the Soviet Union? Unaware of Stassen's questions to AFOAT-1, Inglis probably believed so. For the remainder of the decade, American arms control advocates and treaty negotiators would believe so as well. From AFOAT-1's perspective, however, the AEDS was far from providing such expansive coverage. Indeed, the LRD program had a long way to go.

In March 1955, as Stassen began his appointment as Assistant to the President for Disarmament, the AEC and DoD were conducting new weapons design experiments in a test series codenamed Operation TEAPOT. As promised (perhaps coincidentally and timely in assuaging the public's concern over fallout), the 14 nuclear shots of TEAPOT would not test any devices producing yields in the range of megatons. Indeed, the largest would be Shot Turk, with a yield of 43 kilotons. Nine of the remaining 13 shots produced yields smaller than eight kilotons. For TEAPOT, Los Alamos scientists led the way as they tested "newer, lighter, and more efficient bombs." With all of the TEAPOT shots designed to test tactical weapons, the scientists were quickly becoming experts at miniaturization.[23]

Operation TEAPOT took place between February 18 and May 15, 1955. In total, more than 11,000 DoD personnel participated in the 14 shots as scientists, observers, or soldiers and marines conducting tactical maneuvers. Los Alamos scientists sponsored the series as they sought to test nuclear devices for possible smaller weapons systems, for use in military tactical operations, and for the study of civil defense requirements. However, these test objectives were less useful for LRD experimentation. Consequently, for the first time, this operation did not include a formal LRD Program. Still, AFOAT-1 deployed 25 people to conduct a long-range aerial sampling project with AWS aircraft flying from Bermuda.[24]

The AFOAT-1 airborne collection team in Bermuda utilized the TEAPOT shots to simulate debris collected from hypothetical Soviet detonations. This scenario-driven simulation imposed a collection radius of 2,000 to 4,000 miles from the Nevada test site. The aircrews conducted sampling sorties from 1 to 12 hours after detonation and at altitudes ranging from 10,000 to 25,000 feet. AFOAT-1 established an operations center in Bermuda to coordinate sampling requirements with the AWS squadron and expeditious handling of filter papers. One special sortie collected air samples above 40,000 feet. For this high-altitude mission, the 4926th Test Squadron used an F-84 aircraft equipped with the new "squeegee" equipment.[25]

On May 14, 1955, one day before the final TEAPOT shot (Shot Zucchini), the U.S. Navy led a single-shot operation codenamed WIGWAM. This shot, the first of its kind to be carried out in deep water, occurred about 500 miles southwest of San Diego, California. The Navy's primary objective was to determine the characteristics and lethal ranges of the resulting underwater shockwave, and the effects of radioactivity on naval vessels. In addition to positioning unmanned surface vessels, the Navy suspended three sub-scale submarine-like pressure hulls equipped with measuring instruments underwater at various distances from the device. The nuclear device detonated at a depth of approximately 2,000 feet and produced a yield of about 30kt. The Navy's intent was to eventually develop nuclear depth charges and to determine how much punishment a submarine could absorb.[26]

The AFOAT-1 program of participation in Operation WIGWAM primarily consisted of two missions. The first was the collection of seismic measurements and data to improve knowledge of the travel time and energy of seismic waves as a function of distance from the source. The second objective was to continue air sampling operations, which were now routine for AFOAT-1 in all U.S. weapons tests. AWS WB-29s from McClellan AFB conducted the aerial sampling operations in WIGWAM. Those sorties were short-lived due to the absence of any significant airborne debris.[27]

While AFOAT-1's LRD experimental activities throughout the spring and summer were minimal, AEDS operations picked up speed beginning on July 29, 1955. It had been nine months since the last Soviet test. Now, the first detonation of a new five-shot series served to evaluate the effectiveness of the growing AEDS because of its low yield. Three acoustic stations detected the blast and fixed the location at the Semipalatinsk Proving Ground. AFOAT-1 believed it was a fission device and estimated the yield at 4kt. Four days later, on August 2nd, several AEDS acoustic and seismic stations detected a much larger test from the same location. Whereas the previous test occurred near the earth's surface, this test appeared to have detonated above 20,000 feet. It produced a yield of approximately 30kt.[28]

For the September 21st Soviet test, the young AEDS experienced something new. For this shot, the Soviets used a new test site—the two-island district of Novaya Zemlya—located 450 miles northeast of the city of Murmansk.[29] Six acoustic and 19 seismic stations detected the 20kt detonation. Analysis revealed that the detonation occurred offshore. AFOAT-1 analysts compared the data to the U.S. Operations CROSSROADS data collected from the underwater Shot Baker in July 1946. From the large apparent seismic yield and very low apparent acoustic yield, they concluded it was an underwater test, which produced a yield of approximately 20kt. Because of the close similarities between this detonation and the 1946 Baker shot, analysts estimated a submerged depth of less than 200 feet.[30]

On November 6, 1955, the AEDS performed well as the Soviets conducted another test at Semipalatinsk. Three seismic and 13 acoustic stations detected the detonation. For the first time, the new EMP Q systems recorded the detonation and were able to place the time of the explosion within plus or minus 0.1 second. As usual, seismic data were used to confirm the location as the Semipalatinsk area. Acoustic data determined that the detonation produced a yield of approximately 200kt. The difference in origin time computed from the electromagnetic and seismic signals indicated a burst altitude of 3,500 feet. This event was important to the evolution of the AEDS because it demonstrated AFOAT-1's ability to synthesize data from various LRD techniques to present a more comprehensive and accurate analysis of the test. Still, the fidelity and detection threshold levels needed much improvement. A number of AEDS stations had only recorded the event because it had been a large detonation.[31]

Two weeks later, on November 22nd, the AEDS monitored the largest Soviet detonation at Semipalatinsk to date. Two EMP stations, six seismic stations, and 13 acoustic sites recorded the event. The height of burst was estimated at 10,000 feet. The best estimate of yield of this shot was based on acoustic data. Later analysis of the data and the collected debris estimated the yield at approximately 1,700kt, with the detonation occurring at about 4,500 feet above the terrain. This test caught the world's attention as the H-bomb quickly spread dangerous fallout. By November 25th, Japan reported high levels of radioactivity over three major cities. Within a week, European countries reported increased levels of atmospheric radiation. In the United States, radiation also was detected, but the AEC downplayed its significance due to recent negative publicity. What concerned the Americans the most, however, was the fact that the Soviets had delivered this thermonuclear bomb from an airplane—a capability the U.S. had yet to achieve.[32]

The Soviet Union's delivery of an H-bomb from an aircraft immediately sent shock waves through the AEC and DoD. This demonstrated capability marked a huge advancement in the Soviet nuclear weapons program. Unlike the United States, the Soviet Union now had the capability to produce fusion weapons small enough to fit inside a missile. The primary American concern was of the Soviets now moving ahead in the arms race. This event sprouted into what would soon be known as the "missile gap" and elevated distrust between the two superpowers to an all new level.

At the end of 1955, the United States possessed 2,422 nuclear weapons in its stockpile, while the estimate for the USSR was approximately 200.[33] Suddenly, however, the November 22nd Russian test appeared to diminish that 12:1 superiority. Conceivably, just a handful of fusion weapons, especially if easily deliverable, could negate hundreds of fission weapons. To exacerbate American concerns, only six days after their tests, the Soviets renewed their calls for a total nuclear test ban. They further reinforced that call for a ban on December 13th when Nikita Khrushchev visited India and jointly, with Indian Prime Minister Jawaharlal Nehru, called for total disarmament. As before, the Russian propagandistic proposal omitted any safeguards that would include inspections within the Soviet Union. To make matters worse, the Pope presented a three-part plan for the total elimination of nuclear arms during his 1955 Christmas Eve message.[34]

As the calendar turned to 1956, it was clear the U.S. would continue to insist on an inspection system for any arms control agreement, either for disarmament or for a test ban. With DoD fearing the Soviets had now achieved technological superiority, almost no one in the Eisenhower Administration considered even a test ban feasible at this point. The Soviets had scored a propaganda coup by pressing hard for a ban on the heels of their very successful year of testing. While Harold Stassen and his team were still exploring arms control options in early 1956, the Senate Foreign Relations Committee created a disarmament subcommittee to explore the feasibility of a nuclear test ban. Stassen testified before the subcommittee in late January to emphasize that testing was integral to national security. Other advocates, including Dulles and Strauss, soon followed with even stronger endorsements for testing. During Stassen's hearing, committee member Senator John Pastore pressed Stassen on the issue of long-range detection of nuclear stockpiles and tests:

> **Senator PASTORE.** Don't you think that the condition on whether or not an atomic bomb can be outlawed depends entirely upon developing that kind of [detection] technology?
>
> **Mr. STASSEN.** That is one of the essentials that must be there before you could.
>
> **Senator PASTORE.** Isn't it the essential, I mean, what else could you add to it possibly? If you can't tell by technical facilities whether or not there is the presence of a hidden stockpile then you could never agree with any nation or have any agreement with any nation to outlaw.
>
> **Mr. STASSEN.** That's right.
>
> **Senator PASTORE.** Because you could not trust that nation that it would live up to the agreement, and certainly you are not going to trust the Russians any time with reference to that. Doesn't that place you in a position that we can never hope for the outlawing of an atomic bomb until such time as we develop the technical facilities to detect it?
>
> **Mr. STASSEN.** That is a good statement, but I was going to point out that you not only have to know the scientific method of sure detection but you would also have to have a system set up by which you could move around inside the other countries with that system and check it, so it would require not only a new

breakthrough from the scientists that no one can now describe, but also an agreement as to the application of such a new system that would go beyond anything that anybody has talked about in the inspection field at the present time.

Senator PASTORE. Governor, I am going to ask you 1 or 2 other questions, and if you feel they should not be answered in public hearing, I wish you would just say so and I will be satisfied with your opinion on the matter. Are you satisfied that we are doing enough in developing the techniques and research to detect the presence of hidden atomic stockpiles?

Mr. STASSEN. Yes, we have some of our very ablest scientists constantly exploring and searching in this problem.

Senator PASTORE. Do you think we are devoting sufficient money to that research?

Mr. STASSEN. Yes.

Senator PASTORE. Now you made a statement about not being able to accept the suggestion that was made with reference to the testing. We do have the technical facilities to detect whether or not an agreement in that regard would be violated, have we not?

Mr. STASSEN. No.

Senator PASTORE. You mean that we can't tell now every time that the Russians have an explosion?

Mr. STASSEN. We can tell the big ones under usual circumstances, but you can't necessarily tell every one.[35]

Stassen's testimony clearly contradicted the views of Inglis, whose advocacy for a test ban rested on his assertion that U.S. technology (i e., the AEDS) was adequate to enforce a test ban treaty. While Stassen's assessment was more than correct (having received Northrup's briefings), Inglis' views were gaining increasing public awareness via his writings and speaking engagements. Although the AEDS and any information surrounding the system at that time was treated as classified (Secret or Top Secret), it is surprising the public did not press for more information, especially during an election year in which the Democratic Party adopted a test ban as a primary element of its party platform.

While the AEC had reaped the benefit of miniaturization testing in the TEAPOT shots, the perception of the Soviets' now greatly advanced nuclear weapons program meant that the U.S. would certainly resume testing with a renewed sense of urgency. They did so on May 4, 1956, when Operation REDWING commenced in the Marshall Islands.

REDWING utilized the Bikini and Enewetak Atolls instead of the Nevada site because six of the 17 shots would produce yields in the range of megatons. Such large detonations would have been unacceptable in the United States given the scope of public attention on fallout hazards in early 1956. Importantly, although REDWING experienced large detonations, one of the primary goals of the operation was to test hydrogen weapons, which would produce far less fallout—so-called "clean" weapons. In general, REDWING evaluated the efficiency of new weapons designs and their effects such as blast and shock, nuclear radiation, and thermal radiation. They also tested the ability to set up instruments and fire shots on Bikini and Enewetak Atolls at the same time. The diversity of tests ranged from an airborne delivery of existing stockpile weapons to the successful testing of radically new, experimental mechanisms. More than 10,000 DoD personnel participated in the operation.[36] AFOAT-1's participation in Operation REDWING was extensive. The field element was established on April 9, 1956, at Parry Island, Enewetak Atoll. It consisted of a command section, gas sample handling crews, and electromagnetic project personnel.

The REDWING shots, with different types of bombs, produced a wide range of yields, and presented the LRD researchers with an opportunity to validate and extend previous work. AFOAT-1 collection plans included experiments within close distances (out to approximately 235 miles), and debris tracking extending out to 700 miles. With much data already compiled, LRD researchers wanted to calibrate detection instruments and techniques in order to provide more precise evaluations of the data obtained from earlier Soviet nuclear detonations. Additionally, they utilized REDWING to test new equipment with the aim of improving the effectiveness of the AEDS.

One of AFOAT-1's primary objectives was the collection and analysis of close-in samples of nuclear debris. AFOAT-1 scientists understood nuclear debris collected at long range exhibited a variable isotopic composition depending on a combination of three effects: the composition of the debris at the time of formation, the fractionization that occurred, and the amount and composition of old debris simultaneously collected. In order to establish a basis for making corrections for the latter two effects, they needed to determine the initial composition of the debris. Collection of close-in debris would allow them to make that determination. The collection of short-lived radioactive products could yield valuable data to confirm the explosion and determine the condition of burst. New calibration data was important because such collections were not possible in previous tests. To perform this mission, crews installed gas sampling equipment in F-84 and B-57 aircraft to test a paired compressor system, one bleeding air from the aircraft compressor and the other sampling raw air through a probe system designed to give isokinetic flow. Also, this was the first time the new B-57 Canberra aircraft was tested for use in airborne sampling.[37]

REDWING served to unearth future possibilities for increasing the usefulness of electromagnetic surveillance. While LRD researchers had previously conducted a number of experiments to develop the EMP Q systems, to the point where they were now fully operational in the AEDS, they now wanted to record the electromagnetic pulse close to and at distances from known nuclear detonations. The scientists, having improved their instruments and analytical methods over the last year, wanted to further understand the relationship between the pulse and the bomb that generated it.[38]

Seismic and acoustic long-range measurements for Operation REDWING were supplied by associated agencies. The Navy Electronics Laboratory, NBS, and the Evans Signal Laboratory conducted acoustic measurements. The Houston Technical Laboratories, the Scripps Institution of Oceanography, and the U.S. Coast and Geodetic Survey Team on Guam Island each provided seismic measurements. The Houston Technical Laboratories conducted "deep-hole" experiments throughout the test series, while the latter two organizations provided seismic monitoring teams. Beers and Heroy also conducted seismic experiments to determine where computational equipment and automatic methods could be applied to the seismic system, and to possibly design and construct a device to measure the difference in arrival time between signals recorded at two or more stations.[39]

AFOAT-1 received considerable data from a number of distant stations set up by the Air Force Cambridge Research Center (AFCRC) of ARDC. Their efforts were designed to explore the problem of indirect bomb damage assessments of far-away targets, which AFOAT figured into the SAC/AFOAT-1 war plan (discussed in Annex 2). SAC and AWS flew sampling sorties that provided AFOAT-1 with data to determine the size and trajectories of radioactive clouds resulting from nuclear explosions of various energies. The crews collected data on the time difference between electromagnetic and seismic signals at several thousand feet above the detonations. Analysts utilized the time difference data, recorded at distant stations, to compare with their recordings from the two Soviet tests conducted in November 1955. The reports indicated greater degrees of accuracy in AFOAT-1's capability to estimate height of burst.[40]

As Eisenhower had stated in April 1956, the objective of Operation REDWING was to develop a variety of atomic weapons to meet a wide range of military uses, especially to provide defense against air attack and to perfect devices that reduced fallout hazards. Soon after the last REDWING detonation on July 21, 1956, the AEC announced substantial progress. New devices had proven their operational suitability and were added to the U.S. stockpile, and certain "clean" weapons had achieved predicted effects on the immediate target in producing minimum fallout.

Politically, the most important shot was Shot Cherokee on May 20th. This 3.8MT thermonuclear bomb was the only pure DoD shot of REDWING. Unique, though, was that it was dropped from a B-52 *Stratofortress* of the 4925th Test Group (Atomic) based at Kirtland AFB, New Mexico. Within the nuclear arms race, the United States had just closed an important capabilities gap with the Soviet Union.[41] Operation REDWING ended in July 1956.

REDWING took place as the Soviets were conducting a nine-shot test series between February and December. Their first shot, fired on February 2nd, produced an extremely small detonation. One of AFOAT-1's routine airborne sampling flights picked up the unexpected debris, which radiochemical analysis soon confirmed. Consequently, AFOAT-1 meteorologists backtracked recorded air masses to vector 10 additional sorties into the path of the effluents. While that first Russian shot served to validate the now routine efficiency of the AEDS's "A" technique, it also demonstrated the inability of the other AEDS techniques to detect much smaller tests.[42]

For the remaining eight Soviet shots in 1956, the AEDS performed well. With the exception of the March and December shots, all three pillars of the AEDS (acoustic, seismic and radiochemical) worked in tandem to detect and analyze the detonations. Routinely, acoustic and seismic data gave the location, EMP data indicated the time (within plus or minus .05 seconds), and subsequent radiochemical analysis of collected debris confirmed the detonation. However, by now it was evident that the smaller the shot, the fewer stations in the AEDS were able to detect the event. Northrup and his team of researchers were well aware of the limitations and were beginning to become concerned with external parties involved in the nuclear arms control debates who believed the AEDS could detect any Russian nuclear detonation.[43] Sensing the dangers of this misnomer, Northrup continued to press for more geophysical R&D.

The year 1956 witnessed the continued expansion of LRD systems in other parts of the world. Several more EMP ("Q") sites were added to the AEDS, but with some reservation. On one hand they had proven their worth and would now expand coverage. On the other hand, unlike the acoustic ("I") and seismic ("B") techniques, sufficient empirical calibration data on electromagnetic measurements did not yet exist. As such, AFOAT-1 was hesitant to rely on them at this point in time.[44]

Like the EMP stations, the seismic net also expanded. On October 8, 1956, AFOAT-1 established Detachment 313 at Sonseca, Spain. The previous year, Detachment 421 had also joined the net from its location in Alice Springs, Australia. Plans were in the works of establishing a seismic station on Mindanao in the Philippine Islands. However, AFOAT-1 faced a series of delays as the Philippine government negotiated a base rights agreement with the United States.[45]

In April 1955, Northrup terminated all ground-based particulate sampling operations, the Ground Filter Units (GFUs), within the AEDS. In his view, they were designed to only back up or augment airborne collections, especially given their limited capabilities as stationary systems. AFOAT-1 had procured the first 10 GFUs in August 1949. They were crude devices consisting of a blower forcing air through a filter paper to screen out particulates. Those papers were then processed at AFOAT-1's labs on Guam, or at McClellan or Eielson Air Force Bases. Northrup withdrew the five operating GFUs from the AEDS (including the backup system at McClellan) because the increased number of recent nuclear tests had produced debris difficult to distinguish from one another.[46] However, the ground-based air samplers, most often collocated with the GFUs, remained operational within the AEDS. These B/20-2 systems collected air to look for the presence of xenon. Because of the short half-life of the xenon isotopes (5.3 days), AFOAT-1 personnel shipped the samples by the most expeditious means possible to the Argonne National Laboratory.[47]

By 1956, AFOAT-1's McClellan Central Laboratory had also greatly improved thanks to a close working relationship with Tracerlab. The partnership between the two labs led to important training and operations programs, which continued to build Air Force radiochemical expertise throughout the first half of the decade. Operationally, that collaboration resulted in greater accuracy with measurements.

With only minimal improvements in the geophysical techniques, aerial sampling continued to serve as the workhorse of the AEDS. Throughout 1956, AWS expedited its transition from the WB-29 to the WB-50 as the primary airborne collection aircraft. Begun in August 1955, the conversion was 95 percent completed by the end of 1956. However, AFOAT-1 still struggled with collecting samples at higher altitudes. The WB-29 had a collection ceiling of 25,000 feet, while the WB-50 now added only an additional 5,000 feet to that limitation. The "mushroom" clouds of nuclear detonations were typically rising above 80,000 feet. Over the preceding several years, AFOAT-1 had explored several options to address the problem.

Following earlier awkward tests, balloons now seemed to offer a solution. As early as 1946, the Army Signal Core had taken a strong interest in balloons, initially to carry sonic detection equipment into the stratosphere (33,000 to 66,000 feet) to possibly record the sound of a Russian nuclear blast. For that acoustic experimental program, the Air Force had created Project MOGUL, under the leadership of Colonel Marcellus Duffy. From May to July 1947, Duffy launched a number of experimental balloons carrying low-frequency microphones from the Army Air Force base at Alamogordo, New Mexico. One balloon crashed 75 miles northwest of Roswell, New Mexico, giving rise to conspiracy theories about aliens, which persist to present day. Walt Singlevich, the future head of radiochemistry for AFOAT-1 and AFTAC, saw the potential of balloons for capturing air samples at very high altitudes—elevations no sampling aircraft could reach.[48]

From the time of the "Roswell Incident" in 1947 until 1954, AFOAT-1 experimented with balloons for high-altitude air sampling. As part of MOGUL, the Air Force had tested the sonic-equipped balloons during Operation SANDSTONE in 1948. In fact, AMC's Balloon Sonic Station at Alamogordo reported a positive signal from the SANDSTONE shots at 50,000 feet. AFOAT-1 also used balloon-equipped counters at SANDSTONE to detect fission products up to 60,000 feet. For Operation GREENHOUSE in 1951, AFOAT-1 experimented with balloons to quickly detect atomic bomb debris. The objective of that experiment was to test equipment reliability and to determine the feasibility of measuring at long range the concentration of atomic debris from ground level to 90,000 feet.[49] In 1952, AFOAT-1 contracted with General Mills for the use of balloons to conduct high-altitude sampling under a program codenamed GRABBAG. General Mills had created its aeronautical research laboratory in 1946 and soon developed the Skyhook balloon capable of altitudes above 100,000 feet. In the fall of 1953, General Mills conducted the first successful air sampling flight. Beginning in January 1954, AFOAT-1 flew an average of three collection flights each month in the vicinity of Minneapolis, Minnesota. By mid-1954, the balloon sampling capability was added to the AEDS as the "Db" technique.[50]

As promising as balloons appeared in collecting very high vertical samples, they were no substitute for long-distance airborne sampling. In addition to the 30,000 foot capability of the WB-50s now coming on line, SAC offered up one B-36 to

AFOAT-1 for use in monitoring the 1956 Russian tests. Codenamed SEA FISH, this B-36 flew the Japan-Alaska route at an altitude of 40,000 feet. The initial test of the B-36 occurred in November 1955, with the 6th Bomb Group flying out of Guam. Additionally, the Far East Air Force (FEAF) converted some T-33s from the 6007th Reconnaissance Group to conduct airborne sampling at 40,000 feet. These aircraft, however, were limited by their much shorter ranges as compared to the converted bombers. By late summer of 1956, FEAF had requested nine B-57 Canberra aircraft for one of their reconnaissance units. AFOAT-1 worked closely with FEAF in planning for the conversion of a B-57 to conduct airborne sampling at the plane's 50,000 foot capability.[51]

As the AEDS continued to improve throughout 1956, it was becoming an object of attention at the center of international nuclear diplomacy, despite being a classified system. Northrup was having to explain the limitations more frequently as both scientists and politicians were referencing the AEDS in their debates over arms control. Simultaneously, the workload at AFOAT-1 was expanding at an accelerating tempo as nuclear testing increased amid rising tensions between the United States and the Soviet Union. The LRD researchers were finding themselves under increased pressures to improve techniques and to expand geographical coverages. Both required more personnel, more equipment, and more money—none of which the President was inclined to favor. In fact, despite DoD and AEC fears of Soviet intentions, Eisenhower still hoped to curb spending. However, somewhat paradoxically, his administration was being increasingly viewed as too soft on communism, an accusation that had caused the defeat of Truman's Democratic Party in 1952. Consequently, arms control and nuclear diplomacy took center stage during the presidential election campaign of 1956.

Operation CASTLE – Shot Bravo

On February 28, 1954, the U.S. detonated the largest nuclear device it had ever, or would ever, conduct. It weighed approximately 12 tons and was located inside a small protective building that "resembled a small shack at the end of a dock." It produced a yield of 15MT. Because radioactive fallout was so extensive, Bravo caused widespread fears resulting in world leaders calling for a nuclear test ban.

Photo courtesy of the Department of Energy

Major General
Daniel E Hooks

Commander, 1009th SWS/AFOAT-1 from 1954 to June 1958. Hooks was a pilot and a scientist, with an M.S. in physics from Harvard University. He served on Guam from 1944 to 1946. After the war, Hooks became chief of the Research and Development Field Office for Atomic Energy at Kirtland AFB. Then remaining at Kirtland as the vice commander of the Air Force Weapons Center. Holding the longest tenure in command, Hooks led the critical refinement and expansion of the AEDS as the Cold War intensified.

Photo courtesy of the United States Air Force.

Harold Stassen

In March 1955, Eisenhower appointed former Minnesota Governor Harold Stassen as Assistant to the President for Disarmament. This was a new position and one that carried cabinet level rank. Stassen's appointment signaled Eisenhower strong desire to formulate a nuclear arms control treaty.

Photo courtesy Texas A&M University Archives

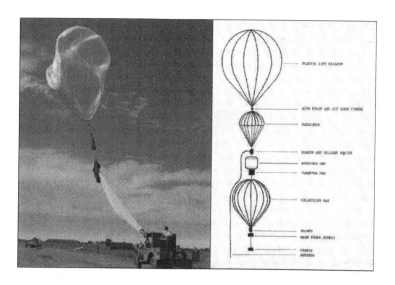

Project GRAB BAG - high altitude air sampling 1953-1959

Unable to obtain air samples above 30,000 feet with the WB-50, AFOAT-1 explored the use of balloons. Balloon-borne air samplers joined the AEDS as the "Db" technique in 1954.

Photo source: AFTAC archives

236

The Mark-9 (AN/AAR-9) Infrared Mapping Set

Used by AFOAT-1 as part of a series of experiments to test the feasibility of President Eisenhower's "open skies" proposal. This prototype, the most advanced IR system available, proved inadequate in locating nuclear facilities. *Photo source: AFTAC archives*

The RB-36 "Peacemaker"

SAC first provided AFOAT-1 with a B-36 during Operation CASTLE in February 1954. The RB-36 could sample at an altitude of 45,000 feet. This was a big improvement over the RB-50 which had a maximum altitude of 30,000 feet. *Photo source AFTAC archives*

The RB-57 Canberra

The RB-57, added a higher altitude capability of 50,000 feet to the AEDS. Was first used in Operation REDWING from May to July 1956.
Photo courtesy of the United States Air Force

237

Notes to Chapter 8

[1] Miller, *Under the Cloud*, 193-194.

[2] Harold Stassen and Marshall Houts, *Eisenhower: Turning the World toward Peace*, St. Paul, Merrill-Magnus Publishing, 1990, 171-172.

[3] Divine, *Blowing on the Wind*, 61-62.

[4] Larson, *Anatomy of Mistrust*, 59.

[5] Gaddis, *We Now Know*, 228. Ironically, Khruschev had now accepted Malenkov's previous views on the mutual destruction inherent in both nations' possession of thermonuclear weapons. Gaddis writes that "Malenkov's heresy thus became Khruschev's orthodoxy: that the common experience of a thermonuclear revolution had left the United States and the Soviet Union no choice but to seek common ground."

[6] Stassen, *Eisenhower*, 339.

[7] Larson, *Anatomy of Mistrust*, 64-65.

[8] Ibid.

[9] Some historians have viewed this development as a serious missed opportunity. As such, historians have remained divided as to whether Eisenhower was morally committed to arms control or whether his pronouncements were staged to assuage public opinion (domestic and worldwide) as the U.S. continued to build up a very large nuclear weapons stockpile.

[10] Gaddis, *We Now Know*, 245.

[11] Divine, *Blowing on the Wind*, 61.

[12] Bio, Major General Daniel E. Hooks, www.af.mil, accessed October 20, 2018.

[13] Technical memo #84, Northrup to Hooks, Subject: "Information Requested for Use by the E. O. Lawrence Study Group," October 6, 1955, AFTAC archives. Also, *History of AFOAT-1*, 1956, 128-129.

[14] Ibid.

[15] Ibid

[16] Ibid.

[17] Walter Singlevich and Robert S. Brundage, "Nuclear Inspection Experiments in Support of Mr. Stassen's Disarmament Planning, Annex C: Radioactive Emanations from Atomic Energy Installations," final report from Northrup to Hooks, July 15, 1956,

AFTAC archives. For WWII reference see Lawren, *The General and the Bomb*, 138.

[18] J. Allen Crocker, "Nuclear Inspection Experiments in Support of Mr. Stassen's Disarmament Planning, Annex B: Fields Associated with Power Transmission Lines," final report from Northrup to Hooks, July 15, 1956, AFTAC archives.

[19] Robert J. Fisher, "Nuclear Inspection Experiments in Support of Mr. Stassen's Disarmament Planning, Annex A: Infrared Emission from Atomic Energy Installations," final report from Northrup to Hooks, July 15, 1956, AFTAC archives.

[20] Divine, *Blowing on the Wind*, 61.

[21] Divine, *Under the Cloud*, 61. Inglis "emerged as a prominent figure in the nuclear disarmament movement. As a founding member of the Federation of American Scientists and participant in the Pugwash conferences during the 1950s and 1960s, Inglis was a prominent voice for rationality in discussions of nuclear weaponry. As early as 1951, he called for the creation of a federal agency for arms control and disarmament (not established for another nine years), and both in congress and before the public, he lobbied steadily for nuclear disarmament and for a partial nuclear test ban (approved in 1963). He also argued against nuclear proliferation and the development of anti-ballistic missiles." The Inglis Papers, University of Massachusetts, http://scua.library.umass.edu/ead/mufs033. Last accessed April 2018.

[22] Technical memo #82, Northrup to Canterbury, Subject: "Estimate of the Long Range Detection Capabilities of the USSR," December 1, 1953, AFTAC archives.

[23] DoE, *Tests*, 5-6. Also Miller, *Under the Cloud*, 213.

[24] Jean Ponton, Carl Maag, Martha Wilkinson, and Robert F. Shepanek, Operation TEAPOT (Washington, D.C.: Defense Nuclear Agency, November 1981), DTIC-A113537; also *History of AFOAT-1, 1955*.

[25] *History of Long Range Detection: 1955*, author unidentified. Washington, D.C.: The Air Force Office of Atomic Energy-One, Headquarters, United States Air Force, 1955. Minimally, the EMP systems of the AEDS also utilized the detected TEAPOT shots to increase the data used in the calibration of the EMP network.

[26] United States Navy, Operation Plan 1-55, "Operation WIGWAM," Task Group 7.3, March 25, 1955. DTIC No. ADA078589.

[27] *History of Long Range Detection:* 1955, 5-11.

[28] Ibid.

[29] The northern island of Severny and the southern island of Yuzhny are separated by the Matochkin Strait.

[30] *History of Long Range Detection:* 1955, 5-11.

[31] Ibid.

[32] Ibid., 5-11. Also Divine, *Blowing on the Wind,* 65.

[33] Robert S. Norris and Hans M. Kristensen, "Global Nuclear Weapons Inventories, 1945-2010," *Bulletin of the Atomic Scientists,* July/August 2010, 81.

[34] Divine, *Blowing on the Wind,* 66.

[35] "Control and Reductions of Armaments," Hearings, before a Subcommittee of the Senate Committee on Foreign Relations, 84th Congress, 2nd Session (Washington, 1957), 21-22. Also cited in Divine, *Blowing on the Wind,* 68.

[36] Martin Blumenson and Hugh D. Hexamer, "A History of Operation Redwing: The Atomic Weapons Teats in the Pacific, 1956" Headquarters, Joint Task Force Seven, Washington , D.C. December 1, 1956. DTIC ADB951592

[37] *History of the Air Force Office for Atomic Energy-One (AFOAT-1): 1 January-31 December 1956,* author unidentified. Washington, D.C.: The Air Force Office of Atomic Energy One, Headquarters, United States Air Force, 1956, 36-37.

[38] Ibid.

[39] Ibid.

[40] Ibid.

[41] Ibid.

[42] Ibid.

[43] Ibid.

[44] Ibid.

[45] Ibid.

[46] *History of AFOAT-1: 1955,* 30.

[47] Ibid., 33.

[48] Richard P. Hallen, *The Roswell Report: Fact vs. Fiction in the New Mexico Desert* (Washington, D.C.: U.S. Government Printing

Office, 1995), 25-29. Also see James Michael Young, "The U.S. Air Force's Long Range Detection Program and Project MOGUL," *Air Power History*, Vol. 67, No. 4, Winter 2020, 25-34.

[49] *History of AFOAT-1: 1947-1953*, 42, 97-98, 112.

[50] *History of AFOAT-1: 1954*, 39-40.

[51] *History of AFOAT-1: 1956*, 22.

CHAPTER 9

The Cold War Becomes Total

In 1956, the capabilities of the AEDS became central to the presidential election campaign. Vying for the Oval Office were the Democratic nominee, Adlai Stevenson, II, and the incumbent Dwight Eisenhower. Throughout the summer, Stevenson made radiation fallout and the test ban the central themes of his campaign. In response, and contrary to previous policy, Eisenhower announced the monitoring (obviously based on AFOAT-1 reports) of the secretive Soviet nuclear tests that occurred on August 24th and 31st, and September 2nd.[1] He did so to illuminate Soviet secrecy surrounding its nuclear weapons program in order to underscore how they could not be trusted, and to discredit previous Soviet calls for a test ban excluding inspections. On September 11th, Premier Bulganin re-issued his proposal for a test ban but was clear any ban would not include inspections. In his letter to Eisenhower, he stated "the present state of science and engineering made it possible to detect any violation at long range."[2] In retrospect, Eisenhower's remarks did not mean he was unequivocally insistent on a test ban. It was general mistrust and the issue of inspections that had erected apparently insurmountable barriers to any agreement. In late September, softening his stance, he requested the NSC to consider Bulganin's recent proposal. Importantly, Stassen and even the hawkish Dulles were suddenly in favor of considering a ban now that disarmament appeared out of reach.

At the same time, Stevenson made his advocacy for a test ban the central theme of the race. He delivered a number of speeches about the destructive power of thermonuclear weapons and the horrific effects of radiation on unborn children that frightened many Americans. It had some impact; support for Stevenson began to increase. At one point, Stevenson even suggested the U.S. should consider a *unilateral* test ban. In referring to the capabilities of the AEDS, he again erroneously stressed that technology could easily detect any Russian violations: "You can't hide the explosion any more than you can hide an earthquake."[3]

On October 15, 1956, Stevenson conducted a telecast from Chicago focusing solely on the nuclear fallout and test ban issues. He paid particular attention to the goal of nuclear disarmament and the prospects of worldwide peace. A great many listeners approved, and Republican leaders suddenly found that Stevenson was gaining momentum just three weeks before the election. Unfortunately for Stevenson, Soviet statements made several days later in support of Stevenson's views torpedoed his campaign. On October 19th, the State Department released another letter from Bulganin to Eisenhower, which again called for a test ban. In his letter, the Russian premier inferred that Stevenson agreed with the Soviet proposal and that inspections were not needed due to "the present state of scientific knowledge." For many Americans, with McCarthyism only in the recent past, Stevenson now appeared too cozy with the Soviets. Bulganin's comments became "a political kiss of death with an electorate hugely mistrustful of Russian intentions."[4] On November 6th, Eisenhower was reelected to a second term. In sweeping 41 states and receiving 58 percent of the popular vote, it was clear the majority of Americans shared the Republicans' mistrust of the Soviets and viewed a strong security policy as the best path toward peace. This choice set the stage for the next five years in addressing the question: What most protects American national security—nuclear superiority or an arms control treaty?

As 1957 began, AFOAT-1 faced its most demanding year to date. In the months ahead, the United States would conduct the most extensive nuclear testing series yet (Operation PLUMBBOB) and, simultaneously, expand the operational capabilities of the AEDS by improving detection instruments and extending the system's geographical coverage. For

Northrup and Hooks, it was clear from the Stevenson-Eisenhower debates over a test ban and international inspections that the capabilities of the AEDS had become an important tool in nuclear diplomacy. Those capabilities, though vastly improved, were not as good as many high-level officials in the government believed. Stevenson's view that they were sufficient led AFOAT-1 to thoroughly assess the current state of the worldwide system in order to dispel growing misperceptions. On October 19th, the same day as Bulganin's last letter to Eisenhower, Northrup delivered a current assessment of the AEDS to Hooks.

Northrup's 18-page report was notable for two reasons. First, it was the first statement ever made that the AEDS should expand in order to cover geographical regions other than the Soviet Union. Second, it was also the first comprehensive report to document the limitations of the four major components (seismic, acoustic, electromagnetic, and nuclear debris collection). The new goals were to detect explosions of 1kt and above anywhere in the USSR, and 100kt and above anywhere in the world.[5]

The report began by detailing capabilities and limitations. In looking at the nine seismic stations in the AEDS, standardization had now been achieved. All stations had operational seismic arrays and were positioned to discriminate against known sources of microseisms. New electronic gear could now increase the signal-to-noise level in the range from 0.5 to 1.0 second where the maximum energy in the earth was found from nuclear explosions. The seismic equipment produced a film record (developed three times every 24 hours), which station personnel scanned and passed on to AFOAT-1 HQs if they found anything significant. AFOAT-1 also had a central analysis station in Laramie, Wyoming, that correlated all of the station reports. From all incoming data, AFOAT-1 could identify the time of a shot to the nearest second, locate the detonation site within a 25 mile radius, and estimate the yield to an order of magnitude.[6]

In operating the acoustic component of the AEDS, the U.S. Army Signal Corps managed 10 stations. Each acoustic station consisted of four condenser microphones positioned at the corners of a six- to 10-mile square. By late 1956, R&D had succeeded in reducing background noises, and the equipment now greatly increased sensitivity to low-yield tests. Station operators scanned the recordings according to AFOAT-1 criteria, looking for indications of a nuclear explosion. If suspected, they encoded the recordings and passed them to the central analysis station at Fort Monmouth, New Jersey, for validation. Validated recordings were then passed to AFOAT-1 HQs. Despite these improvements, AFOAT-1 personnel, in analyzing the acoustic data, could only determine the time of the detonation within 15 to 30 minutes and confirm the location within a 300-mile radius. Yield estimates also included a margin of inaccuracy within 20 percent for large tests and within a factor of two for smaller tests.[7]

The newest component of the AEDS, the electromagnetic or "Q" technique, was currently expanding at the time of Northrup's assessment, with construction of several new sites underway. Though considered mission capable, the "Q" technique was still plagued with background noises. The eight operational Q stations—three in the United States and the other five located overseas—housed oscilloscopes that recorded electromagnetic pulses emanating from the Soviet test sites. Two overseas stations were considered close-in and had just received new experimental "raster"-type recorders capable of recording the broadband wave form of the pulses. Because the "Q" technique was unreliable as an independent means of characterizing a nuclear event, station data were sent to the central analysis station in Duluth, Minnesota, for correlation. The correlated data was then forwarded to AFOAT-1 HQs in Washington, D.C., where an analysis group conducted the final evaluation. The analysis group would only search through the data if seismic or acoustic data indicated the possibility of an explosion within a given time period. The utility of the Q technique was its ability to determine the time of the detonation within 0.1 second. Also, the difference in time between the electromagnetic and seismic times gave a measure of the height of the burst, an observation that had previously been difficult to determine.[8]

The oldest component of the AEDS, nuclear debris collection and radiochemical analysis (the "A" technique), remained the most valued technique. At the end of 1956, the AWS had dedicated two squadrons of WB-29s and WB-50s (based in Japan and Alaska) to support the AEDS. In addition, SAC had recently modified T-33s and B-36s to collect samples as well. AFOAT-1 personnel immediately processed the collected samples in the field laboratories co-located with the squadrons. MCL also analyzed the samples as a redundant capability.[9]

The new, long-term goal of detecting smaller detonations ranging between 1kt and 10kt challenged the current LRD technologies. However, providing a sober assessment was important at this moment in time because it supported Eisenhower's insistence on a thorough inspection system for any arms control agreement, which could conceivably span the immense geography of the Soviet Union. To monitor a 10kt detonation within that geography, the AEDS acoustic, seismic, and electromagnetic techniques would require an additional 16 stations on a 1,000-mile grid *within* the borders of the USSR (obviously unacceptable to the Soviets). As instrumentation improved, Northrup believed that estimate could be reduced by half. A coverage of 1kt detonations would require 32 additional stations deployed on a 500-mile grid for acoustic, and 128 stations on a 250-mile grid for seismic. For the latter, Northrup could not guarantee detection beyond 125 miles. Importantly, Northrup's pessimistic assessment of a monitoring system within the USSR was a direct response to Eisenhower's primary prerequisite for a disarmament treaty.[10]

The second half of the report discussed the additional requirements needed to transition the AEDS from a Soviet orientation to an effective worldwide surveillance network. Noting that such an assessment was in its infancy, a worldwide system should be able to detect detonations of 100kt anywhere on the planet. To do so, the AEDS would require an additional 20 acoustic stations. Most of those would demand careful emplacement in the southern hemisphere and house instruments capable of covering a 2,500-mile detection range. For seismic, the southern hemisphere posed a greater challenge because of problems with the Earth's core shadow. AFOAT-1 believed seismic readings in the southern hemisphere would produce 10 to 20 times more questionable recordings than those already experienced in the northern hemisphere. To reach the goal of detecting 100kt detonations, the AEDS would need five to eight seismic stations in the southern hemisphere. Northrup also predicted problems with weather impacts. High numbers of tropical thunderstorms in the equatorial belt would impose major problems for the electromagnetic technique. Not knowing the details on such impediments, he estimated the AEDS would require anywhere between 20 and 40 more stations. Finally, the most complex problem of expanding the AEDS was with AFOAT-1's most proficient technique—aerial sampling. Extended coverage down to the South Pole would require two additional B-50 squadrons. For pole-to-pole coverage, one at 90 degree west longitude and the other at the Greenwich meridian, several more squadrons were needed, bringing the total to 10 addition squadrons of B-50s.[11]

In agreeing with the hawkish views of the Air Force, the JCS, and DoD in general, Northrup expressed great skepticism of both the idea of on-site inspections and the feasibility of an "open skies" program:

> It is not believed that augmenting the surveillance coverage inside the USSR, expensive as it would appear to be, would accomplish much more than to spread the

trouble over the entire world. There are vast regions of the Pacific, Atlantic, and Indian oceans particularly in the southern hemisphere where atomic tests of relatively large magnitude would go undetected. The very act of putting an extensive network inside the USSR would do little more than drive them to the Arctic, Antarctica, or other remote areas mentioned above for clandestine tests. It is recommended, therefore, that any consideration of an aerial inspection system be based on covering all possible test sites rather than limited surveillance within the USSR itself.[12]

On January 21, 1957, Dwight Eisenhower was sworn in for a second term as president of the United States. Despite the time commitments of the election campaign, he was still wrestling with the ramifications of the recent Hungarian uprising and the Suez Crisis. Both events dealt with the spread of Soviet influence and invoked concern among the West about continuing communist expansionism.

As Americans voted at the polls on November 6, 1956, a violent uprising was occurring in Hungary. What began as student protests against Soviet-imposed policies quickly escalated to a nationwide revolt lasting from October 23rd to November 10th. Armed resistance led to the collapse of the communist government and demands by the protesters for free elections and withdrawal of Hungary from the Warsaw Pact.

Reneging on their promise to negotiate, the Russians invaded the country on November 4th with more than 30,000 troops and 1,100 tanks. As the fighting ended a week later, more than 3,000 civilians were dead and 13,000 injured. Soviet casualties numbered more than 700. Although the U.S.-sponsored Radio Free Europe highly encouraged resistance in radio broadcasts across Hungary, Eisenhower declined to intervene. Much of the West viewed his inaction as hypocritical and contrary to the U.S. foreign policy of supporting freedom throughout the world by stemming communist expansionism.

The president faced another dilemma. Also on November 6th, as Americans were voting for a president and Hungarians were fighting in the streets, British and French paratroopers were jumping onto El Gamil airfield near Port Said in northern Egypt in a joint operation to secure the Suez Canal. Known as the 1956 Suez Crisis, the joint British-French invasion was a result of Egyptian President Gamal Abdel Nasser's nationalization of the canal. The invasion coincided with the Israeli invasion of the Sinai a week earlier (the Second Arab-Israeli War). Over the course of the previous year, Nasser had been developing a close relationship with the Soviets. U.S. attempts since 1953 to form an anti-Soviet alliance of the Near East Arab states (centered on Egypt) were thwarted by Arab hatred of Israel, which Arab states viewed as a greater threat than communism. Concurrently, the British deeply resented American attempts to influence the Egyptian government.

Previously, in September 1955, after failing to procure American arms, Nasser had purchased a large quantity of Soviet weapons through Czechoslovakia. On July 26, 1956, Nasser, in an attempt to assert himself as the leader of the Arab world, sunk several ships to block passage through the canal. When the Israeli-French-British attack occurred to reopen the waterway, Eisenhower asserted effective pressure to stop the operation. He believed any perceived support of the invaders would produce outrage throughout the Arab world, resulting in expanded Soviet influence in the region. He also could not criticize the ongoing Soviet invasion of Hungary while tacitly accepting the Allied incursion. Some historians have written that Eisenhower feared the invasion could spark a third world war.[13] Others have stated that Khruschev, just as Eisenhower had acted to end the Korean War, threatened the use of nuclear weapons in support of Egypt.[14] In any case, Nasser remained in power and only re-opened the canal in March 1957. U.S. relations with Britain and France were severely strained for some time to come.

The Hungarian invasion and the Suez crisis illustrate how international events consistently derailed Eisenhower's sincere intentions to reduce tensions in the nuclear arms race. On one hand, he was compelled to confront Soviet expansionism, especially with the defense establishment and so many Americans believing a war with the Soviet Union was inevitable. On the other hand, Eisenhower believed arms control was certainly possible and critical in lowering international tensions. Inevitably, the president sided with the former as America could not risk letting its defensive guard down. In the aftermath of the Suez crisis, on January 5, 1957, Eisenhower requested authorization from Congress for the authority and flexibility to deploy military forces to the Middle East to assist any nation confronting communist aggression. He also requested $200 million in aid for those countries (2021 = 2 billion). Congress approved both requests in support of the policy, which became known as the Eisenhower Doctrine.[15]

Fear of a Soviet surprise attack continued to grow throughout 1957. At the root of this fear, in all of its manifestations, was the great unknown about the state of Soviet technological progress in developing nuclear capabilities, especially in placing thermonuclear devices on long-range missiles. Eisenhower had already reacted to the fears of a "bomber gap" and a "missile gap" by expediting the development of the B-52 and the Atlas missile. With human intelligence assets sorely lacking or deficient on nuclear programs within the USSR, the president could only rely on surveillance systems for information. To date, the only reliable system was the AEDS. In his view, he needed another capability. So was his reasoning in pushing the development of the U-2 reconnaissance plane, which could fly at altitudes above 70,000 feet. Initially authorized for CIA use only, the first U-2 flight over the Soviet Union occurred on July 4, 1956. Concern that a downing of the highly classified aircraft could spark a war, Eisenhower exercised caution in authorizing U-2 reconnaissance missions.[16] The AEDS now had a partner in the business of long-range detection.

While Eisenhower would continue to pursue his dream of disarmament and reconsider the idea of a test ban in 1957, he reluctantly supported the aggressive test series scheduled for 1957 and 1958. On December 29, 1956, the president approved the AEC request to conduct this series code named Operation PLUMBBOB. Two months later, on February 23, 1957, he authorized the expenditure of plutonium for a final green light on the operation, scheduled to begin on May 28th.

There was little doubt the U.S. would continue aggressive testing, especially given the degree of anxiety generated by fears of tremendous technological progress in the USSR. Just 48 hours before he was inaugurated for his second term, Eisenhower was informed that the Soviets had already begun their 1957 test series—the earliest start date on record. His concern was justified. The first of 15 conducted in 1957, this test was the first atmospheric nuclear explosion fired with a missile launch. The first missile-deliverable warhead, even though only a straightforward fission device yielding 3.5kt, was a breakthrough for the Soviet nuclear weapons program. The next Soviet launch came on March 8th, followed by five more shots in April. Three of the April shots were thermonuclear devices. In total for the year, seven of the fifteen shots were fusion weapons, almost all air dropped. The final test of the series occurred on December 28th.

Considering the yield sizes of the earlier Soviet fission shots, averaging approximately 30kt, the AEDS did not perform well. The electromagnetic technique detected only 6 of the 15 Soviet detonations. Part of the problem was that four stations were under construction, and new personnel had not reached operational proficiencies for most of the year on new experimental equipment. Additionally, the film viewers and cameras experienced significant problems. To economize the number of stations in the AEDS, AFOAT-1 also spent much of the year combining the electromagnetic and the acoustic techniques into a single-station configuration.[17]

The seismic sites detected only seven of the 15 Soviet shots. However, for those seven detonations, the "B" sites provided accurate locations, times, and estimates of yield for larger devices, and for those shots detonating close enough to the earth's surface to trigger the sensors. Still, coverage continued to improve. On April 15, 1957, the seismic station at Sonseca, Spain, became fully operational and was the ninth site to enter the AEDS.

On May 6th, AFOAT-1 began a major upgrade to the seismic network. Called Operation CHANGE OVER, this effort involved installing new equipment and standardizing older equipment at all nine sites to improve operational efficiency and to reduce outage times. AFOAT-1 personnel completed all modernization activities by October 30th. Although many of the new design features immediately improved performance, calibration difficulties continued to plague the instruments. For example, during the months of April, May, and June, calibration problems contributed to 51 percent of the outage times. Despite these problems, AFOAT-1 continued to advance the seismic technique by experimenting with deep-hole systems. AFOAT-1 also expanded the zipagram data transmission system from Team 204 in Fairbanks, Alaska, to AFOAT-1 headquarters in Washington, D.C. This new capability accelerated seismological analysis and evaluation by an average of 18 hours.[18]

The acoustic systems, the "I" Technique, performed better than the other AEDS sensors during the 1957 Russian tests. The 10 stations detected 14 of the 15 Soviet nuclear explosions. The "I" Technique provided reliable estimates of yield, time, and location of nuclear detonations. However, one of the continuing problems of the acoustic technique was differentiating between nuclear explosions and other large nonnuclear events. For example, on November 12, 1957,

AEDS "I" stations in Alaska and Japan detected an event in the Kurile Island area that appeared to be nuclear. However, subsequent analysis of data indicated it was a volcanic eruption.[19]

In February, the AFOAT-1 commander, General Hooks, decided to reactivate an acoustic station in Iran. The Army had operated a station in Iran during 1952 for a short period of time, but political instability had forced its closure. During that brief period of operation, the station experienced a more favorable noise level compared to the average acoustic net station. Because the new station in Iran would be the nearest site to the Semipalatinsk test site, AFOAT-1 expected this close proximity to improve the accuracy of shot times, yields, and locations. During July, Army Signal Corps and AFOAT-1 personnel conducted a site survey for the new station, and Northrup hoped to have the site fully operational sometime in 1958. The Iran station was scheduled to receive the new remote calibrator also installed in the other nine locations throughout 1957. The new instrument increased efficiency in the acoustic system by eliminating bimonthly personnel visits to the outposts.[20]

The Russians had already completed seven shots in their 1957 test series when Operation PLUMBBOB began on May 28th. PLUMBBOB would become the most controversial operation of all nuclear tests conducted in Nevada because it included the testing of a fusion weapon within the continental United States and, consequently, increased fears over the dangers of nuclear fallout. Nationwide anxieties were high in the spring of 1957, not only due to fallout concerns but also as a result of escalating tensions with the Soviet Union. Many still feared a surprise attack and believed a nuclear war between the United States and the Soviet Union was imminent.

Just days before PLUMBBOB began, the president commissioned an extensive study within the Office of Defense Mobilization (ODM) to assess "the relative value of various active and passive measures to protect the civil population in case of nuclear attack and its aftermath."[21] He also asked the panel to "study the deterrent value of our retaliatory forces, and the economic and political consequences of any significant shift of emphasis or direction in defense programs."[22] The former assistant director of the Radiation Laboratory at MIT, Mr. Horace Rowan Gaither, Jr., chaired the panel. Gaither was a powerful administrator of the Ford Foundation and had helped create the Rand Corporation. His group was officially called the Security Resources Panel of the Science Advisory Committee, or more commonly referred to as the Gaither Panel. In the course of the study, more than 90 people contributed to the assessment, to include famous physicists such as Hans Bethe, Isador I. Rabi, and Ernest Lawrence. Other important contributors included Dr. James B. Fisk, Dr. James R. Killian, Jr., and General James H. Doolittle. The Gaither Panel was granted access to intelligence resources, which was used to assess Soviet capabilities.[23]

The creation of the panel was a direct result of widespread concerns across America about surviving a nuclear surprise attack. Fallout shelters sprung up throughout communities in every state. Local governments drafted emergency action plans, and public schools began practicing attack drills. The belief in the inevitability of a nuclear war had spread beyond the boundaries of government and into the homes and lives of many Americans. It would take the panel nine months to complete the study (discussed below). In the interlude, progress toward an arms control treaty proceeded along a very bumpy road.

As many historians have noted, Eisenhower's sincere desire for nuclear disarmament and the sharing of nuclear knowledge with the world for peaceful purposes stood in stark contrast to his authorizations for increasing the nuclear stockpile and advancing nuclear technologies to outdo Soviet advances. The most plausible explanation for the President's contradictory behaviors was voiced by Dr. John Gaddis who stated,

> [T]he Eisenhower administration came to see that even a few Soviet nuclear bombs could produce such devastation as to render any American nuclear advantage meaningless. The United States therefore refrained, for the most part, from exploiting Soviet weaknesses. Far from shifting the status quo, as he had promised to do while seeking office, Eisenhower used nuclear weapons to shore up and stabilize it. In an interesting inversion of military tradition, he saw *superior* capabilities as allowing only *defensive* responses.[24]

The test goals of Operation PLUMBBOB appeared to support Eisenhower's view of "defensive responses." On June 5th, the same day as the third shot of the series, Eisenhower emphasized this view in a press conference: "Our tests in recent years, the last couple of years, have been largely . . . in the defensive type of armament to defend against attack from the air. . . ."[25] Indeed, the tests were used to evaluate the operability of conventional forces on a nuclear battlefield, especially the impact on military tactics, equipment, and training. Several shots evaluated the effects on civilian structures and food supplies, and served to test civil defense emergency preparedness plans. PLUMBBOB also included six safety experiments to ensure no nuclear reaction would occur if the highly explosive components of the device were accidentally detonated during storage or transport. This was an important goal to emphasize because on May 22nd, just six days before PLUMBBOB began, an Air Force B-36 bomber accidentally dropped a 42,000 pound unfused hydrogen bomb (a Mark 7) near Albuquerque, New Mexico.[26]

Many of AFOAT-1's partners contributed to the test series. Northrup and his team extensively relied on the 4926th Test Squadron (Sampling) of the 4950th Test Group (Nuclear) for the collection of nuclear debris.[27] Uniquely, 14 Air National Guard squadrons used many of the shots as training events to augment 4926th missions. Their participation was part of the Strategic Air Command's war plan (see Annex 2). Each squadron flew two T-33 aircraft. Their samples were sent to LASL and UCRL, and used for comparison against the samples collected by the 4926th. For the Q technique, the National Bureau of Standards (NBS), the Defense Research Laboratory (DRL), and the University of California at Los Angeles (UCLA) established temporary electromagnetic sites in Nevada. The U.S. Coast and Geodetic Survey provided close-in seismic measurements of the shots. In conducting acoustic measurements, AFOAT-1 largely relied on the AEDS. During PLUMBBOB, AFOAT-1 deployed 47 people to Camp Mercury, Nevada, where they controlled all AEDS operations from May 2nd through October 7th.[28] It was a busy five months for the team, as the AEDS responded to the many intertwined Russian and American detonations. From then on, the AEDS would serve as both an experimental R&D system and as an operational strategic technological asset that critically undergirded American national security.

The 24 nuclear detonations of Operation PLUMBBOB occurred between May 28, and October 7, 1957, involving more than 18,000 DoD personnel. Compared to previous test series, AFOAT-1's involvement reflected more of an operational evaluation experience rather than experimental R&D. More precisely, the stated goal of AFOAT-1's participation was to "test technical and procedural methods on known devices which might be applied in evaluating foreign tests" and, more specifically, to "calibrate detection equipment and refine techniques for evaluating Soviet nuclear explosions."[29] The one exception was seismic research. For Doyle Northrup, the operation would finally include his long-sought underground test for advancing seismic R&D.[30] PLUMBBOB would also serve to test the operational capacities and fidelity of the AEDS. It proved to be an extremely active year for the AEDS, as the various stations responded to both the 24 U.S. tests, with known schedules, and the 15 Russian tests throughout 1957 that occurred by surprise.

In June, as AFOAT-1 became aware of the shortcomings of the AEDS in detecting the first seven Russian shots of 1957, Northrup felt compelled to draft another AEDS capabilities statement. In all likelihood, he and Hooks were also influenced by the political context generated in the ongoing UN Disarmament Conference at the time (discussed below). Proponents of a test ban had recently made a number of statements in public, asserting the U.S. possessed the technological systems (i.e. the AEDS) to detect any detonation in the Soviet Union. As fears over the dangers of radiation escalated around the world and demands for arms control grew, the president fell under increased pressure to ease tensions with the Soviet Union. For the armed services, however, it was not a time to compromise or accommodate Soviet stipulations for any agreement. Within the U.S. military establishment, there was a genuine belief the Soviets had caught up to the Americans and were quickly surpassing the United States in the arms race.

On June 21, 1957, just three days before Shot Priscilla (the fifth shot of PLUMBBOB), Northrup submitted his assessment to Hooks, succinctly underscoring the purpose and limited capabilities of the AEDS. The purpose, he stated, was singularly clear: "The principal objective of the Atomic Energy Detection System is the detection of nuclear tests within the USSR." He then emphasized the criticality of the synergistic importance of combining the data from the four primary components to reach an accurate evaluation. "Neither the seismic, acoustic, nor electromagnetic components either independently or jointly are capable of unambiguously identifying an explosion as being of nuclear origin." He noted only radiochemical analysis of debris could make that verification. With a tone of sober frankness, he wrote:

> The existing U.S. network for detecting nuclear explosions in the USSR has a 50-100% chance of

258

detecting tests of the order of 20 kilotons or greater detonated between the surface and altitudes of about 35,000 feet with a decreasing capability for lower yields and for tests detonated underground or in the stratosphere. For yields of the order of 1.0 kiloton the capability of the detection system is poor with real possibilities that shots in this range would not be detected. It is estimated that shots including those in the megaton range detonated in the high stratosphere would be difficult, if not impossible, to detect and identify as actual nuclear explosions. Furthermore, deep underground tests would also be difficult to identify.[31]

Northrup's capabilities statement also served to correct any misunderstandings that may have arisen following his previous capabilities statement, which he produced for Hooks on October 19, 1956. The previous report responded to questions about extending AEDS coverage throughout the world, with a goal of eventually detecting detonations of 100kt anywhere on the planet. The same report set a goal of a 1kt threshold for detonations anywhere in the Soviet Union *and its satellites*.[32] To set the record straight, he now summarized that

a. It is not a fact that with existing techniques of long range detection nuclear weapons tests anywhere in the world of one megaton or more can be reliably detected.

b. It is a fact that tests as large as one megaton detonated high in the stratosphere or deep underwater or underground or detonated in geographical locations remote from the existing stations of the Atomic Energy Detection System might not be detected.

c. Tests of either 50 or 20 kilotons fired under conditions mentioned just above would have considerably less chance of detection.[33]

Northrup's distinctive word choice and phrasing reflected a strong concern that policy makers were relying too heavily on the AEDS to monitor and assess Soviet capabilities. He concluded his report by clearly reminding any reader that critical AEDS data was normally passed through the AFOAT-1 Scientific Advisory Committee (i.e., the Bethe Panel) to the JAEIC. The JAEIC, as a committee of the U.S. Intelligence Board under the NSC, should then disseminate AEDS data to a number of agencies involved with estimating Soviet weapons capabilities, especially "in certain areas of the world not well covered by the Atomic Energy Detection System."[34] (See the chart of AFOAT-1's reporting structure in the Introduction).

Despite Northrup's two well-articulated capability assessments of the AEDS and the heavy test schedule of PLUMBBOB, the president was determined more than ever to break the stalemate with the Soviets over a nuclear arms control agreement. Mounting domestic and international pressures to stem the nuclear arms race precipitated a change in Eisenhower's view toward a nuclear test ban. To understand that shift, however, one must examine the confluence of events taking place at the time of Northrup's updated capabilities assessment.

When Northrup released his assessment on June 21, 1957, Stassen was deeply involved in the United Nations Disarmament Subcommittee that comprised 70 sessions held between March 18 and September 6, 1957. The subcommittee, meeting in London, consisted of representatives from the United States, the Soviet Union, the United Kingdom, France, and Canada. The London Disarmament Conference was a result of intensive worldwide concern that the nuclear arms race between the two superpowers was spiraling out of control; that the prospects of a nuclear war were real; and the dangers of fallout posed significant health hazards for everyone.[35]

While politically the time had come for the Eisenhower Administration to show some progress in addressing such widespread fears, the president was determined to reach an agreement with the Soviets for a disarmament treaty. He personally shared his constituent's concerns. "This is a question of survival and we must put our minds at it until we can find some way of making progress."[36] Stassen, who certainly shared the president's views, led the U.S. delegation.

The six months of negotiations began with the re-statement of well-known positions. In short, the Soviets called for an immediate cessation of all nuclear testing and a moratorium on testing not tied to any other topic of disarmament. The U.S. position, unchanged as well, was that the cessation of testing had to be tied to the larger goal of disarmament, which included an inspection system to verify the testing moratorium and, more importantly, a system of international control over the production of fissionable materials.[37]

What frustrated the NATO member delegates the most was the Soviet insistence that a test ban was self-enforcing. During a session in late March, the chief negotiator for the Soviet delegation, Valerian Zorin, stated "separating test cessation would eliminate the difficult problem of control, since nuclear explosions could be easily detected without a complicated control apparatus requiring inspection on the territory of the parties to the treaty."[38] Having visited AFOAT-1 to learn about the capabilities and limitations of the AEDS, Stassen most certainly recognized the fallacy of Zorin's argument. In addition, Stassen was very aware of Northrup's AEDS capabilities statement of October 1956. Although speculative in the absence of any historical record, it is conceivable Northrup's June 1957 update to the AEDS capabilities assessment was a direct result of the discussions then underway in London.

As the talks continued throughout the spring, the real barrier to any progress revealed itself. Stassen had attempted to get Zorin to agree to the establishment of a technical committee to examine the issues of control and inspections. However, Zorin insisted that a political agreement on principles had to precede any technical discussions. This approach, which simply eliminated any technical considerations, frustrated the other member delegates (and foreshadowed the primary Soviet negotiating tactic consistently used in all future negotiations leading to the 1963 signing of the Limited Test Ban Treaty). Zorin continued to argue that existing technologies could detect any evasive testing occurring within the borders of the USSR or the U.S.[39]

Suddenly, on June 14th, Zorin agreed to allow monitoring stations within the Soviet Union as part of a larger system of inspections among the nuclear nations. This was a radical change in direction, which appeared to acquiesce to American demands. He also agreed to suspend nuclear testing for two to three years. In return, Stassen stated the U. S. would suspend testing for a year and perhaps two while the control system was created. For the first time, the Americans agreed to consider separating a test ban from the larger goal of disarmament, an issue that had previously been non-negotiable. However, as talks continued throughout the summer, the Soviets and the Allies could not reach a deal, largely over the issue of conducting a technical conference to assess current LRD capabilities. While the Americans viewed a conference of experts as essential to establishing an effective control system (knowing the AEDS in its current state could not guarantee compliance), the Soviets refused to participate in such a conference. They insisted all parties should first agree on the conditions and the period of time for a test cessation. While the talks dragged on for several more months without reaching an agreement, the idea of holding technical talks gained currency with all parties.[40]

June 1957 was the one moment during the decade when the two superpowers came the closest to formulating a nuclear arms control agreement. Observers of the London Conference have generally concluded that the Soviets achieved more than the western powers in the course of negotiations by receiving a reciprocal pledge to stop testing and the acceptance of a separate test ban treaty. The London Conference set the foundation for the Soviets to pursue this path in the negotiations, which would begin the following year and eventually lead to the 1963 Limited Test Ban Treaty.[41]

The timing of Northrup's revised capabilities assessment of June 21, 1957, raises an interesting question. One week earlier, during the pivotal moment in London when both sides almost came to an agreement, Zorin stated,

> [W]e learn from scientists of countries which are advanced in atomic matters [that] the state of science and technology is now such that nuclear explosions can be detected at long range. It is also a known fact not a single atomic or hydrogen explosion has so far been set off undetected. In view of this, we see no justification for deferring to control.[42]

Did the Russians really believe the AEDS could detect all nuclear explosions within the USSR? It was certainly in their interests to state so publicly, but privately they were surely aware of the technical limitations. AFOAT-1 had already assessed and documented the extent of the Soviet Union's own significant capabilities and limitations in monitoring the U.S. tests in the Pacific and in Nevada (discussed in Chapter 8).[43]

We now know the Russians were far less concerned than the U.S. about developing the capability to monitor nuclear tests from great distances. The Russian version of AFOAT-1, eventually called the Special Monitoring Service (SMS), was young and significantly underdeveloped in 1957. In terms of capability, their LRD system was years behind the AEDS. Although the leader of the Soviet nuclear program, Dr. Igor V. Kurtchatov, had envisioned the need for an LRD program in 1949, the program did not begin until March 17, 1954, when it was created and placed within the Russian armed forces. Unlike the Americans and the British, who began with airborne sampling of nuclear debris, the Russians placed their initial emphasis on the electromagnetic and acoustic techniques. Indeed, these two techniques were the first Russian detections of an American test (Shot Union of

Operation CASTLE) in the Marshall Islands in late April 1954. The airborne technique was organized the following year and first detected the Nevada shots in Operation TEAPOT (February 18 to May 15, 1955). As the program gained momentum, the military established seismic stations at various locations within the USSR.[44]

In 1956, the Russian LRD system began combining analyses from multiple techniques. During Operation REDWING (May-July 1956) in the Marshall Islands, the Soviet system detected 10 of the 17 shots with the airborne technique and 7 with seismic instruments. This meant that at the time of the 1957 London Disarmament Conference, the Soviets were undoubtedly aware their LRD system lacked significant capabilities to monitor the United States' nuclear program. While they recognized the potential of LRD, they appeared in no hurry to bring it to a fully mission-capable status. By late 1957, the Soviet government decided to remove it from the armed forces and, on May 13, 1958, the SMS was established within the Ministry of Defense. Having little or no information on U.S. LRD systems and capabilities, the Russians developed everything on their own; a process that took a long time to evolve.[45]

As seriously as Eisenhower wanted an agreement, he was unwilling to compromise on the issue of international control, especially at a time when American and Soviet R&D were focused on miniaturization. Test yields were getting smaller, certainly falling well below the 20kt threshold of confidence (i.e., the limit of reliability) within the AEDS. Also, it was envisioned that both nations would soon begin conducting all tests underground. Once that happened, airborne effluents would be scarcer as particulates would most likely be contained in the earth. The "B" technique, seismic, would then take on greater importance at a time when distinguishing between earthquakes and detonations remained difficult.

On the morning of October 4, 1957, Americans awoke to shocking news that the Soviet Union had launched the world's first artificial satellite, called *Sputnik I* (meaning "fellow traveler"), into space. Occurring just as the UN nuclear disarmament talks had failed and the two superpowers were detonating nuclear weapons almost weekly, the event created extensive hysteria throughout the United States, primarily fueled by the media. The successful launch, with enough thrust to place a 184-pound payload into orbit, proved the Soviets could now launch nuclear ICBMs across most of the northern hemisphere. They had conducted their first successful test of an ICBM in August, at a time when the first two American ICBM tests had failed (in June and September).[46] The result was increased fear throughout the U.S. that the Soviets had definitely achieved technological superiority and the world was closer than ever to a nuclear confrontation. It did not help that physicists such as Teller were saying *"Sputnik* was a greater defeat for the United States than Pearl Harbor."[47] Four weeks later, the Soviet Union conducted a second successful satellite launch with *Sputnik II.* This time, the satellite carried a living payload — a dog named *Laika.* At that point, "the press became frantic." *Newsweek* wrote, "for the first time in history, the western world finds itself mortally in danger from the East."[48] The "missile gap" was now in full bloom.

On November 7th, four days after the launch of *Sputnik II,* ODM released the long awaited Gaither Report, which Eisenhower had commissioned to examine the defense posture of the nation. The 34-page report was all doom and gloom. Released in the midst of *Sputnik* hysteria, the report fueled the widespread belief the Soviets had achieved significant technological superiority. Using phrases such as "spectacular progress," the authors estimated the Soviets had stockpiled 1,500 fission bombs, and had rapidly manufactured 1,500 B-29 type bombers and 3,000 jet bombers.[49] The report's extensive estimates continued with enormous projected numbers for submarines, jet fighters, air defense systems, and various missiles. In ICBM development, the Soviets "probably surpassed us." The authors speculated the USSR had curtailed the production of bombers and refueling aircraft in order to accelerate ICBM production. Throughout its length, the report made clear the United States was extremely vulnerable to a surprise nuclear attack. In fact, it implied the U.S. lacked an adequate strategic warning system. "By 1959, the USSR may be able to launch an attack with ICBMs carrying megaton warheads against which SAC will be almost completely vulnerable under present programs."[50] As Pulitzer Prize–winning journalist David Halberstam noted, "[w]e had made our great investment in SAC, the bomber attack fleet that was on constant alert, but the men around Eisenhower still feared a surprise attack."[51]

The Gaither Report also heightened the state of fear *Sputnik* generated by detailing the inadequacies of U.S. civil defense preparedness. To "reduce vulnerability to our people and cities," the authors called for a nationwide fallout shelter program. "This seems the only feasible protection for millions of people who will be increasingly exposed to the hazards of radiation." The report went on to emphasize "survival in the aftermath" of a nuclear attack: "It is certain that there must be stockpiling of essential survival items to serve the surviving population for six to 12 months." With an urgent conclusion, the authors stated "the next two years seem to us critical. If we fail to act at once, the risk, in our opinion, will be unacceptable."[52]

Eisenhower largely dismissed the Gaither Report, if for no other reason than its enormous projected cost of $48 billion (2021 = $455 billion). He already viewed the existing defense budget of $35.5 billion to be excessive, especially during the current period of recession and a $1 billion increase over the previous FY56 budget.[53] However, in the days and weeks following the two *Sputnik* launches and the Gaither Report, the executive and legislative branches of government treated the events as a national crisis. Not only had the Soviet Union "caught up" with the United States with its nuclear program, it had leapfrogged ahead with a viable missile capable of carrying a fusion warhead to the borders of America. Persistent fears of a surprise attack and the inevitability of a nuclear war now seemed to be coming true. Still, despite the nationwide hysteria, Eisenhower discounted the report's findings and recommendations, believing the U.S. deterrent posture was adequate.[54]

Within DoD, especially inside the Air Force, the Sputnik crisis led senior leaders to believe the U.S. had placed too much emphasis on long-range bombers at the expense of missile development. Thus the term "missile gap" was now permanently termed. The president and CIA personnel involved in the U-2 reconnaissance flights, however, had already disproved the existence of a "bomber gap" and knew there was also no significant Soviet advantage in missiles.[55] Unfortunately, they could not release that information to the public given the top secret classification of the U-2's existence. Eisenhower appeared more concerned about public anxieties than the Soviet threat. In the weeks and months following the Soviet satellite launches, DoD requested and received moderate increases in defense spending to expedite the development of an ICBM. Eisenhower reluctantly approved the increases primarily to assuage public concerns, although "missile gap" fears would grow and persist through the end of his presidency. In late 1957 and 1958, he was eager to improve his relationship with Khrushchev and was sensitive to any activities or events that might jeopardize a path to an arms control agreement.[56]

Sputnik impacted AFOAT-1 by making the AEDS a "pawn" in the debate over an arms control treaty. As discussed, Northrup had grown increasingly frustrated with so many public pronouncements about the ease of "technology" in detecting any nuclear detonation "anywhere" in the USSR. The Soviets, in propping up their numerous proposals for an immediate moratorium on all testing, certainly implied (but lied) that they could do so against all

U.S. tests. American politicians allied against the Eisenhower administration had routinely asserted the same claim—most apparent in the 1956 presidential election campaign when Adlai Stevenson stated so in several speeches. But the most puzzling use of the "technology can or cannot" justifications ironically occurred within the scientific community, particularly among nuclear physicists.

Some Eisenhower historians have written about the president's respectful embrace of scientific advice to inform political decisions. However, it is now clear he did so minimally and selectively throughout his first term. In fact, he kept a noticeable distance from the prominent physicists of the time, as those scientists strongly debated the morality of developing fusion weapons and when they took different positions during the prosecution of Oppenheimer.[57] Consequently, Eisenhower received almost all of his advice on nuclear matters from the AEC; primarily from Lewis Strauss, who was the AEC Chairman from 1953 to 1958. Strauss had allied himself closely with Lawrence and Teller, all of whom were strongly anti-Russian, anti-disarmament, and pro–thermonuclear weapons. During his years as AEC chairman, Strauss had always succeeded, in one way or another, in diluting the president's momentum toward an agreement. His arguments usually had their roots in East-West mistrust or in the concerns the Soviets were rapidly moving ahead with nuclear technology. For example, he convinced Eisenhower in May 1954 that U.S. monitoring capabilities were inadequate to detect violations of a Soviet moratorium on testing, an accurate understanding that must have come from AFOAT-1.[58]

The pro-disarmament camp included Bethe, Rabi, Pauling, and several other prominent physicists. Although they had lost the fight to prevent the development of the hydrogen bomb, they remained quite vocal in advocating arms control, largely based on strong beliefs that the current escalation of the nuclear arms race could very likely lead to a holocaust. However, they had few avenues to the president's ear due to Strauss's control over information flow to the White House. All of that changed with *Sputnik.*

Eleven days after *Sputnik,* the president called all 14 members of the ODM SAC into a meeting to express his concerns about the pace of Soviet technological advances. He pointedly asked the group if American science was being "outdistanced." Much of the group spoke during the long meeting with a common message—that the Soviets were placing far greater emphasis and resources on science and had tremendous momentum, which could soon overtake the U.S. Rabi, the most outspoken member, argued that the president needed a scientific advisor. Eisenhower agreed such an advisor would be "most helpful."[59]

On October 28th, two weeks after the ODM/SAC meeting, Rabi met again with Eisenhower. He informed the president that Hans Bethe had reported a flaw in the design of the Russian ICBMs. Based on his review of Russian nuclear debris with Northrup (i.e., AFOAT-1's Bethe Panel), Bethe believed incoming thermonuclear missiles could be intercepted and prematurely detonated by exploding a small nuclear device in their path. The president was excited by the idea and met with Rabi and Strauss the following day to explore this concept of an anti-missile defense system.

Strauss downplayed the idea, questioning Rabi's evidence and arguing "the Soviets can always steal our secrets."[60] The discussion then turned to the topic of a test ban. Again, the AEC Chairman and the Nobel laureate espoused opposing views. Rabi believed the proposed "listening posts" inside the USSR, the few the Soviets had accepted, combined with the AEDS, would be enough to deter the Soviets from cheating on a test ban. Strauss strongly disagreed, believing the AEDS was inadequate in ensuring Soviet compliance with a ban. Eisenhower was struck by how diametrically opposed Rabi and Strauss were to one another. He heard, for the first time, the great divide within the nuclear physics community. "The confrontation between Rabi and Strauss had finally awakened the president to the existence of a long and deep ideological split within the scientific community."[61]

Evidently, this "awakening" weighed on the president's mind for another week. That encounter, combined with *Sputnik*, recent demonstrable advancements in Soviet missiles, and setbacks in reaching an agreement with the Soviets, jolted Eisenhower to engage in a broader dialog with more scientists. In a move to reach beyond Strauss, Teller, and Lawrence, Eisenhower pulled the Scientific Advisory Committee away from ODM and elevated it to the Office of the President. On November 7th, during a special nationwide public address, he announced the creation of the President's Science Advisory Committee (PSAC), and named James Killian, Jr., as chairman and the first Special Assistant to the President for Science and Technology. Most importantly, Killian immediately staffed the committee with leading pro-test ban scientists. Most prominent among the 14-member PSAC was Hans Bethe, the one physicist in the world most closely associated with Northrup and AFOAT-1. As the next 18 months would reveal, Bethe and the PSAC "became a rare, technically competent voice of moderation that matched Eisenhower's own political and fiscal conservatism" at a time when "Sputnik helped make the Cold War a total war."[62] Until now, AFOAT-1's AEDS had been a pawn in the pro– or anti– test disarmament debates. However, in 1958, the LRD organization and its leaders would become major pieces on the chessboard of nuclear diplomacy.

Georgy Malenkov

Became Premier of the Soviet Union after Joseph Stalin's death in March 1953. He expressed some willingness to unilaterally ease tensions with the US. He warned the Supreme Soviet that Cold War policies could lead to a new world war and destruction of world civilization. Malenkov was replaced by Nikolai Bulganin in February 1955, before he could propose anything substantial. *Photo courtesy of the Dutch National Archives, The Hague, ANEFO.*

Nikolai Bulganin

Succeeded Malenkov as Premier of the Soviet Union in February 1955. Bulganin made several proposals to the U.S. for a nuclear test ban. However, he insisted that LRD technologies were good enough to police any evasion of a ban and that inspections were not needed. He lost the premiership to Nikita Khrushchev in March 1958. *Photo courtesy of the German National Archives*

Adlai Stephenson, II

Former governor of Illinois from 1949-1953, he was the Democratic presidential candidate in the 1952 and 1956 elections. He made the nuclear test ban central to his campaign in 1956 and asserted that LRD technologies could detect any detonation in the USSR. *Photo courtesy of the Library of Congress*

Nuclear anxieties

Widespread anxieties about the possibility of a nuclear attack and radiation fallout escalated throughout Eisenhower's second term in office. Such concerns, which he also shared, compelled Eisenhower to pursue a nuclear test ban agreement with the Soviets. The issue of monitoring a ban brought national attention to the capabilities of the AEDS and placed AFOAT-1 leaders at the center of international negotiations and diplomacy.

Photo courtesy of the Library of Congress

James Killian, Jr. is sworn-in as PSAC chairman and the first Special Assistant to the President for Science and Technology (November 21, 1957)

Killian, a strong pro-test ban proponent, immediately staffed the committee with leading pro-test ban scientists. Most prominent among the 14 members of PSAC was Hans Bethe, the one physicist in the world most closely associated with Northrup and AFOAT-1.

Photo courtesy of the NASA History Office

Dr. Igor V. Kurtchatov

Head of the USSR's nuclear program, he was the first to envision the need for an LRD program (in 1949). However, the program did not begin until March 1954, when it was created and placed within the Russian Armed Forces. Unlike the Americans, who began with airborne sampling of nuclear debris, the Soviets placed much more emphasis on the electromagnetic and acoustic techniques. On May 13, 1958, the Special Monitoring Service was established within the Ministry of Defense. Having little or no information on U.S. LRD systems, unlike their knowledge of nuclear weapons acquired through espionage, the Soviets developed LRD on their own.

Photo courtesy of the Atomic Heritage Foundation

Dr. Isador I. Rabi

Rabi won the 1944 Nobel Prize in physics for his discovery of nuclear magnetic resonance. He served as chairman of the powerful GAC from 1952 to 1956, when the GAC exercised enormous influence over AFOAT-1's R&D plans and experiments. He was a very strong proponent of nuclear arms control and had been opposed to the H-bomb on moral grounds, siding with Fermi, Oppenheimer, and Bethe. Rabi also served in the ODM/SAC and the PSAC. He was considered a nemesis of Lewis Strauss.

Photo courtesy of the Atomic Heritage Foundation

The 1956 Suez Crisis (October - November 1956)

The tank landing ship HMS *Buttress* disembarks troops and vehicles on the landing beach at Port Said.

Photo courtesy of the Imperial War Museum

Hungarian Revolution (October-November) 1956

Student protests against Soviet-imposed policies escalated to a nationwide revolt and the collapse of the communist government. People demanded free elections and withdrawal from the Warsaw Pact. Reneging on their promise to negotiate, the Soviet forces invaded the country. Despite U.S. support for the resistance, Eisenhower declined to intervene. Events such as a Suez crisis and the Hungarian revolution were not uncommon during the Cold War and served to impede Eisenhower's goal of achieving arms control.

Photo source: public domain.

T-33 configured for air sampling

The configured T-33s first flew sampling sorties during Operation GREENHOUSE in order to collect higher altitude samples (despite their limited range). The 14 Air National Guard (ANG) units that flew the modified T-33 sampler figured heavily into the SAC/AFOAT-1 war plan. Those ANG units actively collected samples during Operation PLUMBBOB to train for their wartime mission.

Photo source: AFTAC archives

The I-2 foil mounted on the T-33 gun deck

Photo source: AFTAC archives

275

The U-2 "Dragon Lady"

With human intelligence assets sorely lacking or deficient in collecting data on the Soviet nuclear weapons program, the president heavily relied on surveillance systems for information. So was his reasoning for pushing the development of the U-2 reconnaissance plane which could fly at altitudes above 70,000 feet. Initially authorized for CIA use only, the first U-2 flights over the Soviet Union occurred on July 4, 1956. The AEDS now had a partner in the business of long range detection. (The U-2 was first modified for air sampling in 1958).

Photo source: AFTAC archives

Equipment used for the acoustic ("I") technique

Photo source AFTAC archives

Sputnik satellite

The USSR launched the world's first satellite on October 4, 1957. The event shocked U.S. leaders and signaled that the Russians had surpassed the Americans in technological achievements. The event immediately led to the creation of the PSAC and encouraged Eisenhower to consult scientists more frequently in order to inform policy. The Soviet launch also prompted Congress to approve the largest U.S. public educational program in the 20th century.

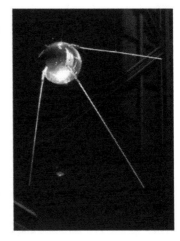

Photo courtesy of the National Museum of the United States Air Force

Notes to Chapter 9

---

[1] The Soviet Union rarely announced their nuclear tests.

[2] *New York Times*, September 15, 1956, as cited in Divine, *Blowing on the Wind*, 85.

[3] Ibid.

[4] *Newsweek* 48 (October 29, 1956) 27, as cited in Divine, *Blowing on the Wind*, 100.

[5] Technical memo #86, Northrup to Hooks, Subject: "AFOAT-1 Capability for Recording Nuclear Explosions," October 19, 1956, AFTAC archives.

[6] Ibid.

[7] Ibid.

[8] Ibid.

[9] Ibid.

[10] Ibid.

[11] Ibid.

[12] Ibid.

[13] Donald Neff, *Warriors at Suez: Eisenhower Takes America into the Middle East* (New York: Linden Press, Simon and Schuster, 1981), 403.

[14] Gaddis, *We Now Know*, 236-237. Gaddis acknowledges that it was a bluff but one that the West could not ignore.

[15] Ibid., 174-175.

[16] Walker, *The Cold War*, 170-172; Ambrose, *Eisenhower*, 227-228.

[17] Glenn B. Fatzinger, *History of the Air Force Office for Atomic Energy-One (AFOAT-1): 1 January- 31 December, 1957*, 4-7, AFTAC archives.

[18] Ibid., 8-10.

[19] Ibid., 11-14.

[20] Ibid.

[21] H. Rowan Gaither, "Deterrence and Survival in the Nuclear Age," Security Resources Panel of the Science Advisory Committee, Washington, D.C., November 7, 1957. AFTAC archives.

[22] Ibid.

[23] Ibid.

[24] Gaddis, *We Now Know*, 222.

[25] As cited in Miller, *Under the Cloud*, 259.

[26] The public was unaware at the time. The bomb fell on University of New Mexico land and created a crater 12 feet deep and 25 feet in diameter. "Accident Revealed after 29 Years: H-Bomb Fell Near Albuquerque in 1957," *The Los Angeles Times,* August 27, 1986.

[27] "4950th Test Group (N) Final Report, Operation Plumbbob," Headquarters, Air Force Special Weapons Center, Kirtland, AFB, New Mexico, November 20, 1957.

[28] Fatzinger, *History of AFOAT-1: 1957,* 26-27.

[29] Ibid., 4, 26.

[30] P.S. Harris, C. Lowery, A. Nelson, S. Obermiller, W.J. Ozeroff, and S.E. Weary, *PLUMBBOB Series 1957* (Alexandria, VA: Defense Nuclear Agency, September 15, 1981); Miller, *Under the Cloud,* 251-293.

[31] Technical memo #91, Northrup to Hooks, Subject: "AFOAT-1 Detection Capability," June 21, 1957, AFTAC archives.

[32] Author's italics. Technical memo #86, Northrup to Hooks, Subject: "AFOAT-1 Capability for Recording Nuclear Explosions," October 19, 1956, AFTAC archives.

[33] Ibid.

[34] Ibid.

[35] For a good summary of the key issues at the London Conference see Divine, *Blowing on the Wind,* Chapter Six.

[36] Eisenhower's comments to Sherman Adams as quoted in Divine, *Blowing on the Wind,* 144.

[37] Jacobson and Eric Stein, *Diplomats, Scientists, and Politicians,* 14-18.

[38] Ciro E. Zcippo, "The Issue of Nuclear Test Cessation at the London Disarmament Conference of 1957: A Study in East-West Negotiation," RAND Corporation, September 1961, DTIC ADA280836.

[39] Ibid., 58-73. If a technical committee had been formed, it would most certainly have included Northrup.

[40] Jacobson and Stein, *Diplomats, Scientists, and Politicians,* 16.

[41] See especially Walter C. Clemens and Franklyn Griffiths, "The Soviet Position on Arms Control and Disarmament - Negotiations and Propaganda, 1954-1964," (Cambridge, MA: Center for International Studies, MIT, February 1, 1965).

[42] UN, Subcommittee of the Disarmament Commission, *Verbatim Record,* No. 121 (June 14, 1957), 11-12., as cited in Ibid., 67.

[43] Technical memo #82, Northrup to Canterbury, Subject: "Estimate of the Long Range Detection Capabilities of the USSR," December 1, 1953, AFTAC archives.

[44] 430 Y. V. Cherepanov, *Born by the Atomic Age: 1958-1998,* Alexey Pavlovich Vasiliev, Ed. (Moscow: 2002), 6-13. Russian geophysicists had experimented with seismographs at Semipalatinsk in 1951, but had only first detected a nuclear detonation against their own test there on September 24, 1951. Then, on October 31, 1952, Russian seismologists detected the 10 MT detonation of the U.S. Shot Mike in Operation IVY. At that point, Grigory A. Gamburtsev, Director of the Geophysical Institute in the USSR Academy of Sciences, recommended a worldwide system of seismic monitoring instruments. His recommendation, however, was ignored until 1954; a year after radiochemists first collected debris samples for analysis and tested against Joe-4.

[45] Ibid.

[46] This was the Atlas Program. The Atlas A failed on takeoff 24 seconds into flight, followed by a second similar failure in September.

[47] As quoted in Divine, *Blowing on the Wind,* 170.

[48] As quoted in Ibid., 172.

[49] The U.S. later learned that the Soviet Union possessed only 660 nuclear weapons at that time.

[50] "Deterrence and Survival in the Nuclear Age," Security Resources Panel of the Science Advisory Committee, Washington, November 7, 1957, 14.

[51] Halberstam, *The Fifties,* 617.

[52] Ibid.

[53] Richard V. Damms, "James Killian, the Technological Capabilities Panel, and the Emergence of President Eisenhower's "Scientific-Technological Elite," *Diplomatic History,* 57-78; 73.

[54] Divine, *Blowing on the Wind,* 170-171.

[55] Halberstam, *The Fifties,* 622.

[56] For a good analysis of Eisenhower's reaction to *Sputnik,* see Rodger A. Payne, "Public Opinion and Foreign Threats:

Eisenhower's Response to Sputnik," *Armed Forces and Society,* Vol. 21, No. 1, Fall 1994, pp. 89-112.

[57] In 1954, Congress summoned Oppenheimer to their hearings concerning his pre-war association with procommunist groups. He was subsequently stripped of his security clearance and largely ostracized from any government activities. Eisenhower sided with the anti-Oppenheimer group led by Strauss. His primary concern was that McCarthyism would elevate the Oppenheimer case to negatively impact his Administration and give "the impression that all scientists are disloyal. 'We've got to handle this so that all our scientists are not made out to be Reds.'" Ambrose, *Eisenhower,* 166. See also Richard V. Damms, "James Killian, the Technological Capabilities Panel, and the Emergence of President Eisenhower's 'Scientific Technological Elite,'" *Diplomatic History,* Vol. 24, Issue 1, January 1, 2000: 57-78, 62. Eisenhower "acquiesced to Strauss's quiet campaign to freeze Oppenheimer out of his advisory positions."

[58] See especially Julia M. MacDonald, "Eisenhower's Scientists: Policy Entrepreneurs and the Test-Ban Debate 1954-1958," *Foreign Policy Analysis* (2015) 11, 1-21, p7. "Strauss' outlook [toward Soviet intentions] bordered on paranoia."

[59] Ambrose, *Eisenhower,* 429-431; also Herken, *Cardinal Choices,* 101-102.

[60] Bethe's discovery and quote in Herken, *Cardinal Choices,* 103.

[61] Ibid., 102-104.

[62] Wang, *In Sputnik's Shadow,* 3.

CHAPTER 10

AFOAT-1 and Geneva

As the year 1958 began, the prospects of any arms control agreement with the Soviet Union looked dim. While the London Conference on Disarmament had come tantalizingly close to an agreement in June 1957, it fizzled out in the fall as the specter of ICBMs introduced a new factor in calculating the Soviet threat. Not only did missiles appear to diminish the dominance of the SAC bombers, *Sputnik* demonstrated the potential of a new Soviet intelligence capability theoretically able to observe any activity in the United States. These rapid advances in technology served to place greater pressures on AFOAT-1 to quickly improve the capabilities of the AEDS. For Northrup, the problem largely resided within the realm of seismology. Fortunately, he had one of the nation's top seismologists on his staff—Dr. Carl Romney.

Romney was first introduced to AFOAT-1 in September, 1949, just as Northrup was evaluating the Joe-1 data. Romney was then starting work at Beers & Heroy (founders of the Geotechnical Corporation) to earn enough money to finish the final year of his doctorate program at the University of California, Berkeley. He had briefly served in the Pacific as a Navy weather officer immediately after the war, and then prepared for a career as a seismologist. AFOAT-1 had contracted with Beers & Heroy (B&H) to develop seismographs capable of detecting nuclear detonations in the Soviet Union.

Romney had led the search for an effective seismograph and had chosen the new Benioff instruments. His team successfully tested the new instruments during Operation GREENHOUSE. The five test sites located in Alaska, Turkey, and Wyoming (three sites), functioned well enough for Northrup to declare them as minimally mission capable for the AEDS. As such, they became known as detachments 204, 301, 141, 142, and 162, respectively. Romney continued his work for AFOAT-1 at B&H by supervising the Laramie Analysis Center in Wyoming from September 1951 to August 1954. While there, he monitored the unexpected detonation of Joe-4. Due to the size of the thermonuclear burst, Romney and his team recorded excellent data, which he described as the "first important success for our seismic [AEDS] system."[1]

Remaining a consultant for B&H, Romney returned to Berkeley in August 1954 to complete his degree under the guidance of his mentor, Dr. Perry Byerly.[2] He graduated a year later and joined AFOAT-1 as a temporary employee ("to keep his options open") but worked full time under J. Allen Crocker, the assistant technical director to Northrup. In 1955 and 1956, Crocker and Romney traveled to Australia and Spain, where they selected and approved the locations of the seismic arrays which became detachments 421 and 313.[3]

When the first underground nuclear test occurred on September 19, 1957, Shot Ranier in Operation PLUMBBOB, Romney began his involvement in the escalating debates over nuclear disarmament and a possible test ban. Along with Leslie Bailie of the United States Coast and Geodetic Survey (USC&GS), he was the primary analyst of the Ranier seismic data, which Northrup had long awaited. Romney and Bailie concluded the existing seismographic instruments used in the AEDS could not detect the 1.7 kiloton detonation from beyond 300 miles. Also, the "tiny wiggles" on the seismogram were "very much like those from an earthquake." In fact, Ranier generated signal sizes "equivalent to those from a small magnitude 4.25 earthquake." Romney noted,

> . . . most seismologists had expected that the explosion would cause the earth to move predominantly back and forth along the path between the explosion and the station, but we discovered that there were also prominent side-to-side motions, perpendicular to the path. The latter motion had been thought to be a definitive characteristic of earthquakes alone.[4]

This was not good news. The findings more than underscored the sober capabilities assessment Northrup had provided Hooks only three months earlier. More than anyone, Northrup and Romney recognized an atmospheric test ban would force all testing to go underground. At that point, with the system heavily reliant on the underdeveloped seismic technique, the AEDS overall would become less effective.[5]

Hans Bethe, the future Nobel laureate and the head of the AFOAT-1 Science Advisory Committee, also knew the seriousness of the seismic problem—more than anyone outside of AFOAT-1. Crocker and Romney had alerted the Bethe Panel to the issue at their meeting on August 27 and 28, 1957. Bethe, by now a well-known proponent of a test ban, had requested an appraisal. The AFOAT-1 seismologists reported on October 1, 1957, that despite extensive work on the problem,

> [u]nfortunately, the recordings from seismic disturbances are considerably more characteristic of the structure of the earth than of properties of the source. A fair summary of our present knowledge is that many natural earthquakes can be distinguished from blasts but blasts cannot be distinguished from some natural earthquakes.[6]

This was an especially critical concern with low-yield detonations. The problem troubled Northrup enough that he issued an additional evaluation of the low-yield issue on October 24th. AFOAT-1 assessed that a Soviet sub-surface test of 1.5kt to 2.5kt stood a 30 to 60 percent chance of being detected by the AEDS. In his conclusion, Northrup cautioned that any future international control system might require "additional stations within the USSR."[7]

On January 9, 1958, Killian, as directed by the NSC, appointed Bethe to head up a working group to examine the complexities of monitoring compliance of a test ban. Officially called the Ad Hoc Working Group on the Technical Feasibility of a Cessation of Nuclear Testing, its task was threefold:

1. A study of the losses to the United States consequent on a total suspension of nuclear tests at specific future dates;

2. A symmetrical study of the losses to the USSR that would accrue from cessation of nuclear testing using the same hypothetical dates; and

3. A study of the technical feasibility of monitoring a test suspension, including the outlines of a surveillance and inspection system.[8]

The first two tasks dealt directly with the issue of "advantage;" that is, a test cessation would freeze the current technological progress of both nations in place. Therefore, with the perception that the Soviets had already reached parity, the Americans would be left wondering who actually held the strategic advantage (i.e., be further ahead in R&D and operational capabilities) on that enactment date. The third task, by far the most complex and debatable, would predominantly focus on the state-of-the-art technologies making up the AEDS (and their limitations).

Three days after Killian appointed the Bethe working group, Eisenhower replied to Bulganin's recent (December 10, 1957), letter in which the Soviet premier recommended a summit to discuss an immediate moratorium on testing lasting two to three years.[9] In his reply to Bulganin on January 12th, the president once again insisted on a technical conference to examine control mechanisms for monitoring a test ban. In the weeks that followed, as the two leaders exchanged letters to reach some common ground, the Senate Committee on Foreign Relations began to hold hearings on disarmament. Chaired by future U.S. Vice President Hubert Humphrey, the Subcommittee on Disarmament called upon a number of experts who represented both sides of the arms control debates. Stassen testified early on and, having just resigned as Special Assistant to the President for Disarmament, voiced his personal opinion, arguing for a test ban separate from disarmament. Given Eisenhower's strong desire for a comprehensive test ban, this was a bold but realistic step forward. In essence, he was supporting the Soviet position that had consistently argued for a clear separation between a test ban and the more complex issue of disarmament. At the time, Bulganin hoped to disrupt the upcoming NATO meeting that would plan for the deployment of nuclear weapons in Europe.

In his remarks, Stassen stated (contrary to his knowledge about the limited capabilities of the AEDS), that the requirements for monitoring controls were not that extensive. He posited that only a dozen or so inspection stations were needed and, therefore, an agreement should be easy to formulate. In contrast and, as expected, DoD and AEC representatives warned that, given the vast geography of the Soviet Union, the risk of secret stockpiles was too great, even with 3,000 to 3,500 inspectors stationed within the USSR. When the subcommittee hearings ended, however, the overall impression aligned with Stassen's view that "control measures needed to police a test ban would not be very extensive."[10]

While Eisenhower continued to call for a conference of experts to discuss the details of an international control system to police a test ban, AFOAT-1 remained busy monitoring the ongoing Soviet test series. That series, which began in January 1957, was still underway in early 1958, making it the longest continuous test series conducted by either nation to date. As discussed, AFOAT-1's workload in late 1957 was especially heavy due to the 15 Soviet shots and the additional 29 U.S. shots conducted in Operation PLUMBBOB. Now, between January 4th, and March 22nd, the Soviets conducted an additional 12 tests. The high operational tempo exercised the AEDS to an unprecedented level.[11]

Overall, the AEDS performed well, although several shortcomings continued to frustrate AFOAT-1 personnel. While the "Q" technique detected all of the Soviet tests to produce time and azimuth data, it was still not capable of distinguishing between the electromagnetic pulse produced by a lightning flash and one generated by a nuclear detonation. Analyzing the electromagnetic data was time-consuming. Teams searched film for specific times furnished by AFOAT-1 and then wired numerical data to the AFOAT-1 analysis center. The teams then forwarded the film to the center, where personnel correlated the telegraphic data and later rechecked the original film to confirm the evaluation used to determine the height of burst of a nuclear explosion.

The seismic stations detected less than half of the Soviet shots. This was due to the fact that 25 of the 27 tests were air dropped detonations. Seismic-derived yield determinations were uncertain as the coupling between the explosion and the Earth was affected both by the height of the burst and the geological formations at ground zero.

AFOAT-1 seismic teams also learned that seasonal conditions affected the fidelity of the systems. During storm activity in wintertime, a general deterioration of the capability for the entire seismic net occurred. Summer storms had similar effects, albeit for shorter periods of time. AFOAT-1 also discovered that a distinct loss of sensitivity occurred at distances between 500 and 1,100 miles, and between 2,000 and 2,700 miles from the explosion. This gap in coverage between 1,100 and 2,000 miles was caused by the Earth's core shadow, also referred to as the "Skip Zone." Since all of the AEDS stations were located beyond 1,000 miles from known Soviet test sites, the seismic detection range limitation for detecting low-yield subsurface shots existed only in the 2,000 to 2,700 mile range.[12]

The acoustic stations in the AEDS detected almost all of the Soviet shots. AFOAT-1 learned there was a marked variation in amplitude of pressure waves caused by lower noise levels at night than at daytime. Stratospheric winds favored transmission of signals toward the east in the winter and toward the west in the summer. Stations north or south of the test sites were not affected by seasons as were stations located east or west of the Soviet test locations. Longer ranges of detection could be obtained in summer and winter than during spring or fall.[13]

As expected, airborne sampling successfully collected against all of the Soviet shots. Relatively small-yield (3kt to 10kt), low-altitude explosions were readily detectable because they created substantial clouds of debris.[14] Ground-based particulate samplers, however, were non-operational. Northrup had removed them from the AEDS in April 1955, because the frequency of nuclear tests was making attribution of samples almost impossible (discussed in Chapter 8).

In observing the extensive test series, U.S. officials believed the Soviet Union was pushing hard to quickly achieve nuclear technological parity or superiority. Those suspicions were confirmed a week after the Soviets concluded their tests. On March 31st, the Soviets announced they would "discontinue the testing of all types of atomic and hydrogen weapons in the Soviet Union."[15] They cautioned the British and the Americans to do the same, or else the USSR reserved the right to resume testing. On April 4th, Khrushchev reiterated the decree in a personal letter to Eisenhower.

The Soviet unilateral moratorium on testing resulted in a propaganda victory, placing the Eisenhower Administration on the defensive. Even the hawkish Dulles suggested the U.S. should follow suit. We now know that the secretary of state had changed his position because he had foreknowledge of the Bethe Panel's report.[16]

Bethe's ad hoc working group delivered the NSC-directed top-secret report to Killian on March 28th. The full report was comprehensive with contributing assessments (annexes) from the various agency stakeholders. The crux of the entire report was strongly undergirded by AFOAT-1's 42-page AEDS capabilities assessment that Northrup provided Bethe on February 5th, and was included in Bethe's final report as Annex A.[17] The other annexes were: Annex B: "Detection of High Altitude Nuclear Tests," authored by Bethe; Annex C: "Concealment and Detection of Nuclear Tests Underground," authored by Bethe and Harold Brown; Annex D: "Chart of Present and Future U.S. Nuclear Warhead Developments," authored by the AEC; and Annex E: Impact of a September 1958 Nuclear Test Moratorium on Soviet Nuclear Weapons Capabilities," authored by the CIA.

As Annex A, Northrup's lengthy report, addressed to Bethe as "Memorandum for the Chairman, Ad Hoc Panel on Nuclear Test Limitation," was entitled "Present and Potential Capabilities and Limitations of the AFOAT-1 Long Range Detection System." The framework of the document was structured around the four technical pillars of the AEDS. It noted the number of stations for each of the four techniques currently in the AEDS and described their basic functions. In assessing the capabilities and limitations of the AEDS, AFOAT-1 drew upon the analyses of the evaluated Soviet tests conducted since Joe-1, and outlined the number of detections that each technique made against the sizes of the yields by category: 3.5-10kt, 10-45kt, 60-500kt, and 750-4300kt. In short, the larger the shot, the easier the detection by more techniques and more stations. However, it was the inverse (i.e., smaller shots) that the remainder of the report addressed. Northrup succinctly noted the parameters of the assessment. The details of these parameters would make future negotiations difficult:

> The Long Range Detection System of AFOAT-1 was primarily designed to detect surface or low air bursts and subsurface bursts detonated within the boundaries of the USSR. In considering its capabilities and limitations as a system for monitoring an international agreement on nuclear test limitations, the following four environments for possible test detonations within the USSR were considered: surface or air bursts below 50,000 ft.; subsurface tests; high altitude tests; and surface or air bursts in remote geographical locations.[18]

The limitations of the AEDS techniques in the aforementioned environments became the focus of discussions at the future Conference of Experts. For example,

Northrup was clear about the limitations of the seismic technique in detecting air bursts, which now characterized almost all Soviet tests. The seismic net, on average, detected approximately 300 seismic disturbances per year in the USSR registering above 100kt. However, 10 percent of those could not be identified as either man-made or natural. The other severe limitation he noted was the AEDS's inability to detect detonations above 50,000 feet. On the heels of *Sputnik*, AFOAT-1 believed it was only a matter of time until both the United States and the Soviet Union would conduct tests at extremely high altitudes or in space.[19]

In addition to the severe shortfalls in detecting subsurface tests, high altitude tests, and possible tests in remote geographical locations, the current limitations of the AEDS under normal conditions should form the criteria of a Geneva system. However, as AFOAT-1 was producing this report for Bethe, the Soviets were finishing their yearlong test series with several detonations that produced extremely small yields (i.e., below 5kt). Smaller yields would seriously challenge the current capabilities of the AEDS. Northrup detailed those challenges in his report:

Summary of Existing Capabilities and Limitations.

Surface or Air Burst below 50 kilofeet within the USSR.

a) The electromagnetic technique may possibly detect shots within the USSR of 25 KT or greater but requires a determination of time independently by some other geophysical technique to take care of the sorting problem.

b) The seismic technique is capable of detecting surface or low air bursts of 100 KT or larger.

c) The acoustic system is capable of detecting tests of 15 KT or larger.

d) The nuclear technique is capable of detecting tests as small as 3 KT as long as the debris does not rise much above 20,000 ft.

e) The overall detection capability for the Atomic Energy Detection System is excellent (90-100%) for surface or air burst tests of 100 KT and greater between the surface and 50 kilofeet.[20]

While there were no seismologists on the Bethe Working Group, the historical record reveals that Northrup worked closely with Romney on this report. On February 1st, he reassured Bethe of this fact (i.e., the challenges of seismic fidelity) in a personal letter:

> The report you requested from me for the Ad Hoc Panel on Test Limitation is about finished. The part that remains concerns the problem of detection of underground shots by seismic means and Dr. Romney is in the process of a final study which may substantially affect our recommendations on the number and distribution of seismic stations required. The basic problem is brought about by a sharp dip in the sensitivity of seismic detection versus distance at ranges between 300 and 900 miles. Recent studies by Carl [Romney] indicate that *this will substantially change the recommended locations and number of seismic stations necessary.*[21]

This last sentence, a "heads-up" to Bethe, foreshadowed the greatest barrier to achieving a test ban during the Eisenhower administration; namely, the number of control stations needed in any future Geneva system.

In early April, Killian asked Bethe to brief a summary of the report to the NSC.[22] Bethe's abbreviated report consisted of a 15-page narrative, which included the key points of the five other annexes. Northrup was well-prepared, having expedited the assessment since early January when Killian had informed the NSC that his PSAC would undertake the study. For Northrup, this was one of the most important reports AFOAT-1 had ever produced. He had known for some time, as had Bethe, that the secretive AEDS would most likely become the foundation of the "Geneva system"—the name that the negotiators used to describe a future international control system which would monitor test ban compliance.

Throughout April, Eisenhower and Khrushchev exchanged several letters that did little to advance any negotiations. Again, the intractable obstacle was the issue of a control system. Eisenhower insisted on a conference of experts to discuss the details and feasibility of an international system as a first step. The American view remained steadfast—technical details had to come first and remain separate from any political decisions. The Soviets, on the other hand, while accepting a system of control in general, viewed the technical and political dimensions as intricately linked.[23]

Having received the Bethe Report from Killian in early April 1958, Eisenhower was faced with the decision of whether to match the Soviet moratorium with a U.S. halt to testing or proceed with the extensive test series codenamed HARDTACK I scheduled to commence on April 28th. Had the Soviets achieved an advantage with the conclusion of their 15-month series? Were they now attempting to freeze everything in place by pressuring the U.S. to stop testing before HARDTACK I? As the president pondered these questions, Killian gathered the members of the PSAC at Ramey AFB in Puerto Rico to consider the necessity of conducting HARDTACK I and whether current LRD technologies were sufficient to support a test ban. Within days they returned with the recommendation to conduct HARDTACK I, but the U.S. should then suspend all testing. Despite the significant list of AEDS limitations so prevalent in Northrup's report to Bethe, the PSAC stated the current system (i.e., the AEDS) was good enough as a model for the design of a Geneva system of control.[24]

Northrup was growing anxious over widespread press reports that an inspection and control system could be created to easily detect any attempt to evade a test suspension agreement. On May 9th, just days after the return of the PSAC from Puerto Rico, Northrup sent a letter to Killian (in essence, disagreeing with Bethe) expressing his concern about the inaccuracies of such reports. He noted there was a significant lack of data on underground tests because only one U.S. underground test (Shot Ranier) was ever conducted. Also, AFOAT-1 possessed no data on high-altitude detonations (although a secret three-shot series codenamed ARGUS would soon include those experiments). Finally, he also told Killian it would take approximately two years to install a Geneva control system. There is no evidence that Killian ever replied to Northrup.[25]

As the first shot of Operation HARDTACK I, Shot Yucca, was detonated in the Pacific on April 28th, Eisenhower delivered another letter to Khrushchev, reiterating his desire for a technical conference. To his surprise, the Soviet premier agreed to a conference of experts in his reply on May 9th. Scheduled for July 1st, this Conference of Experts was charged to consider the details of a Geneva control system, which could effectively monitor compliance with a nuclear test ban. Eisenhower's instructions to the U.S. delegation were clear: any agreements achieved would not commit the United States to a test cessation. Again, in contrast, the Soviets believed that technical and political considerations could not be separated.[26]

In selecting the technical experts for the U.S. delegation, the White House was careful to appoint scientists who took stands both for and against a test ban treaty. Three delegates and 11 scientific advisors were chosen. On June 20th, the public learned that Dr. James B. Fisk, Vice President of Bell Telephone Laboratories, would lead the U.S. contingent. Bacher and Lawrence rounded out the principals.[27] Importantly, in addition to Bethe as a close associate of AFOAT-1, four of the 11 American advisors constituted all of AFOAT-1's top scientists and technical division directors: Northrup, Romney, Crocker, and Olmstead. This large U.S. team spent the latter half of June preparing for the conference. During that time, they reviewed a number of critical reports, including Stassen's material that he had used in the London Conference, and especially Northrup's lengthy assessment that became the heart of Bethe's report to Killian and the president.[28] On June 24th, they also received the seismic study that General Herbert B. Loper, Assistant to the Secretary of Defense for Atomic Energy, had requested on May 28th, along with several other AFOAT-1 studies. These were: "Seismic Detection of Nuclear Explosions" and "Seismic System for Detection of Clandestine Nuclear Tests within the USSR and China" (Romney); "Acoustic Detection of Nuclear Explosions" (Olmstead); and "Electromagnetic Detection of Nuclear Explosions" (Crocker).[29]

The seven-week-long Conference of Experts opened on July 1, 1958, with serious discussions beginning three days later after agreeing to an agenda. The first half of the conference consisted of a review of all techniques that would comprise the international system of control. Particular experts took time to explain the basic science underlying the various methods. Clearly, the American and British delegates were most detailed in their presentations, having worked within the AEDS for some time. The discussions that followed the presentations centered on the capabilities of employed LRD techniques and limitations of the fielded equipment. In regard to the latter, both the western delegates and the Soviets were careful not to divulge the full limitations of their LRD techniques.

Throughout the conference, Northrup and Romney were instrumental in shaping those discussions, and in convincing the Soviet delegates of the accuracy and fidelity of their supporting data. On July 5th, Northrup presented detailed information on the acoustic technique. He included general characteristics of the medium through which the waves are propagated; the mechanism of production of acoustic waves by explosions as well as certain natural phenomena; and the characteristics of the receiving equipment (general capabilities and limitations). Discussions on the acoustic technique continued through July 10th, when all parties drafted an agreement on the acoustic method. The conference was off to a promising start.[30]

This collaborative, positive atmosphere was short-lived. Two days later, Romney began the most important topic discussed at the conference—seismic capabilities and limitations. On July 12th, Romney presented an overview of the basic fundamentals of seismology and the challenges which Earth's geography posed to detection technologies. He explained the importance of P-waves (longitudinal body waves), noting they were the "most useful for detecting and determining the time and location of explosions."[31] In the course of his talk, Romney shared the extensive experiences of the AEDS seismic stations. He indicated that his data was also very recent, having come from the Ranier shot on September 19, 1957, the only underground nuclear test the U.S. had ever conducted. He also emphasized the problems, largely pointing to background noises as particular issues that complicated the interpretation of the data. Importantly, he spoke to the criticality of "first motion" and the serious difficulty in distinguishing between earthquakes and detonations:[32]

> The feature which appears to us to be most useful for this identification process is the direction of first motion in the longitudinal wave group. The first motion criterion is useful because most of the large earthquakes studied to date have been found to produce alternate zones of initial compressions and rarefactions surrounding the source. Blasts, on the other hand, are expected to produce initial compressions in all directions. We may thus assume that the existence of several strong initial rarefactions is a good indication that the recorded signal came from an earthquake source. It must be remembered, however, that the amplitude required for identification by this method is much greater than that required for detection only.[33]

In short, Romney's strongest message to everyone present was that seismologists knew the features that distinguished earthquakes from blasts, but much was still unknown about how to differentiate detonations from earthquakes—a problem Northrup had voiced for years. The Ranier data now provided evidence underscoring this problem. Romney had recently presented the Ranier findings in a paper given at the annual meeting of the Seismological Society of America:

> The seismograms are very similar to those recorded from any small earthquake. This, of course, only confirms what all practicing seismologists know, that is, that the characteristics of seismic waves are largely determined by the structural and mechanical properties of the earth, rather than by the characteristics of the source.[34]

In immediate follow-up discussions, Mr. I. P. Pasechnik, Romney's counterpart in the Soviet delegation, spent much time downplaying the importance of P-waves. He implied that detonations could easily be detected if control stations (i.e., seismic sites) were carefully positioned "in such a way as to record the direct longitudinal P-waves." He argued that other wave forms or factors were equally or more important.[35]

Romney was well-prepared to challenge the Soviet counter arguments. At the conclusion of Pasechnik's presentation, Romney carefully laid out the results of the Ranier shot. With detailed graphs, he illustrated how and why many of the 50 seismic sites could not detect the small 1.7kt detonation from Shot Ranier. Importantly, he noted none of the AEDS stations located outside of North America had detected the detonation. In reviewing the seismograms from some of the 50 stations, the graphs were "all very similar to those recorded from any small earthquake."[36] He displayed the seismogram from the Alaska station, located 2,300 miles from the test site in Nevada, to illustrate how difficult it was to observe first motion. "I think that the seismologists here will all agree that this signal probably is not detectable in the ordinary sense, but, knowing precisely where to look, of course, there it is."[37] Romney concluded that the Ranier shot of 1.7kt was the absolute minimum size that could be detected at distances of 1,250 to 1,875 miles, with good equipment positioned at quiet locations. However, the Earth's core shadow would prevent detection between 700 and 1,200 miles. Within the same parameters, Romney indicated a 1kt nuclear explosion would only be detected out to 250 miles.

The Soviet scientists, Evgeny Federov, Ivan Pasechnik, and Mikhail Sadovski, strongly challenged Romney's conclusions and spent the remaining time during this session rebutting his findings.[38] Most notably, Sadovski disputed the evidence that the Earth's core shadow zone greatly diminished seismic detections beyond usefulness. Although he offered little substantive data, he stated Soviet instruments had detected seismic activity (i.e., by observing other types of waves), which would be important to a detection system within those zonal distances.[39] Western seismologists, however, also stepped forward to politely support Romney. Although Sir Edward Bullard initially tried to close the gap of disagreement by acknowledging the Soviets' assertions that they could record data in the shadow zone, he pointed out that such ideal conditions would not represent the norm of most sites in a Geneva system. Indeed, he cited the Ranier shot as an example of an average conditioned site. He concluded his remarks by suggesting that a better threshold for a monitoring criterion would be 5kt rather than 1kt.[40]

The disagreements continued for several more days as the Soviet delegates sought to de-legitimize Romney's data. Still, western representatives countered each time with data collaborating Romney's analysis. For example, on July 17th, Bethe gave a very technical presentation that addressed the factors of pressure waves, regions in which they become acoustical, and why coupling to the long-period seismic waves enabled long-range detection. Importantly, he noted the significance between TNT explosions and nuclear detonations—the latter causing great vaporization (meaning that almost all of its energy immediately dissipates). In short, chemical explosions (i.e., TNT), which the Soviets had exclusively used in their tests, pushed out energy that endured by not having experienced vaporization. Bethe's calculations clearly showed, as Romney's had, that "the propagation of waves to large distances is essentially a function of the structure of the earth rather than the source." In distinguishing between detonations and earthquakes, Bethe explained that with earthquakes, "most of the energy goes into long waves from the beginning because the source is extended."

On July 18th, seismologist Dr. Frank Press began the day's discussions by using very frank and succinct language to briefly summarize the seismic debates thus far. While agreeing with many of the presenters that seismic instruments would continue to improve, he asserted that everyone

> . . . must consider only what is experimentally realizable at this time. Interpretations must be based on P-wave data alone, recorded at many stations with such clarity as to reveal the direction of first motion. Records from wide-band or long-period seismograms provide useful, but not conclusive, additional information.
>
> Unfortunately, long-period waves are not excited by small blasts. The design of the net must be based on clear recording of the first arriving waves up to the shadow zone, and this should include clear recordings of the direction of first motion. We require this because of our need to have this clear indication of first motion for the smaller blasts.[41]

Press's short address comprehensively set aside all of the Soviets' assertions and illuminated what was required for a Geneva system based on strong evidential science. Although confronted with many probing questions by the Soviets, Press deferred any detailed discussion concerning his short address. "My only purpose was to lay the groundwork, to describe those basic facts which we must take into account in designing a system."[42] By this time, the preponderance of data and analyses surrounding the seismic issues appeared to have strongly shifted the momentum firmly into the West's camp. Romney later noted the reason why:

> I soon speculated that we were dealing primarily with members of Soviet academia, who understood the basic physics of each technique, but had not faced the practical problem of applying the technique operationally to long-range detection of unannounced foreign tests.[43]

At the time of Press's blunt conclusion, Northrup and the other geophysicists from all nations were meeting in an informal working group to draft a compromise agreement on seismic capabilities and threshold criteria for the eventual international control system. On July 18th, Northrup and Sadovski presented an update on their informal meetings at the general session. They reported they had only one main point of disagreement. Northrup stated "the determination of first motion may not be possible beyond 500 kilometers [312 miles] under average conditions or beyond 700 kilometers [437 miles] in favorable conditions;" but under ideal conditions, perhaps 1,000 kilometers (625 miles). Sadovski's viewpoint was that a 1kt underground explosion could be detected out to 1,000 kilometers. "However, the determination of first motion will not be possible in all cases." In other words, Sadovski implied it may be possible in some cases beyond 1,000 kilometers and within the "skip zone" (i.e., the undetectable surface area caused by the Earth's core shadow). Still, everyone appeared committed to achieving a compromise agreement.[44]

However, four days later, on July 22nd, the informal seismic working group was still struggling to reach a consensus. Sadovski informed the general session that a significant amount of contention remained. The Soviets wanted to set the monitoring threshold at a 5kt yield detectable out to 1,000 kilometers. Northrup and Romney, with years of AEDS experience, stuck by their many assessments that current LRD technologies detected approximately 90 percent of all earthquakes.[45] The remaining 10 percent came to "about 300 earth shocks in the range of 1-5 KT [that] could not be distinguished from subsurface nuclear tests." Therefore, the future Geneva system must have a mechanism that enabled the investigation of indeterminable events. Sadovski countered that "Dr. Northrup believes that 90 percent of all earthquakes can be correctly identified and 10 per cent might require additional study.... We believe that such cases will be isolated ones, that in any event, there certainly will be no more than ten or twenty."[46] Northrup replied that the difficulty in estimation was due to the varied geography of the Earth. Earthquakes occurring in mid-continent were significantly different from those originating in coastal areas. Surprisingly, Semenov spoke out strongly against his fellow countryman by criticizing Sadovski's agreement of a 5kt threshold. In bringing up an old, largely disputed argument, Semenov insisted that the first motion of a 5kt detonation could be detected out to 3,700 kilometers (2,300 miles). "There will never be a case where 10 percent of earthquakes will have to be checked."[47]

While the informal seismic working group continued to meet each day, the conference took up discussions about the remaining LRD techniques. As agreements were readily reached on some of those sciences—particularly EMP, as well as acoustic and hydroacoustic methods—serious debates arose over the issue of airborne sampling. At the center of this debate was the issue of overflight.

Sir William Penney had actually initiated the talks on the collection of radioactive debris and contamination on July 9th. At that time, the attendees heard papers on the topics of meteorological and radiochemical issues pertaining to collection operations and how those issues should help determine an international control system. Dr. Federov's paper, entitled "Control Measures through the Utilization of Radioactive Products," was an attempt to seriously downplay the significance of airborne collections. Ever since Eisenhower's "open skies" proposal, the Soviets had always remained seriously guarded about any overflights of their territories. They feared that debris collection could become a "front" for other forms of spying over Soviet territory.[48] Federov strongly argued that ground-based samplers were adequate. Disingenuously, he even claimed the USSR had abandoned the airborne technique in favor of ground-sampling methods.[49] In the end, the two sides worked out a compromise, whereby a system of pre-determined flying routes would be used, and aircrews would include observers from the testing nation.[50]

One other issue arose during the conference that posed a serious challenge for all participants—high-altitude testing. To date, no one had thought much about the issue. However, recent Soviet advances in missile technology and the shock of the two *Sputnik* launches the previous October and November, had awakened both nations to the fact that the dimensions of long-range detection had expanded exponentially. Other environments such as space and other regions of the world, especially the southern hemisphere, suddenly became very important. The only scientist who had explored the issue of high-altitude detection to any depth was Hans Bethe.[51]

On July 22nd, Bethe, apparently comfortable in his role as one of the world's most prominent theoretical physicists, began a very detailed review of his work with an uncomfortable truth that everyone recognized. "In the field of high altitude explosions our knowledge is even smaller than in the case of underground explosions. Experimental data are lacking and we have to rely entirely on theoretical calculations."[52] Bethe then quickly ran through the LRD techniques and eliminated aerial debris collection and seismic monitoring as being capable of detecting high-altitude explosions. That left the acoustic and electromagnetic techniques. Lightning and detonations would be difficult to distinguish with regard to electromagnetic signals, he explained, because both generated frequencies in the same ranges. The acoustic technique presented even greater challenges. In short, Bethe explained that 50 kilometers (31 miles) was the boundary between low- and high-altitude zones because that was the border between two distinct sound channels. Small kiloton tests at high altitudes would be "very difficult to distinguish from natural phenomena and very difficult to observe."[53]

The key to detecting small detonations at high altitudes depended on instruments that could measure additional ionization created by the blast. You could do so, he said, "by observing the absorption of the so-called cosmic noise, that is, the electromagnetic signals arising from the galaxy." Importantly, the acoustic zones located in the low- and high-altitude zones affected detection differently. By his calculations, a 1MT device detonated between 30 and 50 kilometers above the Earth's surface could be detected out to a distance of 200 kilometers (125 miles). A 10kt bomb blast at 70 to 80 kilometer altitude would produce the same limitation (125 miles). He added that a 1MT bomb detonated at 80 kilometers high could be detected out to 500 kilometers (312 miles). Bethe then suggested that radio telescopes pointed in various directions might be able to perform LRD.[54] Finally, he spoke to the potentiality of satellites as the best platforms for high-altitude LRD because they could observe light, and detect gamma rays and neutrons. Bethe concluded his calculations-ridden presentation exactly where he began. "In conclusion, I want to emphasize once more that all the considerations which I have presented are purely theoretical, and that, so far, there is no confirmation by observation."[55]

Despite this qualifying statement, Bethe had already concluded, apparently from his work with AFOAT-1, that LRD techniques were possible — and soon. His presentation at Geneva were largely based on a report he had completed for Northrup on March 8th. His study examined the different ways in which techniques could be developed for the AEDS. These included electromagnetic signals, radio waves, acoustic signals, light, seismic signals, and satellites. He concluded "there are many ways to detect high altitude explosions of nuclear weapons. Since these methods complement each other, one can almost certainly obtain unique identification of a high altitude test without the benefit of the collection of nuclear debris."[56]

Bethe was followed by Ovsei Leipunski. The Soviet physicist complimented Bethe on the thorough presentation and addressed Bethe's recommendation that satellites held the most promise for an LRD platform. He noted his recent experience with the *Sputnik* satellites, and envisioned the practical application of Bethe's theories and calculations. "These calculations of patrolling on the basis of sputniks have been conducted by us, and it is rather interesting to note that Dr. Bethe came to the same conclusions."[57] Leipunski described how six satellites orbiting at 6,000 kilometers (3,750 miles) could provide adequate coverage, especially in observing gamma rays. What he did not contribute was that *Sputnik III*, launched only 10 weeks earlier, was significantly designed to research the upper atmosphere and near space to facilitate such LRD. It carried 12 instruments measuring the pressure and composition of the upper atmosphere, charged particles, photons and heavy nuclei in cosmic rays, magnetic and electrostatic fields, and meteoric particles.[58]

Although additional discussions the next day deeply pondered the complexities of operationalizing the sciences involved in high altitude LRD, the talks and questions demonstrated the best attributes of the entire conference— apolitical scientific collaboration. Federov appeared especially exuberant about the dialog and expressed confidence that both countries were rapidly advancing satellite technologies, which would make a satellite monitoring capability a major component of a Geneva system.

On July 28th, the conference adopted an agreement on high-altitude testing.[59]

While the delegations took satisfaction in quickly arriving at a consensus, the issue of high-altitude testing revealed an extensive gap of knowledge about an immense environment in which a potential treaty violator could hide a nuclear test. As time would soon tell, the Americans viewed this unknown as a potentially serious threat to U.S. national security. Recent Soviet advances in ICBMs and satellites were still fresh in everyone's minds. The Soviet Union now had three satellites in orbit, with the latest specifically designed for geophysical research. The United States only had *Vanguard I* (launched on March 17, 1958), the counterpart to *Sputnik III*, which was simply configured to test the impact of the upper atmosphere on satellite designs. Immediately following the Conference of Experts, the U.S. would focus extensively on the problem of high-altitude testing (discussed below).

Compared to traditional international diplomacy efforts, the first four weeks of the technical conference proceeded rather smoothly. Despite some prolonged debates, the attendees covered much ground. To everyone's pleasant surprise, the seismic working group finally reported to the general session on July 24th that they had arrived at a consensus.[60] With everyone in agreement on the capabilities and limitations of the LRD techniques, the hard work was now at hand—designing the Geneva system.

On July 30th, Federov and Bacher, representing the opening positions of the East and the West, delivered extensive narratives describing their visions of a Geneva system. Federov's remarks were more general than Bacher's and seemed to downplay the complexities involved. "We consider it absolutely necessary that we should propose the simplest possible control system." He went on to assert that every nuclear test ever conducted had been detected by someone. He cited the ongoing U.S. test series, HARDTACK I, as an example, and stated that while the U.S. had not announced every shot, Soviet LRD sites had detected all of them. He asserted a 1kt device detonated anywhere on the Earth in any environment, even "in the area of cosmic space close to the Earth, up to a distance of hundreds of thousands of kilometers from the earth, can be readily recorded by various methods."[61] Federov then discussed the most difficult question in everyone's mind—how many control posts were necessary given the challenges of seismic detection. He arrived at an estimated 100 to 110 needed for the entire planet.[62]

Bacher, taking a similar approach, went into much greater detail on the requirements of a system, paying particular attention to locations, command and control, and the limitations of each LRD method. Whereas Federov had downplayed the need for on-site inspections of suspected events, Bacher specified their importance and composition. He stated that inspection teams had to be readily available, quickly transported to the site, and then provided with the equipment to carry out the inspection. That equipment included radiological survey instruments, drilling machines capable of penetrating the detonation area, and aircraft available for low level, local overflights.[63]

The Soviets immediately tried to pin Bacher down on the number of control posts he anticipated, and the nationality of the post personnel and inspection teams. Bacher and Fisk countered by stating those types of questions reflected political decisions, and the task at hand was to calculate the requirements based on the previous technical discussions. Still, the Soviet delegates, with five now contributing to the conversation, continued to probe for the number of posts they could expect in a Geneva system (obviously curious how many would lie within the borders of the USSR), and the quantity and nationalities of required personnel. Fisk, as chairman of the meeting that day, soon stepped in to try to bring an end to the session. However, Federov continued to press the point. He stated the Soviet concept had given a number of control posts needed, but Bacher "did not indicate concretely such important data as the total number of stations over the whole of the Earth."[64]

Thus began several weeks of debate about the capabilities and limitations of the seismic technique as a prelude to determining the appropriate design of an international control system. In general, the lengthy debates over the seismic technique boiled down to one single factor — interpretation of the data. The Western delegates pointed to extensive sets of data, gained from the numerous AFOAT-1 experiments and AEDS detections, to illustrate the limitations and the general lack of knowledge in the field of seismology. The Soviets, however, used their data to make the case that the LRD seismic capabilities were already sufficient. Many of those conversations frustrated the delegates, as old arguments about first motion and the use of existing stations arose again. The data presented by both sides were used to calculate the spacing between the control posts to determine the number of stations needed. Initially, the West stated that 650 sites were required to monitor 1kt events. The Soviet estimate, however, calculated 100 to 110. The Soviets strongly that argued their existing seismic stations could augment the international system by providing additional confirmation. Their objective, of course, was to minimize the total number of international control posts. The West, predominantly led by Romney, countered that many of the stations in the existing Soviet system were not located in the right locations, and that quite a few were in "noisy" areas such as coastal regions. Most likely, Romney's objective was to minimize any impact to the existing AEDS. That is, he would not want to see any AEDS stations become components of a Geneva system.[65]

The seeds of a solution to this impasse were planted by Penney on August 5th. During this session, he offered a compromise solution, which he had obviously discussed with his allies following the previous day's contentious remarks. Penney proposed a total number of 170 sites, of which 40 would be land-based stations in coastal areas. The sites would be positioned to accurately monitor 5kt events as either explosions or earthquakes. Although the Americans wanted a threshold of 1kt, hence the 650 stations, Penney proffered that under a 5kt detection limit, approximately 90 percent of all earthquakes could be accurately identified. The remaining 10 percent, estimated at about 300 events each year, would be addressed by a random inspection protocol. While this recommendation generated a lot of questions in the following days, all parties warmed to the idea of a randomness feature; one which could reasonably deter a potential violator of a test ban agreement. The delegates thoroughly explored Penney's proposal throughout the remaining sessions.[66]

The Conference of Experts concluded the last session on August 19, 1958, and released the final report the following day. The agreement stressed the criticality of first motion, while acknowledging that seismic stations outside of the Geneva system (i.e., Soviet or American) could contribute to validation. The delegations also recognized the shortcomings of the seismic technique, but expressed optimism that new technologies would soon solve the problem of distinguishing between earthquakes and detonations. Importantly, they all agreed that Shot Ranier occurred in geographically unique conditions; a point the Soviets had previously refused to acknowledge.[67]

Penney's recommendations provided the key to a consensus, especially over the contentious issue of control posts. In concluding that proper spacing could effectively monitor a test ban, the agreement specified 170 posts. The breakout would be: North America: 24, Europe: 6, Asia: 37, Australia: 7, South America: 16, Africa: 16, Antarctica: 4, and other posts on islands (50) and ships (10). All control posts would have standardized equipment. The delegates also agreed that 20 to 100 ambiguous events would, on average, occur each year and would randomly be investigated.[68] For airborne debris collections, the report stated that each signatory country would continue to use its own aerial assets. Aircraft would fly over pre-designated routes and would include multinational observers.[69]

Eisenhower was quick to commend the results of the conference. In a public statement released on August 22nd, the president informed the American people that the experts had all agreed that a system of international control to monitor a test ban was feasible. He then called for an international conference of the three nuclear powers to formulate a treaty based on the results of the Conference of Experts. He recommended October 31st as a start date, the day after the conclusion of Operation HARDTACK II. To highlight his sincerity, he proclaimed a unilateral one-year moratorium on testing to begin on that date as well. On August 30th, Khrushchev agreed to the meeting in Geneva. This was the goal Eisenhower had long sought, and it was the Conference of Experts that had paved the way.[70]

The AFOAT-1 military leaders had no role to play at the Conference of Experts. Hooks had departed on April 12th, leaving his deputy in temporary command. On August 1, 1958, well into the second half of the conference, Brigadier General Jermain F. Rodenhauser assumed command of the 1009th SWS/AFOAT-1. Commissioned out of West Point in 1931, he originally worked in the ordnance career field. After earning a master of science degree from MIT in 1937, Rodenhauser headed the Central Arsenal Plans Division at Picatinny Arsenal in Dover, New Jersey. This was an important assignment because the U.S. was quickly expanding the ammunition and armament industries to meet the huge number of orders from European allies who anticipated the war in Europe. During the first half of the Second World War, he supervised the manufacturing of tanks and ammunition in the Midwest. In 1943 and 1944, Colonel Rodenhauser assisted in the logistics planning for Operation Overlord (the invasion of Europe), and was selected to head the Ordnance and Chemical Warfare portion of the Army

Services of Supply Command. Following the war, Rodenhauser became a founding faculty member of the newly created Air War College in 1946, serving on the faculty for four years and, in his last year, as the academic director for the college. From 1950 to 1954, he attended the National War College and served in the Office of the Air Force Chief of Staff. In this capacity, he was responsible for the Air Force Military Construction Program, for which he annually testified before Congress. He was promoted to Brigadier General in July 1954. Prior to assuming command of the 1009th SWS/AFOAT-1, Rodenhauser served as a member of the U.S. delegation to NATO.[71] Rodenhauser's expertise in logistics and international partnerships brought needed guidance to AFOAT-1 as the engineering requirements for the AEDS rapidly expanded.

Why was the Conference of Experts important in the history of the AEDS? First, by mid-1958, the domestic and international political backlash about radioactive fallout had finally pressured both nations to abandon atmospheric tests — something both East and West clearly desired. All of the discussions over the previous seven weeks about the capabilities and limitations of the LRD techniques were vital in determining operational standards of an international monitoring system; especially so because American and Soviet tests were about to permanently go underground. Also, both sides were entering the ICBM and satellite era. They had reached a stage where miniaturization now allowed low-kilo-tonnage testing (i.e., below 10kt) at their domestic test sites, with little or no impact on the populations. This ensured underground tests would endure well into the future, making the other AEDS techniques less capable or irrelevant, and the fidelity of the seismic technique more critical. In Northrup's mind, seismic was the paramount factor that would either make or break an effective control system. However, in the summer of 1958, there was just too much that remained unknown about the earthquake versus underground detonation problem.

Second, it was clear from the beginning sessions that every American and British presentation was firmly grounded in the experiences gained from a decade of monitoring more than 200 nuclear tests. The well-funded AFOAT-1 experiments conducted against every U.S. test were highly orchestrated to advance LRD very rapidly. Each experiment quickly led to improvements in the AEDS; and with AFOAT-1 possessing full procurement authorization, upgrades to the system occurred frequently and without much bureaucratic impediment. In contrast, the Soviet scientists, though undoubtedly talented in their fields of study, clearly lacked a similar depth of experience in LRD techniques and experiments. We now know that at the time of the conference, the Soviet equivalent to the AEDS was just then becoming fully mission capable.

Third, as time would soon tell, the Soviets were correct that technical discussions could not be entirely divorced from political issues or concerns. This was apparent by the presence of Semen Tsarapkin—one of the Soviet Union's most skilled diplomats. The U.S. delegation had no senior diplomats with them and were under strict orders from Dulles to not become involved in any dialog that could be construed as political. Fisk frequently interjected in the discussions to remind the Eastern delegates of this fact, especially during the sessions where Federov sought a hard estimate of the future staffing of the stations.[72] Federov's request presented a good example of how the technical and political were inseparable. Positioning foreign nationals, numbering in the thousands, within the borders of sovereign nations for extended periods of time as employees of a Geneva system would certainly involve political or diplomatic details. Federov tried to press this point on multiple occasions.

Fourth, as all parties formulated an agreement based on consensus, it was clear there were two gaping holes in the entire concept of a Geneva system. The few conversations on high-altitude testing did little to offer any real LRD solutions other than to anticipate the day when satellites could be utilized. Also, the specter of Ranier hung over the entire conference. The participants were aware that a single underground nuclear test, the first of its kind, was grossly insufficient to help determine the basis of an international control system to monitor a test ban. However, at the time, that was the only data they possessed. The AFOAT-1 geophysicists, Northrup and Romney, were especially anxious about both shortcomings. They eagerly awaited the three high-altitude experiments in Operation ARGUS, set to begin on August 26th, and the seven underground tests scheduled among the 37 shots of HARDTACK II, taking place between September 12th and October 30th.

Fifth, the Conference was deemed a success by all participants and their governments. It marked the first time in the Cold War when East and West could come together to agree on a course of action to curb the escalating arms race. In that regard, the nuclear powers showed the world that mutual understandings could be achieved. As Bethe later remarked,

> . . . not only did we come out with a positive conclusion, but also with one which I believe is technically feasible. And in addition there was quite an occasion for the two sides to get together and see that the other side is also human. . . . It was pleasant to see that in the course of time in every case agreement was achieved concerning the true experimental situation, and the conclusions were drawn accordingly. This is perhaps the best reason for having a conference of technical people rather than a conference of politicians: you can prove something in a technical conference.[73]

Lastly, while the final report stated that a Geneva system was certainly feasible, the details in establishing an international control organization were formidable. That was certainly how it appeared to Northrup when he returned to Washington in late August. During the conference, he had stated that an international control system would probably require two years of planning. While he recognized that the Conference of Experts was only a prologue to the upcoming Geneva political talks (i.e., the Conference on the Discontinuance of Nuclear Weapons Tests, set to begin on October 31st), it was clear to the AFOAT-1 technical director that the AEDS might initially become the backbone of a Geneva system.

Northrup feared such an inevitability. He directed his staff to "determine the feasibility and desirability of incorporating all or a part of the long range detection capability of the Atomic Energy Detection System into an internationally controlled, worldwide system...." AFOAT-1's Operations Division, headed by Colonel Charles E. Collett, completed that effort on September 25th. Entitled "Incorporation of the Long Range Detection Capability of the AEDS into an Internationally Controlled Organization," the 26-page document strongly argued against doing so. The authors outlined how well the AEDS had evolved over the decade into an interdependent system, where the combination of the four main LRD techniques could reasonably detect almost any nuclear tests within the USSR producing yields of 10kt or greater. Any fragmentation of the AEDS would seriously impede the nation's ability to independently monitor tests within the Soviet Union. In essence, the U.S. would likely lose the ability to unilaterally confirm a Soviet test. More importantly, the AEDS was now fully integrated into SAC's war plans (see Annex 2), and it would be "militarily unacceptable" to deconstruct it.[74]

The report provided two other areas of consideration. The narrative portion concluded with a summary of AEDS limitations, especially the challenging environments of underground and high-altitude testing. In much stronger language than the Conference of Experts report, this document stated that the AEDS had "no capability for lower yields [below 10kt] detonated underground, and an unknown capability for tests detonated in the stratosphere."[75]

The most detailed section of the report was AFOAT-1's study of what it would take for the U.S. to establish an international control organization based on the recommendations of the Conference of Experts. The planners estimated it would take 42 months to establish the system and would require 24,000 personnel. Each control post, depending on the location and LRD techniques, would require 39 to 63 personnel. The initial costs and the first operational year would total approximately $2 billion (2021 = $18.6 billion) and the subsequent annual maintenance expense about $375 million (2021 = $3.5 billion).[76] The report concluded with a detailed organizational chart, a budget, and a world map depicting the proposed locations of the control posts.

Between the conclusion of the Conference of Experts on August 21st and the start of the Conference on the Discontinuance of Nuclear Weapon Tests on October 31st (hereafter referred to as the Geneva Conference), AFOAT-1 experienced its heaviest workload to date. In fact, AFOAT-1 issued a memo on July 28th, expressing concern over the resources required to monitor all tests, foreign and domestic. This was the first time in the history of the AEDS when AFOAT-1 could not cover all tests. The document outlined the allocation of resources and prioritized the six different operations which would overlap. Due to growing concerns over high-altitude detonations, shots Teak and Orange in Operation ARGUS took top priority, followed by the UK tests and the underground seismic tests in HARDTACK II.[77]

Killian had recommended the U.S. should conduct its final nuclear tests in anticipation of a moratorium going into effect before year's end. Consequently, both East and West planned a large number of tests for September and October. This was a hectic period of time when all three nuclear powers squeezed in as many tests as possible. The jockeying for last-minute detonations also fed the ongoing propaganda war and the psychological effect of "advantage"—the perception that the party to test last had reached parity or surpassed the other in nuclear weapons development.

In addition to completing the analyses of the previous Soviet test series, AFOAT-1 personnel had an extensive role in HARDTACK I. That U.S. test series had completed its 35th and final shot only hours before the conclusion of the Conference of Experts. Importantly, it included the first LRD experiments designed to determine whether very-high-altitude detonations could be detected by existing geophysical techniques, active radar, and optical surveillance.[78] The two high-altitude shots were conducted from Johnston Island on August 1st (Shot Teak) and August 12th (Shot Orange), and detonated at 250,000 feet and 141,000 feet respectively.

Teak and Orange allowed AFOAT-1 to test the prototype backscatter radar technique used to record ionospheric disturbances. High-altitude nuclear explosions produced changes in the electron density of the ionosphere, which persisted for some time and caused interference with radio communications and radar systems. Ground station instruments could detect changes in the ionosphere, normally at frequencies of 10 to 30 megahertz (MHz). After subsequent modifications, the backscatter radar technique would enter the AEDS as the "R" technique in 1961.[79]

Similar to the backscatter radar technique, AFOAT-1 sought to measure disturbances in the ionosphere, but within a 60-degree cone over a detection facility. The experts specifically targeted perturbations of the "F" layer of the ionosphere. Like the "R" technique, the Vertical Incidence Technique was first explored at shots Teak and Orange, and would enter the AEDS in 1962 as the "V" technique.

Another high-altitude LRD experiment conducted during shots Teak and Orange was designed to record a measurement of nuclear explosion-induced fluorescence in the upper atmosphere. Scientists had observed that x-rays from a high-altitude detonation caused "excitation" of neutral and ionized nitrogen molecules when interacting with low density air, thus creating a bright pulse of visible fluorescence radiation. They believed a ground-based station equipped with a wide-angle lens and a narrowband filter could observe the pulse. Soon, this would prove correct, and the method would enter the AEDS as the Atmospheric Fluoresence Detection technique ("Z") in 1961.

On August 9th, three days prior to HARDTACK's Shot Orange, Rodenhauser deployed AFOAT-1 airmen to the South Atlantic to participate in another U.S. operation codenamed ARGUS. The ARGUS test series, originally code named FLORAL, consisted of three shots designed by the Lawrence Radiation Laboratory to test the theory that high-altitude nuclear detonations would create a radiation belt in the extreme upper atmosphere.[80] U.S. Navy Task Force 88, consisting of nine ships, conducted the test series approximately 1,125 miles southwest of Cape Town, South Africa, near Gough Island. AFOAT-1's involvement "was a hasty effort" because it occurred less than a week after the conclusion of the Conference of Experts. At the time, AFOAT-1 was eager to participate, given the lack of knowledge about high-altitude detection, especially in light of Bethe's conference presentation of the subject on July 22nd. Unfortunately, there had been very little preplanning and no site surveys. Consequently, operational support was extremely limited. Still, shots Teak and Orange of Operation ARGUS served as the initial explorations of high-altitude LRD and justified the last-minute rush to experiment.[81]

Initially, AFOAT-1 largely succeeded in recording close-in electromagnetic measurements of the tests. The recording of those measurements proved successful for shots ARGUS 1 and 2 but failed on ARGUS 3 due to communication problems. The three ARGUS shots occurred at altitudes of approximately 300 miles. The EMP experiments at ARGUS allowed AFOAT-1 to significantly improve equipment to record EMP and optical data. That combination would soon result in new equipment, which could be operated on fixed or mobile platforms (the "F" technique that would enter the AEDS in 1962).

ARGUS also served as a test bed to further explore several other scientific methods which Bethe had hypothesized as potential high-altitude LRD techniques, ultimately producing the "R", "F", "V", and "C" techniques. In addition, AFOAT-1 initiated an exploration into the possibility of detecting high-altitude detonations by measuring phase disturbances occurring between very-low-frequency (VLF) transmitting and receiving stations. In 1961, this capability would result in the VLF Phase Anomaly technique ("U") and be added to the AEDS.

At Geneva, Bethe had spoken at length about the capability to detect the extremely rapid changes of cosmic noise levels resulting from nuclear explosions in or above the ionosphere. The resulting ionization affected radio frequency absorption characteristics of the ionosphere. At the time of ARGUS, technicians had developed a special receiver, called a riometer, which could detect radio noise (i.e., "cosmic noise") coming from great distances in space and passing through the ionosphere. The riometer had a detection range of approximately 10,000 miles for a 1MT detonation.[82]

The final capability that had its beginning at this time was the Magnetotelluric Technique, better known as the Earth Current technique. Shots ARGUS I and ARGUS II produced fluctuations in the Earth's magnetic field. AFOAT-1 produced instruments capable of recording those fluctuations. A nuclear explosion within the Earth's magnetic field produced magnetohydrodynamic waves in this field, with a resulting perturbation of the Earth currents and magnetic field measurable on the surface. This LRD method quickly showed great promise and entered the AEDS as the "H" technique in 1961.[83]

Only six days separated the end of the ARGUS series (Shot ARGUS Ill on September 6th) and the beginning of Operation

HARDTACK II (Shot Otero on September 12th). For AFOAT-1, HARDTACK II would be the most important test series yet conducted. Of the 37 scheduled shots, seven were underground detonations. This was a test environment Northrup had sought since 1952. However, he would need to leave the supervision of those geophysical experiments in Romney's hands. Throughout September and October, Northrup was consumed with the analyses of the HARDTACK I data, the Russian tests, and preparations for the political Geneva Conference.

Going into the Geneva Conference on October 31, 1958, it was evident that a political agreement on a comprehensive test ban would be much more difficult to achieve than the technical agreement formulated at the Conference of Experts. From the start, the level of mistrust was high, and it took several weeks before the parties could begin substantive talks. Eisenhower had been clear in his address on August 22nd that any test ban agreement had to be a step toward disarmament, and that an international control system must be established to monitor compliance. In contrast, the Soviets wanted a test ban agreement up front, independent of any other disarmament considerations, with the details of an international control system to follow. They insisted the Conference of Experts had settled the control issue, to which the Soviets had agreed, and that the details could be filled in later.[84] As political science and law professors Harold K. Jacobson and Eric Stein noted,

> . . . the Soviet response raised a problem which was to become increasingly troublesome for the Western powers during the course of the negotiations. The question concerned the status and significance of the report of the Conference of Experts. To what extent was it the basis for the Geneva Conference? Was it the final formulation of the control system, or merely an initial recommendation? To what extent could the system recommended in the report be supplemented or altered without destroying the basis for the negotiations? It will

be recalled that the Western panel had refused to discuss a number of issues at the Conference of Experts on the ground that they were political. How should these matters now be treated?[85]

All of this must have frustrated Northrup, who was a technical advisor to the American delegation. The U.S. team was headed by James J. Wadsworth, Deputy Representative of the United States to the United Nations. Wadsworth lacked knowledge and experience with the technical details discussed, and confessed so throughout the negotiations. Fortunately, he was assisted by Bacher, who was designated as his deputy. In addition to Northrup, the only other key participants for LRD discussions in the 19-member delegation were Colonel Collett, Chief of AFOAT-1's Operations Division (and author of the recent report on the feasibility of integrating the AEDS into an international system), Hans Bethe, and Bethe's colleague on the PSAC, nuclear physicist Harold Brown.[86] The remaining members were primarily lower-ranking Department of State personnel. In contrast, the Soviet delegation was led by formidable Conference of Experts veterans Tsarapkin and Federov, alongside several other scientists who were with them at the technical conference in Geneva.[87]

Historians of the Geneva Conference have noted that the U.S. delegation was ill-prepared going into the conference. In addition to under-staffing the delegation, Eisenhower failed to heal "the cleavages concerning the wisdom of various courses among the individuals and agencies" of his administration who were stakeholders in formulating arms control policy.[88] While he desperately wanted an arms control agreement, Eisenhower was careful to walk that boundary between the pro- and anti-test ban camps. Strongly opposed still were the AEC and DoD, especially the Air Force. For AFOAT-1, it is noteworthy that there is no evidence of Air Force leaders at the highest levels ever pressuring Northrup and his team to support their opposition to a test ban. Indeed,

Northrup and Romney were the quintessential scientists who consistently remained focused on data and analyses.[89]

With the AFOAT-1 scientists having voiced their concerns over the feasibility of a comprehensive treaty, the possibility of a test ban remained predominantly political (in both the national and international arenas), largely due to H-bombs and radiation fallout concerns. During the first session, Senator Albert Gore, Sr. was in attendance as the congressional advisor to the United States delegation. His observations of the contentious debates left him "profoundly disturbed."[90] In essence, Gore believed the mistrust and insincerity on the part of the Soviets required the U.S. to take a new approach. In mid-November, he proposed to Eisenhower that the nuclear powers agree to an atmospheric test ban as a first, concrete step toward other disarmament measures. Although this was not the first time the idea had arisen, it was the first time an influential member of Congress had taken the issue to a wide audience. On November 16th, Gore wrote an article in the *New York Herald Tribune* outlining his recommendations. Not only would an atmospheric test ban eliminate widespread concerns about nuclear fallout, but the treaty would be safeguarded by the existing technologies that had already been proven effective (i.e., the AEDS).[91]

Of course, the U.S. scientists at the Geneva Conference knew this to be untrue, despite their different views toward establishing a treaty. Bethe and Bacher were proponents of an agreement. Brown represented Livermore, Ernest Lawrence's lab which had strongly opposed any treaty, and the AFOAT-1 leaders (Northrup, Collett, and Romney) who all appeared apolitical. For AFOAT-1, the biggest concern was they lacked data on the seismic issues. They eagerly awaited the underground shots of HARDTACK II.

Romney, exhausted and having lost 10 pounds due to the heavy workload at the Conference of Experts, returned to Washington at the end of August eager to involve himself in the HARDTACK II underground shots. Of the 37 shots in the operation, 7 were underground. AFOAT-1 targeted 5 of those for the, seismic experiments: Tamalpais (October 8th), Neptune (October 14th), Logan (October 16th), Evans (October 29th), and Blanca (October 30th). Blanca was one of the last three shots occurring within hours of the start of the Geneva Conference. These were important tests because three of the detonations produced extremely small yields (55, 72, and 115 *tons*). Romney hoped the data "might go a long way toward settling some of the issues that had been argued inconclusively with the Russians at the conference" [of Experts].[92]

The primary objectives of the close-in and long-range seismic measurements in HARDTACK II were to obtain data on the amplitude–energy relationships for underground and high-altitude bursts, the seismic travel-time curve, scaling laws, and elevation and geologic coupling factors. Throughout September, Romney deployed 18 temporary seismic stations ranging from Nevada to Oklahoma to Maine. He and his teams were careful in calibrating those stations in order to gather accurate data about the amplitudes of the detected waves. They wanted to know how the detonation yields and the distances between the explosion and each seismic site affected the amplitudes.[93] With the underground tests occurring in the last three weeks of the operation, Romney had adequate time to prepare. It would take several weeks for the data to be fully analyzed and compared with Ranier. That new data would be invaluable since neither Northrup nor Romney had felt comfortable about the Ranier data providing the sole basis for the establishment of adequate seismic control posts in a Geneva monitoring system.

Meanwhile, at the Geneva Conference, Gore's publicized recommendations for an atmospheric test ban appeared to have captured Tsarapkin's attention. However, the senior Soviet negotiator ignored Gore's call for an initial atmospheric ban, focusing instead on the senator's harsher remarks about the lack of sincerity on the part of the Soviets. The senior U.S. delegate, Ambassador Wadsworth, used the discussion to make it clear that Gore was not representing the official U.S. position and was only speaking as a private citizen.[94] Interestingly, at this stage of the conference, neither side wanted to suggest or entertain any thought short of a comprehensive test ban.

Three weeks into the conference, the West and the East still struggled to formulate an agenda and remained deadlocked over the sequence of negotiations. Blocking progress was the issue of an agreement on a control system. The atmosphere was nothing like the Conference of Experts. Bickering and petty rebuttals over that one issue kept the conference from getting out of the starting blocks. Watching from the sidelines, Bethe wrote to Bacher on November 25th, pleading the U.S. should agree to the Soviet agenda. Bethe argued the West would not forfeit anything by doing so even if the Soviet Union claimed a propaganda victory from the concession. "Could we not forestall this [propaganda victory] to a large extent by making the acceptance of the Russian agenda the occasion of a big public announcement . . . that we accepted the Russian agenda for the sake of making progress?"[95] Bethe's plea and Gore's remarks nudged the parties toward a position where serious discussions could begin to take place without a formal agenda. Immediately following Gore's publicized opinions (during the last week of November), the parties began, albeit painstakingly slowly, to work on the details of a test ban treaty.

In retrospect, all parties were trying to unravel a decade-long entanglement of mistrust. Simultaneously, another conference was taking place to address the longstanding fears of America's defense establishment, namely, the possibility of a surprise attack. Meeting in Geneva between November 10 and December 18, 1958, the conference was officially called "The Conference of Experts for the Study of Possible Measures Which Might Be Helpful in Preventing Surprise Attack and for the Preparation of a Report thereon to Governments." Due to its long title, it was generally referred to as the Surprise Attack Conference. Eisenhower had pressed Bulganin to establish such a meeting eleven months earlier. DoD and the president harbored serious concerns about Soviet advantages in nuclear missiles, which appeared to make the bombers of the Strategic Air Command less effective. Their hope was the conference would lead to a treaty restricting the proliferation of bombers and missiles, and could be enforced via treaty monitoring.[96] As leadership positions shifted within the Soviet government, Khrushchev agreed to the conference on July 2nd, just as the Conference of Experts began. Khrushchev wanted the Surprise Attack Conference held concurrently with the Geneva Conference in November. Eisenhower had agreed to the request.

Originally conceived as a technical-military conference, Eisenhower reduced the scope to a technical conference largely focused on the perpetual problem of inspections. Like the American and British delegates at the Geneva Conference, their counterparts at the Surprise Attack Conference were meeting strong resistance at the very same time on the very same issue. In sum, both sides were never able to develop a framework for discussion. Indeed, throughout the seven-week conference, East and West spoke past each other. The Soviets only wanted to link any discussion on inspections to disarmament in Europe, especially fearful of West Germany's rearmament and the stationing of tactical nuclear weapons there. They feared a surprise NATO attack of conventional forces (perhaps supported by tactical nuclear weapons). The Americans, however, feared a surprise attack from the Soviet strategic arsenal within the larger geography of the world.[97]

The surprise attack conference held little meaning for AFOAT-1, other than the fact that any inspection system (as nebulously conceived by the American delegates) would ultimately resemble the AEDS. While there is no record of AFOAT-1 scientists advising the American delegation, the senior technical expert was well-respected chemist Dr. George Kistiakowski from Harvard University, who exchanged correspondence with Bethe during this time period. In any case, American fixation on inspections at the Surprise Attack Conference undergirded the many ongoing assertions that the existing LRD techniques were inadequate. Still, the fact that both East and West met to discuss fears of a surprise attack revealed sincerity in addressing a spiraling nuclear arms race.

By mid-December, the distances between the two positions at the Geneva Conference had substantially closed. Details on the control system were addressed despite the strong stance the Soviets had taken that an agreement rather than a control system had to come first. For example, at the December 12th session, the USSR agreed to the composition of a control system that included a governing commission and the authorities of its chief executive officer. This was an important step forward. As the verbatim records reveal, the British and Americans present reacted pleasantly surprised when Tsarapkin, after asking for a clarification, replied:

> It seems to me that after the explanation which the representatives of the United States and the representative of the United Kingdom gave us, namely, that the administrator will be subordinate to the commission and will act upon the commission's directives, as well as in accordance with the provisions to be worked out in the treaty and its annexes, we can agree to accept this text of the article on that understanding.[98]

The following week, it appeared Eisenhower's dream of a disarmament treaty was slowly on its way toward being fulfilled. As Wadsworth recalled in his memoirs," . . . my contention [was] that the Soviet representatives came to the table in 1958 and 1959 with a genuine desire for agreement."[99] Authors Jacobson and Stein wrote, ". . . as East-West conferences go, the Geneva Conference had an unusual record of achievement when the first recess began on December 19, 1958."[100]

It was all almost too good to be true.

Dr. Hans Bethe

Hans Bethe worked closely with AFOAT-1, and understood the capabilities and limitations of the AEDS better than any other scientist outside of the United States Air Force. He served on AFOAT-1s Scientific Advisory Committee, better known as the Beta Panel that reviewed significant LRD R&D issues. The panel met regularly to evaluate the data collected from the Soviet nuclear tests. In 1958 and 1959, Bethe's advocacy for a test ban treaty contributed to placing the AEDS at the center of the international treaty negotiations.

Photo courtesy of Los Alamos National laboratory

Amb. James J. Wadsworth

Wadsworth was the Deputy Representative of the United States to the United Nations and headed the American delegation at the Geneva Conference. He proved to be a superb negotiator with the Soviets. In recognizing his lack of knowledge with the technical details of nuclear science, Wadsworth adroitly utilized the expertise of Bacher, Bethe, Northrup, and Romney. In 1960, Eisenhower appointed Wadsworth as United States ambassador to the United Nations.

Photo courtesy of the Department of State

325

Conference of Experts July 1, 1958

Delegates leaving the first session. From left to right: S. K. Tsarapkin (senior Soviet diplomat), David H. Popper (acting U.S. Representative to the International Organizations in Geneva), Dr. Y. K. Federov (Chairman of the Soviet delegation), and Dr. James B. Fisk (Chairman of the American delegation).

Photo courtesy of Cornell University

Conference of Experts (July 1, 1958)

Eastern delegation is seated on the left and Western delegation on the right. Hans Bethe and Carl Romney are seated in the second and third rows. *Photo courtesy of Cornell University*

Conference of Experts 1958 (date unknown)

United States delegation is seated on the left. Doyle Northrup is seated on the far right of the US delegation.

Photo source: AFTAC archives

Shot Rainier in Operation PLUMBBOB (September 19, 1957)
Ranier was the first underground nuclear test. It was the only data
available at the Conference of Experts for determining the
requirements of seismic monitoring in a nuclear test ban treaty.
Subsequent underground tests would later reveal that the Ranier data
was highly inaccurate, resulting in the derailment of the Geneva
Conference in 1959. *Photo courtesy of Los Alamos national laboratory*

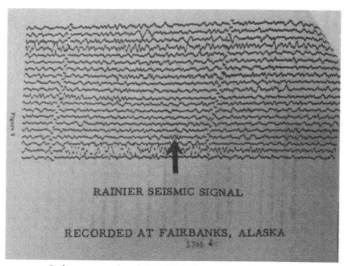

Seismograph of the Rainier seismic signal
The graph of the seismic signal shows the difficulty of identifying first
motion. Romney used this graph to illustrate his discussion at the
Geneva Conference. *Photo source: AFTAC archives*

Notes to Chapter 10

[1] Carl Romney, *Reflections*, (AuthorHouse, Bloomington, IN, 2012), 108-116.

[2] Educated as a physicist, Byerly was a pioneer in the field of seismology. He spent his entire career at Berkeley, from 1925 to 1965. "He played a germinal role in the growth of seismology in the United States during its formative years as a separate discipline." See his extensive bio at http://www.eps.berkley.edu/content/perry-byerly, accessed August 16, 2018.

[3] Carl Romney, *Reflections*, 142-145.

[4] Ibid., 152.

[5] Technical memo #91, Northrup to Hooks, Subject: "AFOAT-1 Detection Capability," June 21, 1957, AFTAC archives.

[6] Underline is in the original. Technical memo #93, Northrup to Hooks, Subject: "Bethe Panel Meeting of August 27-28, 1957," October 1, 1957, AFTAC archives.

[7] Technical memo #94, Northrup to Hooks, Subject: "Additional Evaluation of the Detection Capability of the AEDS in the Very Low Yield Range." October 24, 1957, AFTAC archives.

[8] J.R. Killian, Jr., Chairman, Memorandum for the Special Assistant to the President for National Security Affairs, Subject: "Transmittal of Report," March 28, 1958. In addition to Bethe, members included Doyle Northrup, Harold Brown, Major General Richard Coiner, Herbert Loper, Herbert Scoville, Jr., Roderick Spence, Brigadier General Alfred Starbird, Colonel Lester Woodward, and Herbert York.

[9] Jacobson and Stein, *Diplomats, Scientists, and Politicians*, 40.

[10] Ibid., 43-44. See also United States Congress, Senate, Hearing before a Subcommittee of the Committee on Foreign Relations, "Control and Reduction of Armaments," 84th Congress, 2nd Session, January 25, 1956.

[11] Fatzinger, *History of AFOAT-1: 1958*, 7.

[12] Ibid., 12.

[13] Ibid., 17.

[14] Ibid., 22. They also sent out significant levels of radioactive debris which prompted protests in a number of countries.

[15] U.S. Department of State, *Documents on Disarmament, 1949-1959*, Vol. II, p. 979, as quoted in Jacobson and Stein , *Diplomats, Scientists, and Politicians,* 45

[16] Jacobson and Stein, *Diplomats, Scientists, and Politicians,* 46. He was also aware of Bethe's views from Bethe's testimony before Senator Humphrey's Subcommittee on Disarmament on March 13th. See memo, Humphrey to Bethe, March 6, 1958, Bethe Papers.

[17] Doyle L. Northrup, Memorandum for the Chairman, Ad Hoc Panel on Nuclear Test Limitation, Subject: "Present and Potential Capabilities and Limitations of the AFOAT-1 Long Range Detection System," March 4, 1958.

[18] Ibid., 8.

[19] On March 8, 1958, Bethe produced a technical report for Northrup entitled "Possibility of Detection of Nuclear Tests at High Altitudes." While Bethe provided theoretical calculations that foreshadowed future LRD techniques, he concluded that those LRD instruments "not yet fully developed," would need to be located within the USSR with LRD stations spaced approximately 625 miles (1,000 kilometers) apart. Report, Bethe to Northrup, March 8, 1958, Bethe Papers.

[20] Northrup Memorandum for the Chairman, Ad Hoc Panel on Nuclear Test Limitation, 12-18.

[21] Emphasis added. Letter from Northrup to Bethe, February 1, 1958, AFTAC archives.

[22] Killian, "Transmittal of Report," 1. The other annexes were: Annex B: "Detection of High Altitude Nuclear Tests," authored by Bethe; Annex C: "Concealment and Detection of Nuclear Tests Underground," authored by Bethe and Harold Brown; Annex D: "Chart of Present and Future U.S. Nuclear Warhead Developments," authored by the AEC; and Annex E: Impact of a September 1958 Nuclear Test Moratorium on Soviet Nuclear Weapons Capabilities," authored by the CIA. Also, Bethe's notes outlining his briefing to the NSC dated April 1958. Bethe Papers.

[23] Jacobson and Stein, *Diplomats, Scientists, and Politicians,* 48-49.

[24] Ibid.

[25] Probably because Killian and Bethe were close and in full agreement on achieving a test ban. Fatzinger, *History of AFOAT-1: 1958,* 58-59. On May 28th, General Herbert B. Loper, Assistant to the Secretary of Defense for Atomic Energy, contacted AFOAT-1

that a technical conference was imminent and that AFOAT-1 should prepare a study on the capabilities of the seismic technique and the number of stations required to monitor an international test suspension.

[26] Ibid., 51-52.

[27] Romney, *Reflections*, 156. Also note that Fisk was on the record as being pro- disarmament, Bacher was neutral but leaning toward a treaty, and Lawrence had long partnered with Teller to be extremely vocal against any agreement. Lawrence played almost no role at the conference. He had long suffered from ulcerative colitis. On July 13th, he was bedridden in Geneva and two weeks later flown home to enter the hospital in Berkeley. Ernest Lawrence died during surgery on August 28th.

[28] Ibid., 158-161. Complete list noted in Jacobson and Stein, *Diplomats, Scientists, and Politicians*, 54-55. The other advisors were: Perry Byerly, Director seismographic Stations, University of California (and Romney's mentor); Norman Haskel, Geophysics Research Directorate, Air Force, Cambridge Research Center; Spurgeon M. Kenny, Jr., Office of the Special Assistant to the President for Science and Technology; J. Carson Mark, Director, Theoretical Division, Los Alamos Scientific Laboratory, Herbert Scoville, Jr., PSAC; and Anthony L. Turkevich, Enrico Fermi, Institute for Nuclear Studies, University of Chicago.

[29] Fatzinger, *History of AFOAT-1: 1958*, 60. The Bethe Papers contain several memos and documents that reveal that, throughout June 1958, AFOAT-1 continued to forward technical papers and studies to Bethe. For example, see Memo, Crocker to Bethe, June 13, 1958.

[30] "Report of the Conference of Experts to Study the Methods of Detecting Violations of a Possible Agreement of the Suspension of Nuclear Tests," August 20, 1958, AFTAC archives. Note that this was the final report and not the verbatim records under the same title. See also *Diplomats, Scientists, and Politicians*, Jacobson and Stein, 70-73.

[31] Conference of Experts to Study the Methods of Detecting Violations of a Possible Agreement of the Suspension of Nuclear Tests, Verbatim Record of the Eighth Meeting, July 10, 1958, 26. Hereafter cited as EXP/NUC/PV.#.

[32] In recording an event, P-waves were the first data that appeared on the seismograms, followed by S-waves and then

surface waves. Finding the P-wave in the graph and seeing it point up or down was an indicator of an explosion or an earthquake.

[33] EXP/NUC/PV.11., 21.

[34] Leslie Bailey and Carl F. Romney, "Seismic Waves from the Nevada Underground Explosion of September 19, 1957," Paper presented at the annual meeting of the Seismological Society of America, March 27-29, 1958. Bethe Papers, Cornell University, Box 28, Folder 54.

[35] EXP/NUC/PV.12, 65.

[36] Ibid., 71.

[37] Ibid., 75.

[38] Physicist Evgeny Federov was the Soviet's Chief Delegate and functioned as Fisk's counterpart. At the time, he was one of the world's leading experts in the area of applied geophysics. Federov had been a lieutenant general during the Second World War but was stripped of his rank in 1952. See his extensive biography at https://public.wmo.intlen/aboutus/ awards/international-meteorological-organization- imoprize/evgeny-konstantinovich-fedorov. Pasechnik was a seismologist and Sadovski was a physicist and nuclear weapons expert.

[39] EXP/NUC/PV.13., 46-60. "Personally, I think that Mr. Romney's analysis is slightly pessimistic and is perhaps due to a slight exaggeration of the role which he gives to the disappearance which might occur over an area of 1,000 to 1,900 kilometers."

[40] EXP/NUC/PV.13., 36.

[41] EXP/NUC/PV.15., 11. Press was the co-inventor of the Press-Ewing seismometer that was initially used in the AEDS. At the time of the conference, Press was Director of the Seismological Laboratory at the California Institute of Technology in Pasadena, CA. See https://history.org/phn/116017.html.

[42] Ibid.

[43] Romney, *Reflections,* 162.

[44] EXP/NUC/PV.15., 62-66.

[45] Killian, "Transmittal of Report," Appendix A, 22.

[46] EXP/NUC/PV.17., 6.

[47] Ibid., 11.

[48] Such a fear would soon be validated. Two years later, on May 1, 1960, a Soviet SA-2 missile brought down an American U-2 reconnaissance plane piloted by Gary Powers. The event sparked

an international confrontation that seriously impacted the Geneva test ban negotiations.

[49] EXP/NUC/PV.7., 82-96. We now know that this was not true. While Soviet aerial collection capabilities were not as effective as those in the AEDS, the Soviets were indeed flying such missions.

[50] "Report of the Conference of Experts," 5-6. See also Jacobson and Stein, *Diplomats, Scientists, and Politicians,* 70-73.

[51] Report, Bethe to Northrup, March 8, 1958, Bethe Papers.

[52] EXP/NUC/PV.17., 56.

[53] Ibid., 62.

[54] Ibid., 65.

[55] Ibid., 72.

[56] Hans Bethe, "Possibility of Detection of Nuclear Tests at High Altitudes," Bethe Papers.

[57] EXP/NUC/PV.17., 76.

[58] Asif A. Siddiqi, *Sputnik and the Soviet Space Challenge* (Gainsville: University of Florida Press, 2003), 175-176.

[59] EXP/NUC/PV.21., 3.

[60] EXP/NUC/PV.19., 3. The Soviet scientists were careful not to say anything that could impede future diplomatic talks, hence the constant presence of Tsarapkin. However, as Bethe noticed, "that while they may have started from somewhat naive ideas about some of the detection methods, they were always ultimately convinced by technical arguments; they were real technical people who knew a good experiment when they saw one." Talking papers for Physics Colloquium at Cornell University, September 29, 1958, Bethe Papers.

[61] EXP/NUC/PV.22., 5. At that point, the U.S. had completed the 31st test of HARDTACK I four days earlier (Shot Pine) with four more shots scheduled over the next three weeks. Federov claimed they had detected 32 shots and that the U.S. had only reported 14.

[62] EXP/NUC/PV.22., 16.

[63] Ibid., 36-70.

[64] Ibid., 82.

[65] These debates occurred over a number of sessions in early August. See especially August 4, 5, and 11 (EXP/NUC/PV.25-27). Romney's view of two parallel systems is what we have today with the Comprehensive Test Ban Treaty's International Monitoring

System (IMS) numbering more than 300 stations-and the Air Force's USAEDS.

[66] EXP/NUC/PV.26., 26-65. Note that the 300 unknown figure was not highlighted by Penney in his remarks or follow-up discussions but was the estimate given by Northrup to Bethe in his recent report. Specifically, see Killian, "Transmittal of Report," Appendix A, 22.

[67] "Report of the Conference of Experts," 9-11.

[68] Ibid., 20-24.

[69] "Report of the Conference of Experts," Annex II, 3-4.

[70] For the broadcast of Eisenhower's announcement, see https://www.youtube.com/watch?v=qpZ4nDLvTdQ. Last accessed January 4, 2019.

[71] Fatzinger, *History of AFOAT-1: 1958*, 2. He would earn his second star in May 1960.

[72] See especially EXP/NUC/PV.23., 31-50.

[73] Talking papers for a physics colloquium at Cornell University, September 29, 1958, Bethe Papers.

[74] Report, "Incorporation of the Long Range Detection Capability of the AEDS into an Internationally Controlled Organization," September 25, 1958, AFTAC archives.

[75] Ibid., Enclosure I, 2. Note that the 10kt threshold was twice the size agreed upon at the Conference of Experts.

[76] Ibid., Enclosure II, 2-4.

[77] Technical memo #97, Rock to Acting Chief, AFOAT-1, Subject: "Technical Programs for the monitoring of Nuclear Tests during the Balance of 1958," July 28, 1958, AFTAC archives.

[78] Technical Memo #92, Northrup to Hooks, Subject:"AFOAT-1 Participation in Operation HARDTACK," October 9, 1957, AFTAC archives. Also, Fatzinger, *History of AFOAT-1: 1958*, 37-42.

[79] The "R" technique was an immediate success. The following year (1959), AFOAT-1 began a five-year USAEDS extension/improvement program that called for the construction of 25 backscatter radar stations.

[80] Known as the Christofilos effect. Nicholas C. Christofilos, "The ARGUS Experiment" *Geophysics,* Vol. 45, 1959, 1144-1152. Available at http://www.pnas.org. Last accessed January 4, 2019.

[81] Fatzinger, *History of AFOAT-1: 1958*, 39.

[82] Ibid., 81.

[83] Ibid., 95. Ten stations were in operation by the end of 1962

[84] Jacobson and Stein, *Diplomats, Scientists, and Politicians*, 113-125.

[85] Ibid., 125.

[86] Brown would later serve as Secretary of Defense in the Carter Administration. He had coauthored (with Bethe) the Annex C for Killian's NSC report: "Concealment and Detection of Nuclear Tests Underground."

[87] Jacobson and Stein, *Diplomats, Scientists, and Politicians*, 114. Romney would later attend several sessions as an expert contributor.

[88] Ibid., 115.

[89] Ibid., 116. This must have frustrated all of the AFOAT-1 participants. In preparing for the conference, the U.S. delegation team met from October 20th to October 24th. AFOAT-1 provided the majority of those papers and presentations. "Tentative Agenda, Preparation for Geneva Conference on Nuclear Test Detection, Week of October 20-24, 1958." Bethe papers.

[90] Ibid., 126.

[91] Ibid., 127.

[92] Romney, *Reflections*, 168.

[93] Fatzinger, *History of AFOAT-1: 1958*, 14. Also Ibid., 169.

[94] See the Wadsworth-Tsarapkin interchange in the verbatim records for November 19th. Conference of Three Powers, Verbatim Record of the Tenth Meeting, November 19, 1958, 5-14. Hereafter cited as GEN/DNT/PV.#.

[95] Letter, Bethe to Bacher, November 25, 1958, Bethe papers.

[96] Jeremi Suri, "America's Search for a Technological Solution to the Arms Race: The Surprise Attack Conference of 1958 and a Challenge for 'Eisenhower Revisionists," *Diplomatic History*, Vol. 21, No. 3 (Summer 1997): 417-451. Also, Jacobson and Stein, *Diplomats, Scientists, and Politicians*, 111-113.

[97] Ibid.

[98] GEN/DNT/PV.24., 38.

[99] James J. Wadsworth, *The Price of Peace* (New York: Frederick A. Praeger, 1962), 69.

[100] Conference of Experts July 1, 1958

CHAPTER 11

Metamorphosis

Given the positive momentum achieved in December during the final weeks of the first session of the Geneva Conference, all parties must have looked forward to making further progress when they returned for the second session on January 5, 1959. Despite the fundamental disagreement as to whether a control system or a test ban agreement should come first, the diplomats were slowly working toward an implementation of the Conference of Experts recommendations. The AFOAT-1 participants, however, now had some bad news to share.

The underground nuclear tests in HARDTACK II produced a plethora of geophysical data. As AFOAT-1 poured over that data during the first two weeks of November 1958, Romney became more and more concerned. Immediately following the Thanksgiving holiday, he took the initial analysis and additional unanalyzed data and rushed off to Geneva. Departing on November 26th, he arrived the following day to complete his review of the analysis and reports, and then met with Wadsworth and the lead British delegate, W. David Ormsby-Gore. The new data obtained from HARDTACK II indicated that it was much more difficult to identify underground nuclear explosions than previously estimated by the Geneva Conference of Experts. The method of distinguishing earthquakes from explosions by determining direction of the first motion was less effective than the experts had estimated from the Ranier data. The HARDTACK II detonations had produced seismic signals smaller than anticipated. This meant there were approximately two to four times as many natural earthquakes equivalent to an underground explosion of a given yield than the experts had first estimated.[1]

As the bad news was digested, AFOAT-1 notified Killian and General Loper, Assistant to the Secretary of Defense for

Atomic Energy, who then, presumably, passed the news on to President Eisenhower. Reportedly, Eisenhower was "furious" and "blew his stack." However, the president stated it was important to tell the truth and later expressed confidence that after the other nations examined the data, the real challenges to seismic detection would be universally understood.[2]

On December 12th, Loper called for an immediate ad hoc panel of seismologists to thoroughly examine the HARDTACK II data. Having returned from Geneva, Romney chaired the panel that met from December 16th to 19th. The geophysicists concluded "additional information on the relationship between magnitude and equivalent yield in kilotons indicates that previous estimates of the number of earthquakes per year equivalent to a given yield in kilotons were low by a factor of about two." The report estimated the kiloton equivalent of earthquakes ranging from 3kt to 7kt numbered between 1,100 and 4,400 per year. Using the anticipated international control system envisioned at the Geneva Conference of Experts, only 10 percent of those earthquakes could be identified each year. The report concluded that, if the future control system was designed with the agreed 170 control posts, then the threshold of detection would need to be raised from 5kt to 20kt. The geophysicists recommended further underground tests as seismic detection technologies rapidly improved; meaning much more R&D was required before they could ever approach lower thresholds.[3]

After reviewing the ad hoc panel's report, Killian commissioned a formal review panel to consider whether the international control system designed at Geneva could be improved enough so the 170 control posts could effectively monitor a treaty. The panel would look at two possibilities: first, could existing technology achieve this and, second, if not, could a rapid program of R&D do so. Heading the panel was Lloyd Berkner, president of Associated Universities, Inc. Officially called the Panel on Seismic Improvement, the group soon became known as the Berkner Panel. Importantly, Romney and Bethe were members of the panel.[4] It would take several months for the Berkner Panel to complete the study.

Meanwhile, back in Geneva, the second session reconvened on January 5th. Just hours before the formal opening, Wadsworth informed Tsarapkin of the HARDTACK II results and the report of Romney's ad hoc panel of seismologists. As expected, the Soviets expressed outrage. Wadsworth later wrote that "to withhold [the data] would be keeping them from world opinion and could be construed as deliberate deceit. The presentation of the 'new data' resulted in the most violent reaction imaginable. It spread a pall over the negotiations from which they never completely recovered."[5] Tsarapkin knew immediately the West would want many more control posts within the USSR. Either that or a dramatic improvement in seismic instrumentation was needed. Wadsworth did not, however, inform Tsarapkin of Berkner's charter to assess both courses of action.

The American public became aware of the Berkner Panel on February 11, 1959. However, by then, there was already an enormous amount of debate in assessing the impact of the new data on the Geneva negotiations. PSAC had actually issued a general statement to the public about the new data within hours of Wadsworth's briefing to Tsarapkin on January 5th. In that statement, PSAC was clear that the "method of distinguishing earthquakes from explosions is less effective than had been estimated by the Geneva Conference of Experts." What caught the public's (and especially the politicians') attention was an example of the impact of the new data: "The total number of unidentified seismic events with energy equivalents larger than 5 kilo-tons may be increased 10 times more over the number previously estimated."[6] DoD followed up with more detailed information on January 16th. In short, these debates significantly hardened the pro- and anti-test ban camps, and created a more rigid divide among the public, both at home and abroad. The net result was more people questioning the wisdom of entering a treaty that could not be adequately monitored.

The division also fueled further inquiries from Congress. On January 12th, Senator Gore leveraged the controversy to restate his earlier proposal for the nuclear powers to settle for an atmospheric test ban. On January 20th, Senator Humphrey, a member of the Committee on Foreign Relations, took to the Senate floor to inform his colleagues that his sub-committee hearings were about to begin a thorough study of the controversy. The following week, on January 28th, Romney appeared before Humphrey's Subcommittee on Disarmament along with Fisk and Dr. Leonard Murphy, Chief of the Seismology Branch of the U.S. Coast and Geodetic Society (USCGS). Romney spoke at length about the impact of the new data on the negotiations.[7]

This was an important hearing for several reasons. First, Senator Gore was in attendance and would surely interject his strong views on ending the Geneva Conference and accepting only an atmospheric test ban. Second, the Democrats held the majority in both houses of Congress. With Eisenhower and the Republicans controlling the Executive branch, it was crucial that any forthcoming agreement with the Soviets would be ratified by the Senate. At Geneva, Tsarapkin had already questioned American resolve to formulate an agreement by pointing out criticisms certain Democrats had expressed in the press. Finally, the senators looked to Romney as the definitive expert on the seismic problem. Until now, both Romney and Northrup had always adhered to the scientific facts and had kept a clear distance from the growing politicization of the arms control efforts despite their own service—the U.S. Air Force—being a staunch opponent to Eisenhower's desire for a treaty. To their credit, Northrup and Romney had not joined their close associate, Hans Bethe, in his advocacy of a ban nor sided with Teller and the AEC in their strong opposition.

The senators asked probing questions. Importantly, they wanted to know to what degree the AEDS would be involved in the proposed international control system, implying they would not wish to lose an independent LRD capability. Although Northrup had produced a study on the integration of the AEDS into a Geneva system, he had strongly recommended against it. Fisk reassured Humphrey that "from the point of view of our own country, I should think we would continue to use that system [i.e., the AEDS] very thoroughly and completely."[8]

The committee continued to question Romney and Fisk in an attempt to understand the accuracy of the new data. They wanted to know the impact of the difference between Ranier and the HARDTACK II shots. When Murphy expressed some surprise with Romney's analysis, which differed somewhat from his own experiences at USCGS, the conversation underscored how seismology was a very imprecise science at that time. For the senators, whose ultimate objective was to understand whether a test ban treaty could be effectively monitored, the dialog did little to address that fundamental question. Romney remarked that "seismologists do not claim the estimates are very precise. . . . [Y]ou will notice when a seismologist says there are so many earthquakes of a given size, he is tempted not to say that there are a thousand earthquakes, but there are 500 to 2,000 earthquakes."[9] Several minutes later, Fisk got to the heart of the discussion. "The concept of a hundred percent foolproof system is meaningless. So the question really is how much is enough?"[10]

As the long session continued, the senators pressed Romney on the earthquake versus detonation problem to assess just how much *would* be enough. They were especially interested in how many unidentified seismic events occurred specifically in the USSR and China. Explaining he could not break down the numbers by country, Romney told the senators the new data increased the number of unidentifiable events by a factor of 15 over the previous estimate calculated at the Conference of Experts. At that conference, the experts had concluded there would be between 20 and 100 5kt equivalent events worldwide each year which would require investigation. Now that number would range between 700 and 3,000. He estimated the USSR-China region would experience between 100 and 500 of those.[11]

Bethe followed Romney several days later, on February 2nd, to testify before the Subcommittee on Disarmament. The committee wanted his views on the impact of AFOAT-1's new data and whether he believed the international control system identified at the Conference of Experts could be salvaged. In sum, Bethe's testimony, very cordial, made a positive impression on both the Democratic and Republican members of the committee. He began by bluntly stating that the "statements in the United States press which sound as if the world were coming to an end, and as if the HARDTACK II results were an indication that we could not detect underground explosions at all , or not satisfactory. Well, this is . . . an exaggeration."[12] Bethe stressed such exaggerations and current confusions demonstrated the need for much more research. He then explained the problem of the Earth's core shadow zone and the criticality of first motion in language that everyone on the committee could understand. Importantly, he offered two detection methods that would immediately improve the Geneva system. First, he said, seismometers could be installed in very deep holes, approximately 6,000 feet. This would greatly reduce the problem of noise and may likely achieve fidelities greater than the experts envisioned. Second, the Geneva system should build unmanned seismograph stations and emplace them at 100-mile intervals.[13]

Near the end of Bethe's appearance, a short dialog on the fruitfulness of the Geneva Conference took place between Humphrey and Bethe that surprisingly went unchallenged. When Humphrey asked Bethe whether the Conference of Experts and the current Geneva Conference were productive,

Bethe replied in the affirmative, adding that the ongoing conference "has brought some very solid concessions by the Russians."[14] This was a remarkable statement given that almost all of the stakeholders in the United States believed the Soviets were being unreasonable. It was well known at the time that the Geneva Conference was stalemated on three points over which the Soviets refused to compromise. These were that: (1) the proposed control posts be manned by the host nation with only one or two international observers; (2) each party possess a veto over decisions of the control commission; and (3) an investigative team be allowed to immediately deploy to a questionable location without delay.[15]

Was the U.S. unwilling to compromise, as Bethe implied? Perhaps, but the U.S. was already considering a concession. On January 5th, as Wadsworth was briefing Tsarapkin on the HARDTACK II data in Geneva, Fisk and Killian met with Eisenhower in the White House to discuss the impact of the data issue. Remarkably, the president stated he was "considering severing test suspension from the requirement for progress on disarmament generally, in order to keep the focus on [on-site inspections]."[16] If so, this would be a dramatic departure from the position he had strongly held for years. On January 12th, Eisenhower conducted a meeting with his cabinet heads to discuss the concession. The president reemphasized that a control system was paramount but conceded it would take two years to implement (as Northrup had estimated and reported).[17] Four days later, on January 16th, the U.S. dropped the condition that a test ban treaty must be connected to the issue of disarmament.

Eisenhower's concession served to further harden, not soften, the two camps by making the prospects of an arms control treaty, in some form, now realistically feasible. Previously, almost everyone recognized total disarmament was a pipe dream. Now, Eisenhower's detachment of a test ban from disarmament brought to the forefront Fisk's blunt assertion that "a hundred percent foolproof system" was meaningless and the question of "how much is enough." What was desperately needed was a final assessment from the Berkner Panel, who were taking too long. As all of this pressure built throughout January 1959, the Berkner Panel was asked to provide interim reports on their progress. On January 7th, with Soviet outrage at the new data at its peak, the panel briefed the White House and the State Department that there were "four promising approaches [that] are within the present limits of technology and should be considered." These were: an analysis of long-period surface waves, a network of unmanned seismic stations, larger arrays of seismometers at manned stations, and seismometers in deep holes. Separately, Bethe proposed these improvement methods in his aforementioned appearance before Humphrey's committee three weeks later.[18]

Killian was again asked to provide an update on the Berkner Panel to all key stakeholders at a meeting at the State Department on January 26th. Herter, now serving as acting Secretary of State (as Dulles was dealing with his terminal illness), hosted the meeting, which included 15 participants from DoS, DoD, AEC, CIA, and the White House. The attendees debated the current U.S. negotiating policy and the interim report. As expected, McCone, representing the AEC, advocated for changing the U.S. position at Geneva to accepting an atmospheric ban and to continue underground tests in order to devise an adequate seismic detection system. Colonel Fred Rhea, from DoD, fully backed McCone and stated that Defense "would not support an agreement which would outlaw tests beyond the reach of detection." Like Fisk, Killian countered that a perfect control system could never be designed, but one that could provide a deterrent was within reach. He told everyone "it would be useful to have as soon as possible the quantitative analysis by AFOAT-1 of the effects on the capabilities of the Geneva System of the four approaches recommended by the Panel." Again, McCone shed doubt on current (or soon available) LRD capabilities. He said the AEC estimated a three-year time frame was needed in order to "design a good system for monitoring underground tests." He added "the U.S. would be pursuing a reckless course by relying on techniques, which are based on inadequate appraisal and are too new to be fully understood."[19]

Back in Geneva, despite the U.S. concession on dropping a link to a larger agreement on disarmament, the negotiations proceeded agonizingly slow and were quite contentious. Tsarapkin consistently ignored Wadsworth's repeated calls for a new technical conference, insisting everyone had already agreed to the recommendations from the Conference of Experts.[20] It was clear the Soviets had no intention of examining or even acknowledging AFOAT-1's new data. Tsarapkin remarked, "even the existence of new scientific data would not be sufficient ground for us now to begin revising the conclusions previously reached."[21] Instead, he argued and would continue to argue that rapidly advancing technologies would soon resolve the underground detection problem.

To make matters worse, on January 22, 1959, the PSAC received new information that would exacerbate the underground detection problem further when the committee visited the University of California Radiation Laboratory. Teller, now director of the lab following Lawrence's death, made the issue of "decoupling" his top priority for the visit. The leading scientist working the theory of de-coupling was Dr. Albert Latter, a nuclear physicist at the RAND Corporation. He was a close associate of Teller's, and together they had recently published a book arguing against any test ban agreement.[22] Latter discovered that an underground detonation in a large cavity could greatly reduce the seismic signal. For example, a 300kt yield could conceivably look like a 1kt detonation. Bethe had corresponded with Latter and had made calculations on this theory.[23] He had presented those calculations at the Conference of Experts but had expressed skepticism about the theory's validity. Now, at Livermore, Teller and Latter placed the problem of de-coupling at the top of the agenda in hopes of convincing the PSAC that the proposed international control system was inadequate. Obviously, this was a topic of great interest to Bethe.[24]

Soon after the Livermore visit, Bethe accepted the validity of Latter's theory. Killian concurred and, in late February, extended the Berkner Panel's charter to include the problem of decoupling. Berkner did so immediately by devoting the entire day to the topic at the panel's next meeting on March 5th. Bethe and Latter spent several hours discussing the problem, with Romney offering some supporting data.[25] This was a serious matter because the discovery of de-coupling could lead to effective construction of underground cavities to facilitate cheating. Therefore, Latter's first paper was not disseminated within the government until March 30th and would not be declassified until late October. The PSAC did not release it to the public until December 22, 1959.[26] Too much information became known about de-coupling to keep it classified, especially throughout 1960 and 1961 as the Geneva negotiations continued.

The final event serving to complicate the negotiations in Geneva was the issue of high-altitude detection. As recounted earlier, the Conference of Experts had addressed the issue but was left with an uncomfortable truth: There were no established LRD capabilities to detect high-altitude detonations. Bethe's theoretical work certainly offered great promise, but the practical application of his theories was still in its infancy. Relevant R&D work at AFOAT-1 took on greater importance on January 2, 1959, when the Soviet Union launched *Lunik I*.

Lunik I was one of more than two dozen Soviet spacecraft designed to study the moon. Configured as either orbiters or landers, the *Luna* series of spacecraft conducted experiments to study the moon's chemical composition, gravity, temperature, and radiation. The first three *Luna* spacecraft failed to orbit the moon. *Lunik I* was the fourth spacecraft but only achieved limited success. Due to an incorrect trajectory, it flew past the moon and entered into permanent orbit around the sun, thus becoming the first artificial planet. *Lunik I,* combined with the three successful *Sputniks,* created great consternation within DoD and increased anxiety over the missile gap. This anxiety, more than anything, explained the hard stance DoD and AEC took against almost any agreement with the Soviets, especially a test ban. They believed the USSR was well ahead of the U.S. in developing missile technology, as the *Sputniks* and *Luna* launches demonstrated. DoD badly wanted the resumption of tests to perfect nuclear missiles in order to catch up with the Soviets. At this point in time, they only supported an atmospheric test ban because they were ready to conduct all tests underground.[27]

Many people believed if the Soviets could place a satellite into orbit around the sun, they could certainly test nuclear weapons at high altitudes or in space. Consequently, on the heels of the *Luna I* news, Killian levied PSAC and AFOAT-1 to make the issue of high-altitude LRD a priority. Paralleling the Berkner Panel, he immediately commissioned another panel to study the problem of high-altitude detection. Killian chose Dr. Wolfgang K.H. Panofsky, a professor of physics at Stanford University, to lead the technical group, which became known as the Panofsky Panel. They considered the utility, time schedule, and costs required to conduct nuclear tests in space. More importantly, the panel examined the issues involved in detecting such tests.[28]

By early March, all of this uncertainty, anxiety, and whirlwind of activity about a nuclear test ban—now pervasive throughout domestic and world politics, the public, and the scientific community—was coming to a head. Negotiations in Geneva remained stalemated and contentious. Unlike the first session, this second session had yet to adopt any articles. The Soviets, still angry about AFOAT-1's HARDTACK II data, refused to yield on their three points of contention (i.e., the staffing of control posts, a veto over decisions of the control commission, and rapid deployment of investigative teams). Yet neither the Soviets nor the Americans were willing to break off talks for fear of allowing the other side to claim a propaganda victory.

This left the State Department in a quandary. Pressure was building. Dulles, in an "eyes only" message to Wadsworth on January 30th, wrote

> I think you should be aware that there is growing apprehension in informed private circles in Defense and AEC, in Congress and in the country as a whole less these suspension-of-test negotiations involve us in agreements which are far from being fool proof so far as inspections and controls are concerned.[29]

Wadsworth replied (now to Herter with Dulles gone) on February 16th that he was concerned about a division with the British, who were considering a compromise on the three Russian points. "I have been giving careful thought to the basic question of whether or not it is in our interest to go further with these negotiations. One thing is clear. We have to make up our minds right away."[30] Herter briefed the president the following day, reiterating Wadsworth's suggestion that they break off the talks. Eisenhower did not agree but emphasized "we do not have to have a system better than that agreed upon a year ago [i.e., at the Conference of Experts]. We do, however, have to have the right to go in and inspect any questionable occurrence."[31] On February 25th, the president appeared to modify this statement to Herter. In a private meeting with Killian, Eisenhower stated if the LRD system could detect atmospheric and large detonations underground, then, perhaps, on-site inspections might not be so critical. What he did not want was the right to veto any decision made by the control commission, especially the right to initiate an inspection.[32]

With his most important initiative of any during his two administrations slipping away, Eisenhower began to yield.

However, he needed to buy some time. In an "eyes only" message to Wadsworth on March 10th, Herter stated, "if we do not recess soon, we will be drawn into prolonged inconclusive discussions confusing and obscuring basic issues without making real progress."[33] That same day in Geneva, the U.S. compromised on the issue of the treaty's duration. In essence, the U.S. acquiesced to the Soviet position that the treaty should be of "indeterminate duration."[34] With a recess now arranged for March 19th, both sides adopted a couple of minor articles on the last day of the session in order to keep the conference alive. The first was an article defining the future relationship between the control organization and the United Nations. The other article concerned the issue of annual reviews of the control system to assess its effectiveness.[35]

As the conference went into recess on March 19, 1959, the White House and Congress needed the final reports from the Berkner and Panofsky panels since the core of every argument about a test ban treaty was the problem of a detection threshold. Again, as Fisk had told Humphrey, "the concept of a hundred percent foolproof system is meaningless. So the question really is how much is enough?"[36] Wadsworth, Killian, and others still hoped the Soviets could be persuaded to convene a technical conference to really examine the new data. Wadsworth had consistently pressed the issue throughout the session, but Tsarapkin remained steadfast that the Conference of Expert conclusions were definitive.

On March 31st, two weeks into the recess, the Berkner Panel finally delivered its report, which included more than a dozen annexes. The summary report, entitled "The Need for Fundamental Research in Seismology," laid out the details behind the previous interim reports, which advocated for an analysis of long-period surface waves, a network of unmanned seismic stations, larger arrays of seismometers at manned stations, and seismometers in deep holes. In sum, the panel recommended a broad program of basic research in seismic wave generation, propagation, and detection. They also recommended a program to develop a prototype seismic detection system such as the Conference of Experts had envisioned. This prototype would provide the technical requirements for an international control commission to monitor a nuclear test ban. The report recommended the most advanced seismographic instruments be first installed in the AEDS "and to establish new stations since the existing distribution is not optimum."[37] In its entirety, the report was an unabashed outcry for an extensive R&D effort. In blunt terms, the Berkner Panel report revealed that the U.S. had underappreciated and virtually ignored geophysical research throughout the decade. Consequently, the nation had now positioned itself in an untenable position at Geneva because it had relied on the data from only one underground test (i.e., Ranier). Now, at the eleventh hour, only two of the HARDTACK II underground shots (over 1kt) were being considered. In short, three data sets were far from adequate.

Unfortunately, the Berkner Panel report did little to help determine "how much is enough" or to bolster the U.S. negotiating position at Geneva. Although the Panofsky Panel would not deliver its final report until several weeks later, Bethe had already completed his work on the panel. It, too, did little to help inform the U.S. negotiating position. Bethe provided an assessment of detection capabilities (none ready for the AEDS) for tests conducted at various altitudes in the atmosphere and space. He addressed the challenges posed by the use of ring shields on the weapons to obscure or degrade detectable radiation. He considered the propulsions necessary to launch various weights of the test platforms and concluded that yields in the range of megatons could not go undetected. In terms of Soviet capabilities to conduct high altitude tests, he concluded,

> ... the USSR would, at least as far as weight carrying capacity is concerned, be able to carry out sophisticated tests before the detection system is capable to detect

such tests. Even after solar satellites are launched, the detection apparatus will have to contend with concealment by ring shields, and will thus only be able to detect explosions from 100 to 300 kt.[38]

Despite Bethe's sober assessment, the Panofsky report concluded that no nation was likely to test at high altitude in the near future, if for no other reason than the expense involved.

Where did all of this uncertainty leave Northrup and AFOAT-1? And what did it mean for the future of the AEDS? By the time the third session of the Geneva Conference reconvened on April 13, 1959, it was clear to Northrup and Rodenhauser that, although never mentioned by name due to classification restrictions, AFOAT-1 and the AEDS had become a political football. Almost all of the capabilities and limitations of the AEDS had been revealed in the course of the Conference of Experts and the Geneva Conference, and in the press. Everything about the envisioned Geneva system reflected the research and experiences of AFOAT-1 in creating the AEDS. For Northrup, there was genuine concern the AEDS could likely become an integral component of a future international control system. In sum, AFOAT-1 was trapped between opposing camps throughout the executive and legislative branches of the U.S. government, the scientific community, the domestic and international media, and public opinion at home and abroad.

Hans Bethe, the one person outside of AFOAT-1 who knew the most about the AEDS, was partly responsible, albeit for honorable reasons. Much of his work found its way into official reports in the PSAC and into international conference sessions. He derived his knowledge from his close association with AFOAT-1 and his work with the AFOAT-1 scientists on the Bethe Panel. Bethe, a leader within the pro-test ban group, had become an influential voice in the government and in the scientific community. As discussed above, he, Killian, Bacher, and others had shown how science could directly influence and shape American foreign policy.[39]

As the Geneva Conference stumbled along throughout the first half of 1959, many remarkable changes were underway that would determine the future of the test ban negotiations, and AFOAT-1 and its AEDS. The launching of the two *Sputniks* in late 1957 had jumpstarted those changes. One effect was the U.S. immediately accelerating its missile program. A much larger effect, however, stemming from the perception the Soviets had surpassed the U.S. in science and technology, was the enacting of the National Defense Education Act of 1957. To alleviate a national shortage of scientists and mathematicians, the act allocated more than $1 billion (2021 = $9.3 billion) to advance those subjects in higher education. The act was also meant to attract quality professionals into government positions. The shortage was pervasive throughout DoD; Northrup, especially, had faced this problem throughout the decade. Consequently, AFOAT-1 had been forced to contract out much of its work since the early days of Tracerlab. In addition to Northrup's largely unsuccessful attempts to recruit scientists, AFOAT-1's re-enlistment rate was especially low throughout this time period, as private industries or college science programs lured the talented airmen away.

With significant impact on AFOAT-1 and the future of the AEDS, all of these aforementioned forces at play came to a head in the reports from the Berkner and Panofsky panels. Both studies were highly critical of the government's lack of adequate support for R&D. The Berkner report in particular implied the U.S. predicament at Geneva, caused by a poor or inadequate understanding of seismology, was a direct result of underfunding geophysical research. The Soviets, who had appeared less knowledgeable about seismology at both Geneva conferences, were now assessed to actually be far ahead of the Americans in seismic R&D. In citing an example of crustal studies (i.e., using small explosions to sound the

Earth's crust), the Soviets were spending 10 times the amount that American seismologists were spending. "Comparison of recent U.S.S.R. and U.S.A. publications in this field clearly reflect the resulting higher level of work."[40] The report criticized the meager amount the U.S. spent in total on seismological research—only $700,000 per year (2021 = $6.5 million).

The Berkner Panel report served as a call to arms for massive increases in geophysical R&D. The report emphasized that DoD (essentially AFOAT-1) could not go it alone: "A seismological research program of the magnitude discussed in this report must draw upon the facilities of the universities, the government, and industry for implementation." The report developed a two-year budget for the required individual research projects and systems development. Excluding implementation costs, the budget totaled $52.8 million (2021 = $491 million). The authors cautioned it would take time to address the problem of earthquakes versus underground nuclear detonations:

> It is the opinion of the Panel that the research studies described in this document will certainly improve detection capabilities of underground nuclear detonations. However, the improvements are not likely to be evaluated adequately for proper assessment or value in a detection system before one year of research activity at best. Most of them will undoubtedly require more time, perhaps three years. Thus, it is important to conceive of the detection system as one which will gradually evolve with time and reach a high level of detection capability only after several years.[41]

With Romney focused on the Berkner Panel, Northrup and his team paid particular attention to the Panofsky Panel report as a basis for an extensive study about the impact of high-altitude testing on the AEDS. The objective of Northrup's study was "to improve the capability for detection and identification of nuclear tests in the atmosphere and in cislunar space out to distances of 100,000 kilometers [62,500 miles] from the Earth."[42] The most striking aspect of the study's 11-page report, released on June 25, 1959, was Northrup's anticipation of an international monitoring system and the Soviets' ability to cheat on an agreement by conducting clandestine tests outside of the Soviet Union. Therefore, the study advocated for the expansion of the AEDS throughout other areas of the world.

The analysis was based on several facts. In addition to the possibility of a test ban treaty, the study acknowledged the advanced state of Soviet missile technology and the Soviets' capability to test at high altitudes or in space. Therefore, "the installation of an international control system will force a potential violator to find other regions than presently used for conducting tests on a clandestine basis. . . ." Northrup had the USSR in mind:

> The U.S. and Great Britain can be expected to adhere to the terms of the treaty, whereas the Soviet Union can be suspected of attempting to conduct clandestine tests under conditions difficult for the control system to detect. . . . [That is why the AEDS must continue to improve as quickly as possible]. . . . The performance of an international control system will require checking by the U.S.A.E.D.S. for a period of from five to ten years.

Given these assumptions, the study summarized the current capabilities and limitations of the AEDS. As of mid-1959, the AEDS had:

> • a good capability for detecting and identifying nuclear explosions in the atmosphere up to 10 kilometers [6.25 miles] in the U.S.S.R. and a fair to good capability in China;

> • no capability for *identifying* [i.e., beyond just *detecting*] underground nuclear explosions in the U.S.S.R. and China and no promise of a capability for several years until research programs presently planned have been executed;

- little capability for detecting explosions high in the atmosphere and no capability for detecting explosions in cislunar space; and

- a very limited capability for detecting air bursts any place in the Southern Hemisphere.[43]

The remainder of the report identified the objectives of the technical programs designed to improve and expand the AEDS. Northrup provided estimates on the increases needed for adequate coverage of the USSR-China region and those required for coverage of the southern hemisphere. In total, the AEDS would need 42 new stations: 18 for the "I" technique (acoustic); 21 for the "Q" technique (EMP); and three for the "B" technique (seismic—but only in the Northern Hemisphere). Importantly, Northrup underscored the fact he could not estimate the number of seismic stations needed in the southern hemisphere because of the immaturity of the current science and instrumentation:

> In view of the fact that present techniques for identifying underground explosions are under investigation, no additional seismic coverage is recommended in other parts of the world at this time. . . . [I]t is expected that a period of three to five years of research will be required before this additional requirement can be specified.[44]

The most important information contained in Northrup's study was the information on the progress achieved thus far in developing new techniques to monitor high-altitude detonations. In citing the Panofsky Panel's recommendations (in actuality, Bethe's work), Northrup indicated there were a number of methods or mediums which could be operationalized. These were: direct visible light, fluorescence of the atmosphere, backscatter radar, direct radar reflection from the fireball, ionosphere disturbance, and magnetic methods. Northrup summarized the science of each method, its potential in contributing to the identification of events, the number of systems needed within the AEDS, and the cost per system. All of the methods were in various states of testing at the time. In retrospect, the number of AFOAT-1's innovations was impressive. These methods would soon join the AEDS. The study of "ionosphere disturbance" would lead to the Cosmic Noise Absorption Technique ("C") and the Vertical Incidence Technique ("V") in 1962; and the Surface-Based Resonance Scatter Technique ("E") in 1963. The Back-Scatter Radar Technique would join the AEDS on September 1, 1961. "Magnetic methods" resulted in the creation of the Magnetotelluric (Earth Current) Technique ("H") and would be added to the AEDS in April 1961; "Fluorescence of the atmosphere" became the Atmospheric Fluorescence Detection Technique ("Z") in 1961 as well.[45]

The final sections of Northrup's study dealt with the future of nuclear debris collection, the radiochemistry laboratories, and the potential for a hydroacoustic detection system. For airborne collections, Northrup recognized the expense of dedicated aircraft to cover the southern hemisphere as prohibitive, estimated at $20 million per year (2021 = $185 million). However, he insisted "there is now a vital need for AFOAT-1 to take the initiative in sponsoring a study of the air-sampling requirements of the future." He expressed appreciation for the recent availability of U-2 aircraft for high altitude collection but believed the study should determine how other high-altitude aircraft located throughout the Air Force could be utilized to provide AFOAT-1 "with a special 'vector' possibility in all parts of the world." In looking at the issue of detecting underwater explosions, Northrup agreed that current hydroacoustic systems held great potential but noted much more research was still needed. He pointed out the type of signals generated by a nuclear detonation were unknown, and the current system had never detected an underwater explosion. He added, like the seismic challenge overall, "the background problems due to T-phase earthquakes which will be recorded on hydroacoustic stations is poorly known." He estimated a total of 25 hydroacoustic stations were needed to cover the major oceans of the world.[46]

By early summer of 1959, it was clear AFOAT-1 was under enormous pressure to mature the AEDS as quickly as possible. In retrospect, all roads of debate led back to the singular problem of "threshold" and Fisk's question: "How much is enough?" It was one thing to tackle the requirements of monitoring the Soviet Union; it was quite another to adapt the AEDS to a much more complex world. Until now, AFOAT-1 had to concern itself with only two countries (the USSR and the UK) and three testing environments (the atmosphere, the Earth's surface, and, minimally, underwater). Now, however, the environments extended to underground, underwater, high altitude, and space. In addition, the U.S. expected the French to test at any time and suspected the Chinese were well on their way to developing nuclear weapons. So how much was enough? The answer to that question would be determined in Geneva.

During the recess, concern grew in Congress and the White House over the possibility that the conference could fail. Wadsworth, back in the U.S. during the interlude, testified before Humphrey's Subcommittee on Disarmament on March 25th. The senators wanted to hear from the chief negotiator his assessment of the stalemate and his prognosis for success. Wadsworth carefully explained why the West could not concede to the three Soviet demands. The most fundamental requirement of all, he stated, was an effective international control system. In this regard, the greatest obstacle was the Soviets' insistence on a veto. This was strongly unacceptable. "The United States and the United Kingdom maintain that this across-the-board application of the veto power would render the control system meaningless and ineffective."[47] He explained that, should a suspicious event occur in the Soviet Union, the deployment of an inspection team could be seriously delayed by a required vote and subsequent veto. What was needed was a cadre of highly trained mobile inspection teams, composed of multiple nationalities, who could immediately deploy to the site. The Soviets feared these teams would provide a cover for espionage. Humphrey then discussed how the U.S. could overcome the Soviet objections, suggesting the president should write a letter to Khrushchev about the many counter-proposals the West had offered and then make that letter public. Wadsworth agreed it would help. Toward the end of Wadsworth's lengthy appearance, Senator Church asked if just an atmospheric ban could be enforced with current LRD systems, and if such a treaty would eliminate the problem of mobile teams conducting on-site inspections. Wadsworth replied, "it would do that."[48]

Whether Humphrey and Church's questions precipitated it or not, the president did send a letter to Khrushchev on April 13th, which marked a major course change in the American position at Geneva. The timing was intentional as this was the first day of the reconvened negotiations. Eisenhower emphasized the sincerity of the U.S. in achieving an arms control agreement:

> In my view, these negotiations must not be permitted completely to fail. . . . Could we not, Mr. Chairman, put the agreement into effect in phases beginning with a prohibition of nuclear weapons tests in the atmosphere? A simplified control system for atmospheric tests up to fifty-kilometers could be readily derived from the Geneva experts' report [and] would not require the automatic on-site inspection which has created the major stumbling block in the negotiations so far.[49]

This was a pragmatic concession. By this time, Eisenhower had most likely accepted the fact that a total disarmament treaty was beyond reach, at least during his remaining time in office. While it is true that he wanted his greatest legacy to be something that could curb a spiraling nuclear arms race, an atmospheric ban would allow some level of agreement, and appease the AEC-DoD opposition who wanted to continue testing underground. The public, as well, could then rest more easily about nuclear fallout.

Khrushchev replied to Eisenhower and British Prime Minister M. Harold Macmillan on April 23rd. In short, he rejected a ban on atmospheric tests but said he would consider dropping the veto in return for a quota system for on-site inspections. The British Government, under enormous pressure from their citizens to do something about fallout and the threat of nuclear war, had suggested a quota system, whereby each nation would be allotted a fixed number of inspections per year. The U.S. did not immediately respond.

Meanwhile, the Geneva Conference dragged on into early May with no signs of progress. Throughout that time, Wadsworth continued to stress the need for another technical conference. However, the Soviets kept insisting their baseline was the technical agreement reached at the Conference of Experts.

Finally, on May 5th, just three days before the current session of the Geneva Conference recessed, the president held another meeting in the White House with the principal stakeholders. Again, AEC Commissioner McCone stuck to his strong opposing position, while Secretary of Defense Quarles, though agreeing with McCone, suggested the quota proposal be explored. Eisenhower stated public opinion would soon force the U.S. to unilaterally declare a moratorium on atmospheric tests. "We must find a reasonable and decent way to do this [i.e., an atmospheric ban] by agreement if possible, even if the arrangement is not necessarily a perfect one." If not, the U.S. would find itself in "a renewed testing race in which we might break ourselves by over-insurance."[50]

As the Geneva Conference recessed on May 8th, the delegates had not adopted any additional articles. However, there was enough potential movement to keep the conference alive. The West was hopeful the upcoming Foreign Ministers Conference, scheduled for May 11th, would result in dislodging the logjam over the issue of on-site inspections.[51]

At the Foreign Minister's Conference, a hopeful sign appeared. On May 14th, Khrushchev agreed to a short technical conference, which would only consider the problem of high-altitude testing and detection. The official Soviet position remained that the seismic problem had already been discussed at the Conference of Experts. The ministers agreed that the next session of the Geneva Conference would work out the details of the new technical conference—called the Technical Working Group I (TWG 1).

When the Geneva Conference reconvened on June 9, 1959, the delegates spent the first day arguing over the issue of on-site inspections.[52] The following day, however, Wadsworth redirected the conference towards establishing the TWG. He read Khrushchev's statements supporting a technical discussion into the record.[53] On June 12th, Tsarapkin began several days of discussion to narrow the scope of the TWG. That afternoon, Wadsworth released the Berkner Panel report to the conference, with the hope that the TWG would consider the seismic problem as well.[54] Again, the Soviets refused to consider any new seismic data beyond what was discussed at the Conference of Experts.

The delegates gave TWG I a June 22nd starting date and only one week to return with an interim report.[55] The core of the U.S. delegation to TWG I were several former members of the Panofsky Panel, including Panofsky who served as the U.S. lead. Surprisingly, neither Bethe nor anyone from AFOAT-1 were included.

TWG I presented its final report to the conference on July 10th. With the Secretary General of the United Nations present, Panofsky and his counterparts (Federov and H. R. Hulme from the UK) each offered an overview of TWG I's work and working relationship. They reported the group had identified 10 techniques applicable to high-altitude LRD and had agreed nine should be incorporated into the future international control system. Federov reported that the Soviets had rejected the backscatter radar technique (the future "R" technique that would enter the AEDS in 1961) because it could be used for espionage. He stated such systems located at control posts "could enable alien staff to spy upon military objectives in the territory of sovereign states. The first radar stations which provided backscatter intelligence during the Second World War operated on the very frequency range now contemplated."[56] The 10 techniques which TWG I considered were the result of all of AFOAT-1's expedited R&D work on the problem of detecting high-altitude detonations. Following the TWG report on July 10th, the Geneva Conference dragged on for the remainder of 1959 with very little progress.

What did these frustrating political machinations occurring between late-April through early July actually mean to AFOAT-1? In total, they conveyed the fact that it was difficult to assess whether the world was better off with an imperfect test ban treaty, which could be monitored with an imperfect international LRD system, or accept the status quo of a nuclear arms race with the possibility of an atmospheric test ban. All roads of discussion led to the threshold problem of the under-developed state of seismic R&D. This, of course, placed enormous pressure on AFOAT-1 to mature the AEDS at an ever-increasing rate. Fortunately, behind the scenes of the politics and diplomacy, the United States had begun to enable AFOAT-1 to do so.

The "missile gap" and the *Sputniks* had not only compelled the U.S. to seriously change the American education system to emphasize science and technology, but the sense of urgency now extended throughout various organizations in the government. The AEC and DoD, of course, wanted to regain the lead over the Soviets in missile and space technologies. To achieve that objective, they advocated for more nuclear tests. In recent years, this reasoning had been the basis of their strong opposition to Eisenhower's disarmament goals. Now, the AEDS became central to their plans because the AEDS provided the primary means of ascertaining the state of the

Soviet Union's nuclear weapons testing program. With all testing soon to go underground, the AEC and DoD now recognized that the seismic stations were the most critical component of the AEDS.

For example, in a meeting at the White House on June 5th to discuss whether to release the Berkner Panel report to the Geneva Conference, McCone voiced his harshest opposition to a test ban yet. The AEC, he said, would not agree to any agreement which could not be safeguarded:

> It is our opinion, as a result of intensive and careful study of this question that an agreement to suspend underground shots cannot be properly and safely safeguarded by known technology. Therefore, we continue to advocate research, experimentation and testing either by the United States alone or jointly with the Soviets and British or through the United Nations as a means of developing this technology. We believe means of properly safeguarding can be developed through this experimentation . . . Dr. Northrup had stated that from three to five years would be required for such experimentation and, furthermore, it was possible even then results would not bring forth a satisfactory system.[57]

Fortunately for Northrup and AFOAT-1, they now had the means to finally advance geophysical research.

On April 23rd, the same day Khrushchev rejected Eisenhower's suggestion of an atmospheric ban as a first step toward a comprehensive test ban, the president met with Prime Minister Macmillan, Killian, McCone, and Quarles to initiate an enormous research program based on the recommendations of the Berkner and Panofsky panels. He assigned primary responsibility to DoD. In turn, DoD later chartered the Advanced Research Projects Agency (ARPA) with the "responsibility for research, experimentation, and systems development leading to a system for the detection of nuclear explosions underground and at high altitude."[58] DoD named the program VELA. VELA would soon split into three primary sub-programs codenamed VELA UNIFORM, VELA SIERRA, and VELA HOTEL. VELA UNIFORM would conduct R&D to address the underground testing problem, focused extensively on seismic R&D. VELA SIERRA included R&D to develop ground systems that could monitor high altitude tests. VELA HOTEL managed R&D for satellite systems. The scope and significance of the VELA program cannot be understated. Northrup and his organization would essentially be the sole beneficiary of the vast majority of VELA funding. ARPA provided that funding and oversight, shielding Northrup and his organization from political influence.[59]

By the summer of 1959, with VELA coming to fruition, the two camps supporting and opposing a test ban treaty could all agree on one thing: an effective LRD monitoring system required much more research. By now, most realized AFOAT-1 could no longer be totally responsible for advancing LRD R&D. The Berkner Panel report had clearly stated that fact:

> The Panel believes that the research program [VELA] can best be carried out by various existing private, university, and government laboratories. . . . A seismological research program of the magnitude in this report must draw upon the facilities of universities, the government, and industry for implementation. . . . [It] would likely sub-contract with private industry for much, or perhaps all, of the specific hardware development and procurement.[60]

AFOAT-1, however, was a highly classified organization, and the AEDS a classified system.[61] It would be almost impossible to partner with so many non-governmental organizations and still apply classification protections to the LRD mission and its monitoring system. The best solution was for the Air Force to create a new organization to provide stewardship over the AEDS. That way, Northrup could more easily detach unclassified techniques, most notably seismic, from those requiring continued security protections.

On July 6, 1959, the Air Force inactivated the 1009th SWS/AFOAT-1 organization. The following day, it activated a new unit to provide stewardship over LRD and the AEDS. Leadership of the new organization called for a major general who would serve as the commander of the new unit designated the 1035th Field Activities Group (FAG) and as director of the Air Force Technical Applications Center (AFTAC).[62] [62]

On July 2nd, four days before the inactivation of AFOAT-1, Rodenhauser sent a lengthy memo to the Assistant Secretary of Defense for Atomic Energy (Loper), which outlined his plan to move rapidly ahead with seismic research, soon to be VELA UNIFORM. In his memo, Rodenhauser informed Loper of the reorganization and his explanation of the rationale:

> The scope of the geophysical experiments involved, together with the very nature of the basic research problems, have convinced us that this program should be conducted in large measure on an unclassified basis. By this means, the best talent among the earth's scientists available in this country can be readily stimulated to work on the projects in view of their known national and international importance and of the unprecedented opportunities for research and subsequent publication of results provided by the proposed program. Toward this objective action has been taken to [create a new organization] as an Air Force Technical Applications Center. This re-designation should permit the staff to carry on both classified and unclassified projects with the same effectiveness and without compromise of classified work. It is expected that wide interest both in the government and in the scientific community at large, will continue in connection with this program. It is believed that the press in the United States will certainly make repeated requests for information on the programming of various

phases of this research as well as in the results that are obtained from the research. To a large extent this interest can be satisfied by the publication in scientific literature of results of the research in the normal fashion. However, it is expected that there will be occasions when an overall review of progress should be prepared for official release. [The Air Force] believes that this fact of life should be recognized from the very beginning and adequate plans made for public information.[63]

The 1035th Field Activities Group/AFTAC was born in the middle of a long-term effort to achieve a comprehensive nuclear test ban treaty. The new organization, however, would no longer be handcuffed by the AEC nor bantered about as a political football. Under the auspices of ARPA, AFTAC would receive virtually unlimited funding, and be free to improve and expand the AEDS into a worldwide LRD treaty monitoring system. With the Cold War growing ever colder, AFTAC's tasks would be extensive, especially as other nations acquired nuclear weapons. The decade of the 1960s was only a few months away and, with technology advancing at a rapid rate, this new era promised to pose new challenges Northrup and Rodenhauser could only imagine as the 1009th SWS/AFOAT-1 colors were permanently encased on July 6, 1959.

Lloyd V. Berkner

Berkner was a physicist and engineer who helped invent measuring devices later used to measure the ionosphere. He was a member of the PSAC when he led the Panel on Seismic Improvement. Known as the Berkner panel, the group produced the Berkner Report in 1959, which was instrumental in promoting geophysical research and the creation of the VELA UNIFORM program. *Photo courtesy of the U.S. National Science Foundation*

Dr. Wolfgang K. H. Panofsky

Panofsky was a professor of physics at Stanford University. News of the Russian *Luna I* launch sparked the appointment of Panofsky to lead a technical group to examine the problem of high altitude nuclear detection. The group became known as the Panofsky Panel. The Panofsky Report, largely reflected the work of Hans Bethe, and articulated the R&D required to develop the necessary LRD techniques.
Photo courtesy of the U.S. Department of Energy

John A. McCone

McCone first served at the national level as Undersecretary of the Air Force from 1950 to 1951. In 1958, he succeeded Lewis Strauss as Chairman of the AEC. Like Strauss, McCone strongly opposed Eisenhower's attempts to formulate a test ban treaty. He stood with DoD to argue that LRD technologies, especially seismic, were not adequate to monitor a ban. McCone later served as Director of the CIA (1961 to 1965). *Photo courtesy of the U.S. Central Intelligence Agency*

366

Operation HARDTACK II - Shot Blanca (October 30, 1958)

Blanca was the last and largest of five underground nuclear tests in operation HARDTACK II. Data derived from Blanca seriously disrupted the test ban treaty negotiations at Genva in January 1959. The data, largely negating the Ranier data, indicated that underground nuclear detonations were much more difficult to identify than previously believed. *Photo courtesy of the National Nuclear Science Administration*

Lunik I

The USSR launched *Lunik I* on January 2, 1959. Designed to study the moon, the spacecraft flew past the moon and entered into permanent orbit around the sun, thus becoming the first artificial planet. *Lunik I,* combined with the three successful *Sputnik's,* created great consternation within DoD and increased anxiety over the missile gap. This anxiety, more than anything, explained the hard stance that DoD and the AEC took against almost any agreement with the Soviets on nuclear arms control. DoD and the AEC supported

an atmospheric test ban only because they wanted to continue testing underground to perfect tactical nuclear weapons and nuclear missiles. *Photo courtesy of NASA*

Notes to Chapter 11

[1] Fatzinger, *History of AFOAT-1: 1958*, 61-62. See also Romney, *Reflections*, 170; and Jacobson and Stein, *Diplomats, Scientists, and Politicians*, 136.

[2] C.J.V. Murphy, "Nuclear Inspections: A Near Miss," *Fortune*, Vol. LIX, No. 3 (March 1959), p. 122, as quoted in Jacobson and Stein, *Diplomats, Scientists, and Politicians*, 155.

[3] Letter and report, Northrup to Bethe, December 22, 1958. Bethe papers. Members of the ad hoc panel were: Carl Romney, AFOAT-1, Chairman; Billy G. Brooks, Chief Seismologist of the Geotechnical Corporation; Perry Byerly, Director of the Seismographic Stations, University of California; Dean S. Carder, Chief Seismologist, U.S. Coast and Geodetic Survey; Frank Press, Director of the Seismological Laboratory, California Institute of Technology; Jack Oliver, Professor of Geophysics, Columbia University; and James T. Wilson, Chairman of the Department of Geology, University of Michigan.

[4] Jacobson and Stein, *Diplomats, Scientists, and Politicians*, 150-151. Berkner, an expert on the upper atmosphere and space, was a physicist who invented a device that measured the height and electron density of the ionosphere. In addition to Romney and Bethe, the other members of the panel were: Professor Hugo Benioff, California Institute of Technology; W. Maurice Ewing, Columbia University; John Gerrard, Texas Instruments, Inc.; David T. Griggs, University of California at Los Angeles; Jack H. Hamilton, the Geotechnical Corporation; Julius P. Molnar, Sandia Corporation; Walter H. Munk, Scripps Institute of Oceanography; Jack E. Oliver, Columbia University; Frank Press, California Institute of Technology; Kenneth Street, Jr., Lawrence Radiation Laboratory, University of California; and John W. Tukey, Princeton University.

[5] Wadsworth, *The Price of Peace*, 24. Also Jacobson and Stein, *Diplomats, Scientists, and Politicians*, 136-137.

[6] "Statement by the President's Science Advisory Committee Regarding Detection of Underground Nuclear Tests, January 5, 1959," in U.S. Department of State, Historical Office, Bureau of Public Affairs, Documents on Disarmament, 1945-1959, 2 vols., GPO, 1960, p. 1335.

[7] U.S. Congress, Senate, *Hearings: Disarmament and Foreign Policy,* 86th Congress, 1st Session, GPO, 1959.

[8] Ibid., 11.

[9] Ibid., 20.

[10] Ibid., 24.

[11] Ibid., 29. Also, Jacobson and Stein, *Diplomats,* Scientists, *and Politicians,* 150.

[12] Senate, *Hearings: Disarmament and Foreign Policy,* 86th Congress, 1st Session, February 2, 1959, 166.

[13] Ibid., 175-177, 195.

[14] Ibid., 185. Humphrey agreed. "So far we have not made too many concessions of any appreciable degree, have we?" Bethe replied: "we have not."

[15] See all of the Geneva Conference verbatim records for January and February. Summary also provided in Senate, *Hearings: Disarmament and Foreign Policy,* 86th Congress, 1st Session, January 28, 1959, 35.

[16] *Foreign Relations of the United States, 1958-1960,* National Security Policy, Arms Control and Disarmament, Volume 3, eds. Edward C. Keefer and Donald W. Mabon (Washington: Government Printing Office, 1996), Document 190. https://history.state.gov/historicaldocuments/frus1958-60v03/d190 [accessed February 4, 2019].

[17] Ibid., Document 191. https://history.state.gov/historicaldocuments/frus1958- 60v03/d191 [accessed February 4, 2019].

[18] Ibid., Document 191. https://history. state.gov/historicaldocuments/frus 1958-60v03/d 194 [accessed February 4, 2019].

[19] Ibid.

[20] See verbatim records, especially GEN/DNT/PV.29, p. 38; GEN/DNT/PV.30, p. 21; and GEN/DNT/PV.31, pp. 7, 35.

[21] GEN/DNT/PV.31, p. 29.

[22] Edward Teller and Albert Latter, *Our Nuclear Future: Facts, Dangers, and Opportunities* (New York: Criterion Books, 1958).

[23] Bethe also corresponded with AFOAT-1's J. Allen Crocker in requesting an earlier study on the elasticity of waves generated

from a detonation in an underground cavity. Letter from Crocker to Bethe, June 13, 1958, Bethe Papers.

[24] Letter from Bethe to Teller, January 14, 1959, Bethe Papers. Bethe writes that "I am just as anxious as you to discuss the generation of seismic waves in various materials. Please give this a prominent place in the agenda on Friday." Note that the topic of de-coupling was classified at this time thus the benign wording. See also Jacobson and Stein, *Diplomats, Scientists, and Politicians,* 151-154.

[25] "Tentative Agenda: Panel on Seismic Improvement," March 5-6, 1959, Bethe Papers.

[26] Jacobson and Stein, *Diplomats, Scientists, and Politicians,* 154.

[27] Bethe agreed as well. In his personal notes he wrote: "We had a late start . . . a year or more behind." Notes dated April 15, 1958. Bethe Papers.

[28] Fatzinger, *History of AFOAT-1: 1959,* 25.

[29] *Foreign Relations of the United States, 1958-1960,* https://history.state.gov/historicaldocuments/frus1958-60v03/d196 [accessed February 4, 2019].

[30] Ibid., https://history.state.gov/historicaldocuments/frus1958-60v03/d200 [accessed February 4, 2019].

[31] Ibid., https://history.state.gov/historicaldocuments/frus 1958-60v03/d202 [accessed February 4, 2019].

[32] Ibid., https://history.state.gov/historicaldocuments/frus1958-60v03/d203 [accessed February 4, 2019].

[33] https://history.state.gov/historicaldocuments/frus1958-60v03/d204 [accessed February 4, 2019].

[34] Jacobson and Stein, *Diplomats, Scientists, and Politicians,* 143.

[35] GEN/DNT/PV.72, p. 34. Also ibid.

[36] U.S. Congress, Senate, *Hearings: Disarmament and Foreign Policy,* 86th Congress, 1st Session, GPO, 1959, 24.

[37] Lloyd V. Berkner, "The Need for Fundamental Research in Seismology," Bethe Papers. Note that Bethe significantly contributed to the report, especially in examining the issue of concealment. See especially his appendices to the report entitled "Concealment of Underground Explosions," and "Theory of Seismic Coupling," dated March 24, 1959, 8. See also Fatzinger, *History of AFOAT-1: 1959,* 24.

[38] "Draft of Section E: Discussion of Report on High Altitude Detection," February 20, 1959. Bethe Papers.

[39] There is much scholarship on the role of scientists in policy making during the Cold War. Some scholars believe that scientists "sold out" to government control and influence in return for funding. Others believe that scientists in the service of government risked being ostracized or faced the subjugation of their ethical standards.

[40] Berkner, "The Need for Fundamental Research in Seismology," Bethe Papers, 2. For crustal studies, the Soviets were spending one million dollars per year as opposed to the $10,000 in the U.S.

[41] Ibid., 16.

[42] Memorandum from Northrup to Rodenhauser, "Improvement and Expansion of the A.E.D.S.," Technical Memo #99, June 25, 1959.

[43] Ibid.

[44] Ibid.

[45] Ibid.

[46] Ibid. When seismic energy radiated from earthquakes below the ocean floor enters the water, it can generate an acoustic wave called the T phase.

[47] Senate, *Hearings: Disarmament and Foreign Policy,* 86th Congress, 1st Session, March 25, 1959, 4-6.

[48] Ibid., 13.

[49] "Letter from President Eisenhower to the Soviet Premier (Khrushchev), April 13, 1959," in U.S. Department of State, Historical Office, Bureau of Public Affairs, Documents on Disarmament, 1945-1959, 2 vols., GPO, 1960, 1392-1393.

[50] https://history.state.gov/historicaldocuments/frus1958-60v03/d216 [accessed February 4, 2019].

[51] Jacobson and Stein, *Diplomats, Scientists, and Politicians,* 183-184.

[52] GEN/DNT/PV.91 , p. 5-10.

[53] GEN/DNT/PV.92, p. 4.

[54] GEN/DNT/PV.94, p. 25-30.

[55] GEN/DNT/PV.95, p. 12.

[56] GEN/DNT/PV.109, p. 10.

[57] https://history.state.gov/historicaldocuments/frus1958-60v03/d220 [accessed February 4, 2019].

[58] U.S. Congress, *Hearings before the Joint Committee on Atomic Energy: Developments in the Field of Detection and the Identification of Nuclear Explosions,* 87th Congress, First Session, July 25, 26, and 27, 1961, 7-8. DoD assigned responsibility to ARPA on September 2, 1959.

[59] Jacobson and Stein, *Diplomats, Scientists, and Politicians,* 178.

[60] *Berkner Report,* 13-14.

[61] The AEDS would be de-classified in 1976.

[62] General Orders #39, Headquarters Command, Bolling AFB, July 8, 1959. Also, Fatzinger, *History of AFOAT-1: 1959,* 2.

[63] Memorandum from Rodenhauser to Loper, "Proposed Program for Seismic Improvement," July 2, 1959, AFTAC archives. Widespread coverage of the AEDS techniques and connection to AFOAT-1 in the media prompted Rodenhauser to launch a staff study on potential reorganization in late January 1959. That study began in earnest in March but only gained serious traction after the Berkner and Panofsky reports. "Redesignation Study," 1959, AFTAC archives.

CHAPTER 12

Conclusion

The 1009th SWS/AFOAT-1 served the nation for 10 years, 10 months, and 8 days. During that time, the inventors of long-range detection and the Atomic Energy Detection System created a legacy that is unique in the annals of U.S. Air Force history. Over the course of the decade largely overlapping the two Eisenhower administrations, the pioneers of LRD took a nebulous concept from idea to full fruition in order to enhance the national security of the United States.

When the 1009th cased its colors on July 6, 1959, the AEDS had become a true worldwide system. In total, it included the following sites: 13 seismic (B), one balloon station (Db), 11 ground-based air samplers (Ds), 11 acoustic (I), eight EMP (Q), four labs, and 10 airfields capable of providing support for airborne sampling (A). The personnel assigned to the organization (authorized for 946 people) numbered 256 officers, 614 Airmen, and 76 civilians. Such maturity over the course of a decade was impressive.

Talented people, of course, were the most important factors in the success of the LRD mission. At the time of its birth in August 1948, the new AFOAT-1 consisted of 22 officers, 63 Airmen and 43 civilians.[1] The officers and the enlisted corps personnel were certainly special. Initially, the Air Force sought people with advanced skill sets or very high aptitude scores. The first group of officers began training on laboratory techniques in mid-June 1949, just two months prior to the first Soviet nuclear test. Tracerlab, AFOAT-1's invaluable partner, provided the equipment and the instructors. After Tracerlab helped to establish the AFOAT-1 lab at McClellan AFB in 1950, on-the-job training became routine as nuclear tests occurred more frequently. In those early years, young officers with college degrees in chemistry and physics were involuntarily assigned to AFOAT-1 to work in the lab. Training on the operation and maintenance of the other LRD techniques took place at Keesler AFB, Mississippi; Brockley AFB, Alabama; and Fort Sill, Oklahoma.

As Hegenberger and Northrup sought to establish the Interim Net (i.e., the AEDS) as quickly as possible, the first several teams deployed without much training on the seismic equipment. AFOAT-1 located the first two stations in late 1950 and early 1951 at College, Alaska, and near Seoul, Korea. Those teams became the nucleus of detachments 207 and 405. By the end of 1953, dedicated AEDS seismic sites were in operation in Wyoming, Turkey, Germany, and Greenland.

As the initial detachments were established, however, AFOAT-1 faced a serious challenge in retaining a highly trained and specialized workforce. For example, in 1951, the first-term airmen reenlistment rate was 6.75 per cent compared to an overall USAF average of 23.7 percent. Leaders attributed the low reenlistment rate to AF Regulation 39-14, which permitted early discharges to attend college. This problem mirrored Northrup's primary concern at that time about recruiting trained scientists.

Formal training for AFOAT-1 personnel began in June 1954, when Air Training Command consolidated the LRD courses from the four separate bases, which had established rudimentary training facilities for the initial AEDS techniques. The new training center was located at Lowry AFB, Colorado. The Air Force chose Lowry because of its optimum geography for seismic instruments. At that time, the Air Force activated the 3454th Technical Training Squadron (TTS) to formalize LRD training and designated the school as the Special Instruments Training Branch (SPINSTRA) under the Department of Special Weapons. However, because of management and security issues, the 3415th TTS realigned SPINSTRA as a stand-alone unit in April 1955.

Training during the first two years focused on the airborne, laboratory, and seismic techniques, with the student load ranging from 50 to 75 students. Facilities were limited to two temporary buildings on Lowry AFB and a "B" technique field station on Genesee Mountain, 26 miles west of Denver. Formal training on the electromagnetic instruments began in 1956, and a "Q" technique field station was activated at the Rocky Mountain Arsenal in 1958. AFOAT-1 airmen who graduated from the specialty courses received an Air Force Specialty Code (AFSC) of 99125.[2]

By the time of AFOAT-1's inactivation in July 1959, SPINSTRA had formalized several courses to focus more intensely on the advancing technologies that made up the expanding AEDS. For example, in 1959 the school created Course ABR 99125FL, Special Weapons Maintenance Technician, to train AFTAC airmen in the chemical and nuclear counting specialties. The duration of the course was extended from 22 to 31 weeks to include additional laboratory training and fundamental electronic training, which was provided in other technical courses.

Other indications of the prominence of AFOAT-1 and its importance to the nation was the large allocation of funds to build two new major facilities. The first was the construction of a new laboratory at McClellan AFB in 1957. Also serving as the headquarters for AFOAT-1's Western Field Office, Building 628 housed the new lab, which was capable of fulfilling all mission requirements as the workload dramatically increased throughout 1957 and 1958. The second facility was a new headquarters building in Hybla Valley, Virginia (later Telegraph Road in Alexandria) built in 1958.

For the men and women of AFOAT-1, it had been a long, stressful journey. As they learned new LRD technologies, they operated the AEDS during a time when significant fears over nuclear fallout and the prospects of a possible nuclear war dominated the concerns of American society. As the nation's primary "eyes" on Soviet nuclear activities during most of this time, they essentially operated on the front lines of the Cold War. This fact became apparent as AFOAT-1 leaders wrestled with Strategic Air Command's war planners to incorporate the AEDS into SAC's war plan. General Hooks had made this point clear when he told SAC by way of the AFOAT-1 war plan (WP 1-55) in mid-1955, that the AEDS would go to war "as is;" that, in many ways, it was already at war (see Annex 2).

Beginning with Operation CROSSROADS in 1946, AFOAT-1 found itself being dragged, slowly at first, into the center of the Cold War ideological confrontation between communism and capitalism. The international climate of mistrust between the two superpowers compelled U.S. senior leaders to lean heavily on AFOAT-1 to advance LRD technologies as quickly as possible. That was not the case initially. Prior to the Soviets' initial nuclear test, the Joe-1 event of August 1949, President Truman's austerity measures threatened to gut the fledgling LRD organization of its vital R&D funding. While the shock of Joe-1 quickly reversed that attitude, AFOAT-1's technical directors Ellis Johnson and Doyle Northrup struggled to acquire financial support for geophysical research, the technique that would prove the most vital (albeit inadequate) by the end of the 1950s. Indeed, it led to Johnson's resignation even before Joe-1 and Northrup's decade-long frustration over a lack of funding.

The awkward birth of AFOAT-1 also occurred during an intense period of serious inter-service rivalries made even more severe as the U.S. demobilized the large Second World War military force that consisted of 11 million servicemen and women in 1945. In the immediate years following the war, the American military underwent the most significant reformation in its history. It took two acts—the National Security Act of 1947 and its amendment two years later—to provide a stabilized environment in which the new LRD organization could grow and develop. If not for the persistence of Generals Hoyt Vandenberg and Curtis LeMay, the 1009th SWS/AFOAT-1 may never have come into existence, at least not as an Air Force unit.

Today, we remember the 1950s as a time when Americans viewed technology as a panacea for almost any problem or challenge. This belief was both a blessing and a curse for AFOAT-1. On one hand, the young LRD organization was given the freedom to experiment and innovate, albeit with some limitations. The AEC allowed LRD experiments in the nuclear tests, but only as an adjunct to the primary purpose of those operations, namely weapons design. Without the experiments in Nevada and in the Pacific, LRD would not have been possible. Fortunately, after Joe-1, Northrup and his commanders began to receive most of the funding they needed. On the other hand, more and more people soon assumed that the emerging LRD technologies were much more capable than they really were. Over time, these beliefs pulled AFOAT-1 into contentious debates and, eventually, onto the national stage. AFOAT-1 captured so much attention not as an identifiable organization (it remained classified), but as a *capability*. During the latter half of the decade, many people—politicians, scientists, the media, national and foreign diplomats—professed knowledge about the capabilities of the LRD techniques. They utilized that alleged knowledge to promote their causes, such as Adlai Stevenson's bid for the White House in 1956, or in arguing for or against a test ban treaty.

Four strong forces brought LRD and the AEDS into the national spotlight. The first were the members of the scientific community, of whom most had become closely acquainted during the wartime Manhattan Project. These nuclear physicists, chemists, mathematicians, and geophysicists were well-known to the public. Some had become celebrities, such as Oppenheimer. Northrup and his scientists consulted with many of these scientists as the LRD experiments progressed. In the case of Hans Bethe, AFOAT-1 contracted with the prominent theoretical physicist to lead the evaluation of all of the Soviet tests throughout most of the decade (the Bethe Panel). In the process, Bethe had become extremely knowledgeable about the capabilities and limitations of the AEDS.

By the time of the Geneva Conference, Bethe had used that knowledge for several years to support his advocacy for a test ban treaty. His opinions appeared frequently in newspapers, magazines, and interviews. As his personal papers reveal, his strong advocacy for a treaty sometimes put him at odds with the AEC. AEC officials often redacted or restricted his writings, citing the classification of the information as restricted data (RD). In retrospect, he was always careful in protecting information. Clearly, though, the AEC did not welcome his views, especially as his commentaries gained momentum in the press alongside the views of his pro–test ban colleagues such as Conant, Oppenheimer, Rabi, and Bacher.

Why did Bethe, who was extremely knowledgeable about the limitations of the AEDS, push so hard in 1958 and 1959 for a test ban and an international control system? Clearly, he was a man of conscience. As one of his biographers noted, "how to act morally has been one of Bethe's constant concerns. The boundary between the moral and the political have never been sharp for him."[3] Indeed, he demonstrated that fact as he remained close to Oppenheimer from the beginning of the decade, when the famous scientist decried the development of fusion weapons based on moral grounds, and later when Oppenheimer fell from grace. For Bethe, the path was always clear. It was better for humanity to have a flawed international control system than to live in a world without a treaty that could more easily witness a nuclear war.[4]

When the debates about whether the LRD techniques were adequate or not came to the forefront in 1956 and remained a hot topic beyond the inactivation of AFOAT-1 in 1959, Northrup and Romney were careful in distancing themselves from the politicization of the AEDS capabilities. While there is no evidence the Air Force ever "leaned" on Northrup to support the service's strong opposition to a treaty, it is clear that Northrup was always driven by an obligation to protect the integrity of the data. As one of his former administrative assistants noted, "Dr. Northrup was always a serious man. He constantly emphasized the paramount necessity to *always* be absolutely correct in its assessments, even when the evidence contradicted a preferred or popular position."[5] By all appearances, Northrup succeeded in letting the facts speak for themselves, as his use of the new seismic data from HARDTACK II demonstrated. By 1958, he understood the LRD techniques held enormous potential in monitoring the safeguards of an international test ban treaty, but only if more R&D funding for geophysical research was available. By the end of the decade, AFOAT-1 and the capabilities of the AEDS were close to doing so but not quite close enough.

The second force that elevated the visibility of LRD and the AEDS was the opposition from the AEC and DoD. Their hard stance against Eisenhower's persistent attempts to formulate a treaty with the Soviet Union almost always contained statements that the LRD techniques were inadequate. The AEC and the DoD were consistently close partners on this issue throughout the decade. For the AEC, it was the dominance of Chairman Lewis Strauss that fueled the opposition. Until November 1957, when Eisenhower created PSAC and named James Killian, Jr., as chairman and the first Special Assistant to the President for Science and Technology, Strauss was practically Eisenhower's only conduit to the nuclear science community. His strong alliance with Teller and Lawrence created an anti-treaty barrier that was not lowered until Killian, Bethe, and the majority of the PSAC had gained access to the president. For DoD, predominantly the Air Force, it was the fear of Soviet technological advantages that urged their opposition, first over the perceived "bomber gap" and then the "missile gap." By the time of the Geneva Conference, the Air Force supported an atmospheric test ban only because they and the AEC no longer had a requirement to conduct tests in the atmosphere. They still wanted many more tests, especially in light of the missile gap, but they could conduct those tests permanently underground.

The third powerful force that placed the capabilities of the AEDS in the national spotlight was the media. The media's influence began with growing concerns over nuclear fallout generated by the 15MT Shot Bravo in Operation CASTLE on February 28, 1954. Anxieties about the dangers of radioactivity became more vocal five weeks later, when the government released photographs of the 1952 Shot Mike in Operation IVY that created a 10MT detonation.[6] TV stations, major magazines and newspapers, in using the images of the large mushroom cloud, spread fears to a much larger audience in illustrating the enormous power of thermonuclear weapons. Those fears then fueled the science fiction community, as books and movies fictionalized the horrific effects of radiation. Politicians then adopted publicized stands on the issue while running for various offices at all levels of government. By late 1957, as the fallout issue compelled Eisenhower to seek an arms control treaty more actively and became more inclined to consider an atmospheric ban, Strauss, Teller, and Lawrence convinced the president that they could develop "clean" hydrogen weapons. That news (not entirely factual) quickly spread around the world and hardened all of the factions that either supported or opposed a test ban treaty between the two superpowers. After the world learned of the new HARDTACK II data and the ensuing stalemate at the Geneva Conference, the public debates over the capabilities of the AEDS received constant media coverage—so much so, that AFOAT-1 found it difficult to safeguard its mission and capabilities.

The fourth force that elevated AFOAT-1's visibility were the Soviets. Beginning with the agreements coming out of the Conference of Experts in the summer of 1958, they consistently insisted the LRD techniques were more than adequate in monitoring compliance with a test ban. They frequently stated so in public and achieved propaganda victories in the press, where they appeared to favor a test ban more than the Western powers. During the Geneva Conference the Soviets kept close tabs on the publicized debates occurring in the United States, such as Senator Gore's outspoken views, and highlighted them during the formal sessions of the conference. Soviet negotiators leveraged this tactic well into the new decade, as the Geneva Conference extended into the Kennedy Administration.

The elevation of AFOAT-1's visibility, as the LRD capabilities were battered about by so many, had one very positive effect. The most significant historical milestone AFOAT-1 created was its central role in underscoring the need for science and technology to inform political decision-making. Prior to Eisenhower's second term in office, neither he nor Truman really relied on the input or advice from scientists in their decision-making processes. However, as we have seen, technology and politics became tightly intertwined ever more frequently as the decade progressed. When Eisenhower finally turned away from Strauss to seek technological advice from the scientific community via Killian and the PSAC, he jumped to the other extreme by totally excluding advice from those scientists who remained opposed to a test ban. Because of the rapid advancement in various weapons technologies, he quickly adopted the "mistaken assumption that America could formulate a technological solution to the arms race."[7] In doing so, however, he built a great barrier between politics and technology. In retrospect, the Soviets had always been correct—the two were inseparable. Eisenhower was, perhaps, the only senior official to be blind to this truth. The technology versus politics conundrum was always at the center of discussions related to arms control. Nothing illustrated this fact more than the topic of onsite inspections in an international control system. The president failed to achieve his most important goal, in part, because he awoke to this fact too late. To be sure, technology would be key to any international control system charged with the responsibility to monitor compliance of a nuclear arms control treaty. However, as the subsequent decades would demonstrate, international treaty negotiations require much more than just a consensus on technological capabilities.

AFOAT-1 left a tremendous legacy for the Air Force. The creation of the AEDS and its criticality to national security cannot be understated. None of that would have been possible without exceptional talent. That talent ushered in an era of remarkable R&D and created a culture of innovation that has continued to present day. In the process, the Air Force scientists and technical experts established enormous credibility with the scientific community at large. By the late 1950s, that level of credibility extended throughout the legislative and executive branches of government. AFOAT-1's *bona fides* as LRD experts went unquestioned by diplomats and senior politicians across the political spectrum. It was the people who made that possible. Northrup and Romney would continue to influence nuclear diplomacy and the formulation of American foreign policy during the Cold War for years to come. They were not alone. Their colleagues, such as Singlevich, Urry, Crocker, Olmstead, and many others continued to improve the AEDS. The Air Force recruited and selected the most talented airmen, and because of their dedicated service and vigilance, the nation was ever more secure.

The first AEDS stations

Detachments 405 in Korea and 207 in Alaska (early 1950s)

Photo source: AFTAC archives

McClellan Central Laboratory (MCL)
Building 628 (constructed in 1957)

Photo source: AFTAC archives

1009th SWS/AFOAT-1 Headquarters
(constructed in 1958)

Photo source: AFTAC archives

Doyle Northrup and Carl Romney

Of all the talented people who worked at AFOAT-1 or AFTAC over the course of seven decades, Northrup and Romney had the greatest influence in advancing long range detection. Both the executive and legislative branches of government held the two scientists in high regard and relied on their expertise to help inform foreign policy decisions. Northrup received the USAF Exceptional Civilian Service Award in 1950 and 1954, and the Distinguished Civilian Service Award by the Secretary of Defense in 1958. President Eisenhower awarded him the President's Award for Distinguished Federal Civilian Service in 1960. Romney received the USAF Exceptional Civilian Service Award in 1960 and the President's Award for Distinguished Federal Civilian Service in 1967. *Photo source: AFTAC archives*

Notes for Chapter 12

[1] The officers: 1 major general, 1 brigadier general, 2 colonels, 3 lieutenant colonels, 5 majors, 7 captains, 4 first lieutenants, and 1 chief warrant officer.

[2] *The Monitor,* AFTAC newsletter, October 1988, 7.

[3] Silvan S. Schwerber, *In the Shadow of the Bomb: Bethe, Oppenheimer, and the Moral Responsibility of the Scientist* (Princeton, NJ: Princeton University Press, 2000), 149.

[4] Bethe supported Oppenheimer through the congressional hearings that eventually led to Oppenheimer's ostracism and revocation of his security clearance. Bethe regularly corresponded with Oppenheimer until Oppenheimer's death in 1967.

[5] Interview with Ms. Judy Milam Henderson, January 2019.

[6] With less concern about world opinion, the Soviets would continue to spew fallout around the world until their declared moratorium in 1958.

[7] Suri, "America's Search for a Technological Solution to the Arms Race," 432.

ANNEX 1

Allied Cooperation in the AEDS

On September 19, 1949, when General Vandenberg gathered the leading nuclear scientists from the United States and the United Kingdom together at AFOAT-1 headquarters to validate the analysis of Joe-1, it became immediately clear that British concurrence was essential in reaching an unquestionable conclusion (see Chapter 4). That meeting also served as a reunion of sorts of the allied alumni from the wartime Manhattan Project. Indeed, the majority of the 29 attendees had worked closely together to develop the first nuclear weapon, and many had become close friends. Yet, as the meeting transcript revealed, such close collaboration now had significant limitations and restrictions. In fact, early in the meeting, Chairman Vannevar Bush had reminded his American colleagues to adhere to "Area 5" parameters in their discussions.

Area 5 was a category of information control, which was formulated in law under the Atomic Energy Act of 1946. That act was the culmination of several legislative actions recognizing the need to control nuclear energy after the war. The primary bill leading to the act was sponsored by Democratic senator Brien McMahon from Connecticut. Signed into law on August 1, 1946 by President Truman, the act went into effect on January 1, 1947. The primary purpose of the law was to determine who would control nuclear energy—the military, as an extension of the Manhattan Project, or a new civilian agency. In choosing the latter, the act is best known for its creation of the Atomic Energy Commission (AEC).

Among several other important provisions, the new law appeared to stifle scientific research by creating a new information classification level called "Restricted Data" (still in effect today). As one journalist noted in 2005, it was "a permanent gag order affecting all public discussion of an entire subject matter. There is nothing like it anywhere else in American law." This was a radical new approach to information control because it stipulated that certain types of information are "born secret." That is, the nature of the information automatically makes it classified (as opposed to the previous practice, in which information was deemed classified after a review). In essence, almost any type of information pertaining to the science of nuclear weapons fell under this new classification category of "Restricted Data" (RD). The new AEC and its five commissioners exercised legal authorities over RD. As such, the AEC acquired supervisory oversight of all atomic energy research and development. This explains why the Special Weapons Group (General Kepner) and then AFOAT-1 (General Hegenberger) initially faced so many obstacles in funding LRD R&D between November 1947 and September 1949 (see Chapter 1). At a time when intense inter-service rivalries were interlocked over severe budget battles, the AEC formalized its new authority over DoD via the Military Liaison Committee (MLC), a creation of the act as well. The real controls over AFOAT-1's R&D occurred in the AEC's General Advisory Committee (GAC), which was stipulated in Section 2 of the act. The act also allowed for congressional oversight via the Joint Committee on Atomic Energy (JCAE). All of these oversight bodies placed AFOAT-1 under a microscope during the entire decade of the 1950s (see graphic entitled "AFOAT-1/AEDS Data/Information Flow" in the Introduction).[1]

As many historians have noted, the Atomic Energy Act of 1946 greatly restricted the sharing of atomic energy information with U.S. allies, specifically Canada and the UK. As one British historian noted, "the act severed all channels through which technical information could cross the Atlantic."[2] Literally overnight, the collaborative relationships formed during the war came to a halt. America's two allies felt betrayed; they viewed the new law as a violation of the Quebec Agreement, signed by Prime Minister Winston

Churchill and President Franklin Roosevelt on 19 August 1943 during the Second World War. That agreement created the joint Combined Policy Committee that outlined the terms for the coordinated development of the science and engineering projects, which in turn led to the accomplishments achieved in the Manhattan Project.

This agreement was followed up a year later, in September 1944, with the Hyde Park Aide-Memoire. Section two of that agreement stated: "Full collaboration between the United States and the British Government in developing Tube Alloys [atomic energy] for military and commercial purposes should continue after the defeat of Japan unless and until terminated by joint agreement."[2] The Atomic Energy Act of 1946 clearly reneged on those agreements.[3]

The U.S. did not terminate those relationships simply out of hand. What initially prompted the extreme restrictions during the formulation of the McMahon Act was the February 1946 arrest of 22 people in Canada suspected of passing atomic secrets to the Soviets. Involved in various aspects of the Manhattan Project, their names were revealed by a defecting Soviet embassy cypher clerk by the name of Igor Gouzenko. The discovery of this spy ring coincided with several FBI investigations looking into suspected espionage activities in the United States.[4]

Fortunately, the Atomic Energy Act had not disestablished the Combined Policy Committee. In the fall of 1947, the American members of the committee (Marshall, Forrestal, Bush, and Lilienthal) agreed to reopen negotiations with the United Kingdom and Canada to discuss a wide range of atomic energy topics while not violating legal restrictions. A subcommittee identified nine potential areas of discussion. The fifth one—"Area 5"—directly pertained to the new LRD mission. It was defined as "the detection of a distant nuclear explosion including: meteorological and geophysical data; instruments (e.g., seismographs, microbarographs); air sampling techniques and analysis; [and] new methods of possible detection." While these specified caveats did little to aid real, substantial information sharing with the Canadians and British, they greatly benefitted AFOAT-1 in the establishment of the AEDS.[5] As the Special Weapons Group (soon AFOAT-1) began to create the AEDS in the spring of 1948, Johnson and Hegenberger immediately recognized that the UK and Canada would have control or influence over geographical areas of the world that would tremendously facilitate the establishment of AEDS stations much closer to the USSR than the U.S. could ever obtain. On May 13, 1948, Hegenberger wrote to the Senior Air Member, Joint Board on Defense Canada-United States, requesting a meeting with leading nuclear physicists from the three nations. The intent was to bring the best minds together in order to design an "interim network of long range detection stations."[6]

On July 2, 1948, mindful of his new responsibilities under AEC oversight, Hegenberger informed the AEC chairman (Lilienthal) that the initial challenges to establishing the AEDS could be rapidly overcome by including the two allies in its development. He was careful to define the objectives. His two key points were that British and Canadian scientists could possibly provide unanticipated solutions to LRD, and that the U.S. would gain access to critical geography. In regard to the latter, he was frank that the U.S., in return, would need to give the partners the results of the radiochemical analysis of Project FITZWILLIAM, the recent LRD experiments in the Operation SANDSTONE nuclear tests (see Chapter 2). He asked Lilienthal to provide security and legal guidance on the proposed agenda. He also noted that time was an issue because Sir William Penney and several of the British scientists would be in Washington on September 2nd and 3rd.[7]

On August 24, 1948, Major General David M. Schlatter, Deputy Chief of Staff, Operations for Atomic Energy (AFOAT-1's higher headquarters), presented Hegenberger's request to the MLC. After much discussion over the advantages and disadvantages, the MLC, chaired by Donald Carpenter, agreed to recommend approval to the AEC as long as AFOAT-1 closely adhered to Area 5 restrictions. However, the AEC commissioners were not scheduled to meet until September 2nd, the same day as the proposed joint conference. With some risk, Carpenter, assuming Sumner Pike (one of the five AEC commissioners) would vote in favor, notified Schlatter on August 30th that AFOAT-1 was cleared to conduct the conference within the defined restrictions.[8]

Representatives from the three allied countries met at AFOAT-1 headquarters on September 2 and 3, 1948. In addition to the senior military and civilian staff from AFOAT-1, American attendees included Navy chemist Dr. Edward S. Gilfillan, Jr., who had served as the Technical Director of the Operations CROSSROADS Committee of DoD, and Mr. Roland J. Beers, co-founder of the Geotechnical Corporation, which provided seismic instruments to AFOAT-1. Three members of the AEC were also present: Dr. Walter F. Colby, Dr. Paul G. Fine, and Dr. Spofford English. English had been a group leader in the Manhattan Project and was now chief of the chemistry branch in the AEC's Research Division. Colby's presence was key. A former professor of physics at the University of Michigan, Colby had developed a personal relationship with Eric Welsh when they conducted the ALSOS mission in Europe at the end of the war. As discussed in Chapter 4, Commander Eric Welsh headed the British Atomic Energy Intelligence Unit (AEIU) within the Ministry of Supply. Because the AEIU held stewardship over British LRD, Welsh would soon become the key figure in fostering a close working relationship with AFOAT-1 going forward.[9]

In addition to Welsh, the UK contingent included Penney and W. Gregory Marley, Penney's radiological expert and a former section leader at Los Alamos during the Manhattan Project. In the 1950s, Marley became well known for establishing radiation safety standards.[10] The two Canadians present were Dr. W. Bennett Lewis and Dr. Edward G. Bullard. Lewis was director of the Chalk River Laboratory, a position he had assumed in the wake of the Canadian espionage revelations at that facility in February 1946 (and which had prompted the more severe restrictions in the Atomic Energy Act of 1946).[11] Bullard was a professor of geophysics at the University of Toronto.

As Hegenberger was absent with an illness, Schlatter presided over the meeting. Succinctly, he stated the agenda was limited to discussions on AFOAT-1's LRD R&D program and speculations about the potentiality of a Soviet nuclear test.[12] With Technical Director Ellis Johnson's resignation only days earlier and his successor's (Dr. George Shortley) imminent departure, Northrup facilitated the technical discussions.[13]

Considering it was the first meeting of these former allied scientists since the war, and despite the tensions caused by the American restrictions on information sharing, the two-day conference resulted in a rich information flow between the participants. In general, the Americans understood they were giving much more than they were receiving in this initial stage of the relationship. This was the price to be paid for later access to other geographical areas of the world needed to expand the AEDS. In short, Northrup provided his colleagues with detailed data from operations CROSSROADS and SANDSTONE. This data covered six areas of scientific work: AFOAT-1's experiments in the previous two test series, the current status of seismic and acoustic studies, magnetic and radio methods of study (ionospheric phenomena), studies of the properties of fission clouds, and the failed experiment to detect light reflection from the dark face of the moon. Northrup held nothing back. He presented, in detail, the full range of capabilities, successes, and failures of the initial AEDS techniques under development.[14]

In the conversations about airborne sampling, Northrup learned for the first time that the British program of aerial debris collection had detected the SANDSTONE shots, and that the radiochemical analysis, which the British readily shared, lined up very closely with the Tracerlab results. This was an important comparison because it showed that AFOAT-1's "A" technique was effective, especially since the British were never informed of the SANDSTONE testing dates and times. In discussing future programs for seismic and acoustic detection, the Canadians suggested they might be able to contribute some station locations, but those stations must remain under Canadian control. Northrup welcomed the offer and expressed special interest in seismic coverage within the Arctic region.[15]

Near the end of the conference, all parties presented their information "wish lists." AFOAT-1 requested help from Penney on mathematical equations for calculating cloud heights. Northrup specifically asked for Marley's final report on seismic activity measurements Marley had compiled off the European coast. Northrup also requested a copy of Bullard's estimate of the fraction of the energy of various explosions converted into seismic waves. British geophysical research was of paramount interest to Northrup. At the time, he was fighting to prevent the loss of geophysical R&D funding; a fight he had inherited from Johnson and the issue over which Johnson had just resigned. Northrup also asked Penney for any spare munitions AFOAT-1 could procure for R&D testing. Within American channels, Johnson and Northrup had been unable to obtain any explosives for seismic research (see Chapter 2).[16]

British and Canadian requests were confined to geophysical and meteorological studies. They wanted all available data on seismic and acoustic research programs, the seismic and acoustic reports from CROSSROADS, and any reports on the frequency of Soviet earthquakes. Penney also requested all air movement diagrams from the three SANDSTONE shots. In retrospect, their modest requests reflected their acknowledgement of Area 5 sensitivities. Clearly, they wanted the relationship to grow.

In anticipating more meetings, the British and Canadian representatives suggested future topics of discussion. Recommendations included additional meteorological data affecting the dissemination of radioactive clouds, improved methods of air filter sampling, more speculative discussions about when the Soviets would test, and the scientific factors which might enable clandestine testing.[17] Unfortunately, immediately following the successful conference, AFOAT-1 began fighting a year-long battle for the future of LRD (the focus of Chapter 3). Consequently, the allies would not meet again until September 19, 1949, when they validated the analysis of Joe-1 (see Chapter 4).

Following Joe-1, the group reconvened on November 28 and December 2, 1949, to discuss the various results and reports on the Joe-1 detonation.[18] After completing the examination of Joe-1, the British suggested a new conference for early 1950. However, AFOAT-1 believed that, within the parameters of Area 5, there was little more to be gained from additional inter-Allied meetings. In addition, Northrup and his team were consumed with the planning for Operation GREENHOUSE, only 16 months away, and with meeting the high expectations to field an AEDS by mid-1951 (see Chapter 5).

Most likely, General Nelson suspended any further meetings due to the investigation into the Klaus Fuchs espionage case. On September 22, 1949, three days after the Joe-1 validation meeting, the FBI officially opened a case file on Fuchs, who had been linked to several other spy cases for some time. Fuchs was a theoretical physicist who was well-known to the scientists at AFOAT-1. He had participated in the Manhattan Project as a member of the British contingent, where he produced significant theoretical calculations for the first bombs, and worked closely with Teller's early research on the hydrogen bomb. By September 1949, he headed the Theoretical Physics Division of the secret British nuclear weapons program. Immediately following the passage of the

1946 Atomic Energy Act into law (that August), which the British viewed as "a declaration of nuclear rivalry," the UK had embarked on its own nuclear weapons program under the direction of Penney.[19] While the U.S. was aware as early as May 1947 that the British were interested in developing nuclear power, it was not until May 1948 that the AEC learned reactors at Harwell were producing plutonium. Ever the Anglophobe, Strauss reacted harshly to the news and called for additional restrictions on Allied collaboration. By the time of the FBI investigation, Fuchs had already participated in joint American-Canadian-British talks about declassifying nuclear information the British had produced at Los Alamos during the war. That work, classified as RD and confiscated from the British as they departed Los Alamos, further frosted the U.S.-UK relationship.[20]

On February 2, 1950, British authorities placed Fuchs under arrest on charges of espionage. On March 1st, the court found him guilty on four counts of violating the Official Secrets Act and sentenced him to 14 years in prison. From his confession, it was clear Fuchs had passed significant information on fusion weapons research to the Soviet Union.[21] Strauss was livid. Within days of Fuchs's arrest, the AEC instructed AFOAT-1 that "no new areas of collaboration were to be opened."[22] The AEC and Los Alamos scientists immediately conducted a damage assessment, which they completed and passed to the FBI in May. Mr. Robert F. LeBaron, assistant to the Secretary of Defense for Atomic Energy (and who also served as the chairman of the MLC), also commissioned an assessment. The investigation and subsequent report, conducted by Major General Kenneth Nichols and Brigadier General Herbert Loper, concluded that "if [the Soviets] had accepted anything [Fuchs has provided] and taken action they could very well be ahead of us in the development of the hydrogen bomb."[23]

All of these events, occurring within a relatively short period of time, paralyzed any significant relaxation of Area 5 restrictions. Both sides regretted the situation. Gordon

Arneson, Undersecretary of State for Atomic Energy, remarked "we were getting very close to really going into bed with the British, with a new agreement. Then the Fuchs affair hit the fan and that was the end of it."[24] A fully, cooperative and unrestrictive relationship would not occur until 1958, when a new Atomic Energy Act was enacted.

In the fall of 1950, Nelson and Northrup appeared to have a change of heart in dealing with the Allies. In October, they met with members of the British embassy staff in Washington to explore the possibility of resuming the scientific conferences. The allies, pleasantly surprised, were quick to agree; and on November 17, 1950, Area 5 discussions resumed. In addition to the previous attendees, representatives from the State Department and the MLC were also present. While the Fuchs affair was still fresh in everyone's mind, Northrup grounded the participants in discussions around LRD R&D. As the meeting concluded, all agreed to share more information on airborne and ground-based sampling, meteorological studies of nuclear clouds, physical studies of particle debris, and the radiochemical analysis of samples. State Department representatives were there because AFOAT-1 had experienced only limited success in getting access to other geographical regions for AEDS stations. The MLC was present as a function of their responsibilities to coordinate with the AEC on all policy, programming, and funding of any military organization involved with any aspect of atomic energy.[25] The AEC was there to ensure compliance with Area 5 restrictions.

The November 1950 conference propelled inter-allied cooperation forward. The British hosted subsequent meetings with AFOAT-1 in April, May, and June 1951, in order to formalize a new agreement. On July 2, 1951, a briefing about the three meetings was presented to members of the MLC in LeBaron's office. The briefing summarized the recent history of AFOAT-1, and the previous relationship between AFOAT-1 and the allies as constrained under the Area 5 restrictions. In essence, the MLC members were informed the AEDS was rapidly expanding and the current date (July 1951) had been the target date for the system to be "fully" operational. While that date had now arrived, the current expansion could not achieve full functionality without the assistance of the British (and the Canadians to some extent). LeBaron was told current plans called for the AEDS to soon consist of 177 teams located in 82 worldwide locations. In sum, AFOAT-1 was in a position to offer new reports and data from the recent RANGER and GREENHOUSE operations (discussed in Chapter 5) to the British in exchange for access to important geographic regions. In addition, AFOAT-1 would receive validation of its analyses from the other nations' scientists to improve ongoing R&D programs. AFOAT-1 offered assurances to LeBaron that the detailed information exchanges would remain within the confines of Area 5.[26]

On August 3, 1951, LeBaron signed the bi-lateral agreement. It took an additional eight months to do the same with Canada. The joint U.S.-Canadian agreement was signed on April 15, 1952. In June 1952, the president approved the minor modifications to the Atomic Energy Act permitting the new scope of information sharing. That modification, however, did little to impact Area 5, an irritant that continued to irk the British. Still, the agreements facilitated the rapid expansion of the AEDS. AFOAT-1's basic techniques—airborne and ground-based debris sampling, geophysical, and laboratory—were all shared to expand coverage of the Soviet Union. AFOAT-1, being so far ahead in developing detection instrumentation, offered to provide the Allies with LRD equipment. Consequently, a remarkable degree of standardization occurred as the Canadians and, especially, the British developed their own LRD programs with American instruments.[27]

Throughout 1952 and 1953, as the three partners continued to share LRD information, the AEC kept a close watch on the relationship primarily through the GAC, which exercised oversight and approval authority. On numerous occasions, the AEC reviewers prohibited the exchange of data by classifying the information as RD. Their rationale was that because the LRD experiments were conducted against U.S. nuclear weapons tests, subsequent analyses and shared data might inadvertently divulge American weapons design information. The AEC also objected to any discussions speculating on Soviet weapons primarily for the same reason—that any comparisons could reveal American designs. The latter disapprovals were especially frustrating to Northrup and his international colleagues because, up until August 1953, the Allies were gravely concerned about when the Soviets would conduct a thermonuclear test (see Chapter 7). Throughout 1954, the AEC appeared even more restrictive. For example, the AEC prohibited AFOAT-1 from sharing any acoustic or seismic data with the British, simply stating the data was RD. This made absolutely no sense to Northrup. The British had already agreed to man and operate two U.S. equipped acoustic stations in the AEDS, yet now AFOAT-1 could not provide typical acoustic data to assist the British in evaluating their own records.

On September 10, 1954, Northrup documented his concerns in a memo to General Hooks, in which he outlined how the extreme restrictions were impeding progress. AFOAT-1 was losing out because the British had much to share. More importantly, he stated, their closer access to the Soviet test site now made it possible for the detection of Soviet detonations well under yields of 10kt. Northrup also pointed out their extensive work in the field of low-frequency radio propagation, which had been "more extensive than that in the United States." Northrup asserted there was also no reason to deny lab data and reports on Soviet tests. While he acknowledged AEC's cautions about sharing radiochemical analyses of U.S. tests, Northrup argued the U.S. would not really be divulging any critical information. Such data and reports would only reveal the yield, date, and place of the tests as well as some basic diagnostic information from the debris samples.[28]

Just days earlier, President Eisenhower signed the Atomic Energy Act of 1954 into law. The act, an amendment to its predecessor, essentially codified how nuclear materials were handled. It was a direct result of Eisenhower's previous "Atoms for Peace" proposal (see Chapter 7). In essence, it loosened some of the RD restrictions to enable the future use of atomic energy for peaceful purposes. It did little, however, to eliminate the strong barriers of Area 5, a frustration that continued to fester with the British and the Canadians.

As discussed in Chapter 4, AFOAT-1's counterpart in the UK was Eric Welsh's Atomic Energy Intelligence Unit (AEIU) within the Ministry of Supply. While both AFOAT-1 and the AEIU conducted LRD, the units were dissimilar. In the United States, "atomic intelligence" was performed by specific intelligence organizations located in the CIA and the AEC. Kept separate, AFOAT-1 was an Air Force technical surveillance unit. In contrast, during 1953 and 1954, Welsh came under great pressure to merge his organization into the Directorate of Scientific Intelligence (DSI). If this occurred, all atomic and non-nuclear information and intelligence would be consolidated. Those pressures increased in 1954, when the AEIU was moved out of the Ministry of Supply and into the Ministry of Defense (MoD). At this point, the Americans interfered in what was essentially an internal dispute in the British Government and strongly insisted the AEIU remain independent. AFOAT-1 was adamant that such a reorganization would impede cooperation by eliminating important boundaries; an important separation given the strong differentiations between U.S. intelligence organizations and AFOAT-1's LRD surveillance mission. In retrospect, the importance of Welsh's relationship with Northrup cannot be understated. Northrup's insistence on the separation helped enable Welsh to retain his full authority over the AEIU. In the end, the British recognized the "indispensable component of the American [LRD] process."[29]

The British had long given up on relying on America's nuclear arsenal as a dependable deterrent in their national security, even within the collective agreement of NATO. While the Atomic Energy Act of 1946 had compelled the British to go their own way, their desires for nuclear weapons were strongly rooted in a sense of vulnerability to a Soviet surprise attack. Geography dictated this vulnerability. Great Britain was already within range of Soviet bombers whereas the United States still enjoyed the benefit of distance. National pride and a real concern for national security led the British to develop their own nuclear weapons—with no assistance from the Americans.

By late 1954, as U.S.-UK LRD talks were making progress—albeit within the parameters of Area 5—the UK had already conducted three nuclear tests. The first, codenamed Operation HURRICANE, occurred on October 3, 1952, in the Monte Bello Islands off the western coast of Australia. The British government deliberately kept the test of their plutonium implosion fission device secret from the Americans. Like Operations CROSSROADS and SANDSTONE in the American nuclear program, HURRICANE was a first step in weapons development. As with AFOAT-1, it also afforded an opportunity for Welsh to begin testing his expanding LRD program.

The British counterpart to AFOAT-1's McClellan Central Laboratory (and TracerLab) was the Atomic Weapons Research Establishment (AWRE) located at Aldermaston, 50 miles west of London. The leading radiochemist in the British LRD program was Dr. Frank Morgan, the counterpart of AFOAT-1's Walt Singlevich. Unlike AFOAT-1's participation in U.S. tests, Morgan's debris sampling experiments for HURRICANE were given the highest priority. As an angry reaction to the American restrictions, Morgan ensured the Americans (or the Soviets) could not collect viable debris to obtain accurate radiochemical analysis. He did so by placing radioactive samples procured from the chemical separation facility in Risley, England, on the test site in order to "make the radiochemistry impossible for everybody else except ourselves." This decision, however, gave him pause. AFOAT-1 had previously notified the British of the U.S. tests so they could conduct debris sampling. Somewhat vindictive, the implemented decision was a result of damaged pride.[30]

The British returned to Australia in October 1953 to conduct two more tests. This time they utilized a new test site at Woomera (Emu Field) in southern Australia. Codenamed Operation TOTEM, the two detonations produced yields of 10 and 8 kilotons respectively. With relations beginning to warm between the two allies, the British now notified the U.S. of their test dates and invited AFOAT-1 to collect aerial debris. Northrup welcomed the invitation, knowing that the British would be testing a new EMP technique for the first time. Although aware that the Atomic Energy Act prevented AFOAT-1 from sharing recent EMP data from Operation UPSHOT-KNOTHOLE (discussed in Chapter 6), the British still graciously welcomed the presence of AFOAT-1 observers.[31]

To get around some of the restrictions, AFOAT-1 "piggy-backed" on a new agreement signed on June 15, 1955, between the two allies, entitled the "Agreement between the Government of the United States of America and the Government of the United Kingdom of Great Britain and Northern Ireland for Cooperation Regarding Atomic Information for Mutual Defense Purposes." The agreement provided for a NATO-type exchange of atomic weapons information for defensive purposes. Northrup was able to add several paragraphs to the document that would, in essence, relax some of the Area 5 constraints. Those additions were:

(10)   Information, including samples, on the amount and geographical distribution of radioactive debris resulting from past and future nuclear explosions.

(11) Acoustic, seismic, electromagnetic, and other geophysical data from nuclear explosions, provided

they reveal only general characteristics of weapons or devices.

(12) Time and locations of scheduled nuclear explosions.[32]

The new leeway was timely as the British began to establish several critical stations for the AEDS, giving the allies a greater monitoring capability over the Soviet proving ground near Semipalatinsk. The expansion also coincided with the British stations receiving American geophysical LRD equipment. Those AEDS instruments enabled an important level of standardization critical to maintenance support, as well as data integrity and speed of transmission. The British elements of the AEDS provided valuable data in the detection of the five Soviet tests in 1955. While Strauss and the AEC kept a close eye on the relationship to ensure compliance with Area 5, the *depth* of cooperation within the law immediately increased. Following the signing, the British provided AFOAT-1 detailed information about their scheduled 1956 tests. Likewise, Northrup furnished the British with post-shot data from Operation TEAPOT (February to May 1955), such as time of event, location, test conditions, and trajectories of radioactive debris clouds.[33]

In May 1956, the British returned to Montebello Island to conduct operations BUFFALO and MOSAIC. BUFFALO consisted of four tests (May to October) that produced yields ranging from 4 to 15kt.[34] The two shots of MOSAIC (October 11th and 21st) created yields of 5kt. and 30kt.[35] Adequate planning time allowed AFOAT-1 to conduct several geophysical experiments and airborne sampling. After the tests, the British provided AFOAT-1 with close-in radioactive debris samples in return for analytical results from U.S. laboratories. Northrup worked with the AEC to ensure that the transmittal of the UK analytical data was properly classified. AFOAT-1's Western Field Office (WFO) at McClellan AFB arranged for two C-118 aircraft to courier the samples back to the McClellan Central Laboratory (MCL).

After each shot, the USAF Military Air Transport Service (MATS) successfully returned samples to MCL within an allotted minimum time period to ensure adequate analysis on important short-lived isotopes.[36]

The Montebello tests also served to exercise the AEDS. One seismic and three acoustic stations detected the May 16th shot, and subsequent radiochemical analysis correctly determined it was a tower detonation. Two acoustic stations in the Pacific region accurately estimated the yield at 25 kt. However, no seismic yield estimates were calculated due to rapid variances in seismic wave amplitudes in the vicinity of the test site. Similar results were obtained from the subsequent shots on June 19th, September 27th, and October 4th.[37]

One week after the conclusion of Operation BUFFALO, the British initiated Operation MOSAIC. By this time, the new permanent nuclear test site at Maralinga was ready for operations. The British abandoned Emu Field after HURRICANE due to its geographic isolation, and instead built a new test site at Maralinga, situated 140 miles further south near the coast. Both sides agreed to continue the same level of cooperation. The UK agreed to deliver the debris samples to the Changi Royal Air Force Station at Singapore, Malaya. As during the previous test, MATS furnished two aircraft to courier the samples to MCL. Both deliveries took approximately five days, well within the time frame needed for accurate radiochemical analysis.

As expected, the AEDS was less successful in detecting the two shots. The October 11th air shot, which produced a 3kt yield, was detected by only one acoustic station in the Philippine Islands and the in-country seismic station at Alice Springs. The larger shot on October 21st produced a respectable yield of 30kt; however, it was still only detected by the few AEDS stations monitoring the BUFFALO shots.[38] The nine British tests in Australia between 1952 and 1956 served to underscore the concern Northrup expressed in his October 19, 1956 AEDS capabilities assessment about the difficulties in extending the AEDS into the southern hemisphere (discussed in Chapter 9).[39]

While AFOAT-1 was focused on the British partnership during 1955 and 1956, Northrup was careful to simultaneously cultivate his relationship with the Canadians. Although the Canadians were not developing nuclear weapons, they possessed a radiochemistry laboratory and were building an LRD capability consisting of both ground and aerial filtering operations. They also provided a back-up sampling capability to augment American airborne operations in Alaska. Importantly, AFOAT-1 gave the Canadians the newly-developed filter paper recently manufactured by the International Paper Company. The new paper allowed standardization of analytical procedures between the laboratories of the two countries. The Canadians provided AFOAT-1 with data from both ground and airborne collections as well as radiochemical analyses of debris samples AFOAT-1 had collected and shared. They also contributed independent analytical results for use in evaluating Soviet bomb debris. The Canadians were especially useful in providing analysis on the two Soviet shots that occurred in November 1955.[40]

The limited agreement between the governments of Canada and the U.S. restricted some of AFOAT-1's direct communications authority. For example, the AEC permitted the advanced notification of Operation TEAPOT (February to May 1955), but only through State Department channels. However, the AEC authorized some exchange of subsequent information on individual events. Under this agreement, AFOAT-1 could keep the Canadian liaison officer apprised of routine schedule changes in the TEAPOT series, and could also transmit post-detonation information relating to time, location, and meteorological conditions. However, AEC's strict interpretation of the Atomic Energy Act prohibited AFOAT-1 from providing information on methods used to calibrate yields.[41]

While the November 1956 Suez Crisis seriously strained U.S.-UK political and diplomatic relations, the crisis appeared to have little or no effect on allied LRD relationships. In fact, all partners built upon their initial successes from the previous year. During the course of 1957, the world's three nuclear powers detonated a total of 53 nuclear devices. It was the busiest year to date (see Chapter 9) and the willingness to share information, albeit within Area 5 restrictions, was impressive.

The U.S. and UK test series both began in May. For the Americans, Operation PLUMBBOB included 32 shots. The UK conducted seven nuclear tests during two separate operations. The first operation, codenamed ANTLER, was a huge step forward for the British nuclear weapons program. Much like Operation GREENHOUSE in 1951, the ANTLER shots tested "trigger" devices required for the production of a hydrogen bomb. The shots occurred at the Maralinga test site and produced yields ranging from 1 to 26 kt.[42] Acquiring a thermonuclear weapon had become a high-priority endeavor, especially after Winston Churchill became prime minister again in October 1951. Lamenting the loss of great power status, Britain's Defence Policy Committee, chaired by Churchill, viewed the acquisition of a thermonuclear weapon as a means of regaining that position in the world, alongside the United States and the Soviet Union. After much debate, they concluded in June 1954 that a hydrogen bomb was necessary not only as a real deterrent against Soviet aggression but to regain their international stature. "More than ever the aim of United Kingdom defence policy must be to prevent war. To this end we must maintain and strengthen our position as a world power so that her Majesty's Government can exercise a powerful influence in the counsels of the world."[43]

Britain's second operation, codenamed GRAPPLE, was critical to achieving such a goal. Penney and his team were determined to demonstrate they could harness the science of nuclear fusion without American assistance. Those scientists on Penney's team who had been at Los Alamos in 1946 and had had their research confiscated the day after the enactment of the McMahan Act had much to prove.

Because Australian law forbade the detonation of megaton devices, the UK had selected Christmas Island, located almost 1,000 miles northwest of Australia in the Indian Ocean, as the test site for Operation GRAPPLE. Having no knowledge of the successful American design, the British scientists and engineers built three different devices, all planned as air drops. The first three shots failed to produce a thermonuclear reaction. However, GRAPPLE X, conducted on November 8, 1957, succeeded with a yield of almost 2MT.[44]

With adequate notification from the British, the U.S. was able to monitor the shots to a greater extent than the previous tests in Australia. AFOAT-1 emplaced special electromagnetic, seismic, and acoustic stations on Palmyra Island, located about 400 miles from the test site. The special acoustic and seismic stations were well positioned to monitor all of the detonations; a good test of new equipment given that the first shot on September 14th produced a yield of only 1kt. Plans for RB-57 samplers from Kirtland AFB, New Mexico, were cancelled when the UK agreed to collect debris samples. The RAF and SAC worked well together. The RAF's Vickers *Valiants* and SAC's B-36s simultaneously collected debris and gas samples. This was the first real operational use of the B-36 (the F-36 foil tore apart on the first sortie but a new type of screen was immediately installed in the aircraft). In an unprecedented gesture, British Foreign Minister Selwyn Lloyd sent a letter of appreciation to U.S. Secretary of State John Foster Dulles, thanking him for U.S. participation.[45]

On July 2, 1958, Congress amended the Atomic Energy Act of 1954 to permit greater information sharing with U.S. allies. To a large degree, the amendment was necessary in order to facilitate some nuclear-related distribution of systems earmarked for NATO defense in Europe. The following day, with the amendment as a foundation, the United States and the United Kingdom signed the Agreement of Cooperation. In essence, it permitted a much greater level of nuclear-related information sharing than the earlier amendment allowed. The new amendment was the culmination of a meeting between Eisenhower and Macmillan on October 25, 1957, when the president promised the British prime minister that he would request Congress to alter the 1954 act to permit a "close and fruitful collaboration of scientists and engineers of Great Britain, the United States, and other friendly countries."[46] Arguably, the greatest beneficiaries of the act were the American and British LRD programs.

Greater levels of cooperation between AFOAT-1 and the AEIU were immediately apparent. With the Conference of Experts beginning just hours before the agreement was signed, the American and British experts were able to coordinate their negotiating positions to the maximum extent possible (see Chapter 10). The new freedom continued with the second of five British nuclear tests on August 22, 1958. The final three events followed soon thereafter on September 2nd, 11th, and 23rd. Because the RB-57 aircraft were committed to HARDTACK I, the British provided debris samples to AFOAT-1 from all of their tests.

AFOAT-1 also provided the UK with certain meteorological information to enable their aircraft to sample Soviet debris. After the debris was analyzed, AFOAT-1 and the AEIU compared their reports to confirm the Soviet nuclear detonations and to evaluate Soviet capabilities.[47]

To illustrate the new level of trust, on November 3rd the UK raised the security classification of all analytical data on UK tests from secret to top secret. However, AFOAT-I was permitted to maintain the test data at the secret level. At the same time, Americans provided their allied counterparts with time, location, and debris trajectory information on U.S. nuclear tests.[48]

The relationship with the Canadians remained restricted, however, until December 1959, when an agreement similar to the British agreement was signed. Still, throughout 1958 and

1959, the information exchange between AFOAT-1 and the northern neighbors remained quite adequate. Canadian cooperation with AFOAT-1 primarily involved debris sampling. AFOAT-1 provided meteorological information for Canadian T-33 and DC-3 aircraft to sample debris from Soviet nuclear tests. MCL then analyzed those samples. AFOAT-1 also provided the Canadian Government with the time and approximate yield of the HARDTACK II detonations. On May 12, 1959, just two weeks prior to the inactivation of AFOAT-1, the Canadians established a seismic station in support of the AEDS.[49]

After AFOAT-1 was inactivated in July 1959, the three allies continued to improve those important relationships for many years to come. Their collaboration would prove invaluable almost immediately as the Geneva Conference struggled on to eventually achieve an atmospheric test ban treaty in 1963. In time, New Zealand and Australia would join this tight-knit club. For AFOAT-1 and Doyle Northrup, however, the AEDS would never have achieved operational efficiencies to such a high degree without the assistance of their British counterparts.

Notes for Annex 1

[1] Howard Morland, "Born Secret," in *Cardozo Law Review* (vol. 26, No.4, March 2005, pp. 1401-1408), 1401. Morland is best known as an activist against nuclear weapons. Goodman, Michael E. *Spying on the Nuclear Bear*, 212.

[2] Copy of memo in AFTAC archives. Also found at http://www.atomicheritage.org/ Last accessed October 4, 2018.

[3] The U.S. rebutted British and Canadian objections by claiming that the Quebec Agreement was only an "executive" pact within the Roosevelt administration because Congress had never reviewed or approved it. The Hyde Park memo was "lost" among the President's papers and only discovered after his death.

[4] Rhodes, *Dark Sun,* Chapter Ten. For details on the scope of the various FBI spy investigations, see especially Herken, *Brotherhood of the Bomb.*

[5] *History of Long Range Detection: 1947-1953,* Volume Two, Appendix X, author unidentified (Washington, D.C.: The Air Force Office of Atomic Energy- One, Headquarters, United States Air Force, 1953), Tab X, AFTAC archives. See also Goodman, *Spying on the Nuclear Bear,* 44-45.

[6] Ibid.

[7] Ibid.

[8] Memo from Carpenter to Schlatter, Subject: "Proposed Procedure for Effecting Technical Cooperation in Approved Areas of Exchange," August 30, 1948, AFTAC archives.

[9] List of conference attendees included in Memo, Nelson to Kenneth Gettman (AEC), Subject: "Report of Discussions between Representatives of AFOAT-1 and Representatives of the British and Canadian Atomic Energy Groups," September 28, 1948, AFTAC archives.

[10] Marley was also an inventor of high speed photography that enabled important research at Los Alamos.

[11] Chalk River had participated in the Manhattan Project as a producer of heavy water (deuterium oxide). The defection of Soviet cypher clerk Igor Gouzenko revealed the spy ring. 39 people were arrested of which 18 received convictions. See Rhodes, *Dark Sun,* 126-127.

[12] Agenda, "Conference between AFOAT~1 and United Kingdom and Canadian Scientific Missions Dated September 3, 1948 (Area 5)," AFTAC archives.

[13] Dr. George Shortley was Johnson's designated successor but at the time of the conference was preparing to resign as well in order to accept a position at the Ohio State University (discussed in Chapter 2).

[14] Report, "Conference between AFOAT-1 and United Kingdom and Canadian Scientific Missions Dated September 3, 1948 (Area 5)," AFTAC archives.

[15] Ibid.

[16] Ibid.

[17] Ibid.

[18] Memo, Nelson to Chairman, MLC, Subject: "The Presentation on 23 November 1949 of Additional Results Obtained by the United Kingdom on Operation Joe," January 24, 1950; and memo, Nelson to Chairman, MLC, Subject: "U .S. and U.K. Meeting of 2 December 1949 for the Discussion of Results on Operation Joe," January 24, 1950, AFTAC archives.

[19] Williams, *Klaus Fuchs*, 95. Williams also quotes Foreign Secretary Ernest Bevin's comment that "we simply could not acquiesce in an American monopoly on this development." 93.

[20] Ibid., Chapter Three. Williams writes that "the Americans feared that the British bomb program would assist the Russians, both because British security was lax and because in the event of war the Soviet Union might invade and occupy Great Britain." 98.

[21] Fuchs was active at Los Alamos in attending many of the seminars (the most important being Oppenheimer's three-day secret conference in April 1946 on the hydrogen bomb and future courses of research into a fusion weapon). See especially Rhodes, *Dark Sun*, 252-253, and Anne Fitzpatrick, dissertation, *Igniting the Light Elements: The Los Alamos Thermonuclear Weapon Project, 1942-1952* (Washington, D.C.: George Washington University, 1999).

[22] Goodman, *Spying on the Nuclear Bear*, 64.

[23] Ibid., 63.

[24] "Interview with Arneson by Neil M. Johnson, June 21, 1989 (Washington, DC)." Harry S. Truman Presidential Library: Papers of R. Gordon Arneson,

http:l/www.trumanlibrary.org/oralhist/arneson .htm accessed
October 15, 2018; also quoted in Ibid., 64.

[25] *History of Long Range Detection: 1947-1953*, Volume Two,
Appendix X.

[26] Memo, Lieutenant Colonel McClanahan to Le Baron, Subject:
"Joint US-UK AEDS Operational Planning Progress," June 29, 1951,
AFTAC archives. Note that the oral presentation in LeBaron's office
was based on this memo written three days earlier.

[27] *History of Long Range Detection: 1947-1953*, Volume Two,
Appendix X.

[28] Technical Memo #83, Northrup to Hooks, Subject:
"International Cooperation in Atomic Energy," September 10, 1954.
AFTAC archives.

[29] Goodman, *Spying on the Nuclear Bear*, 131.

[30] Ibid., 91. Quote attributed to William Penney.

[31] Ibid., 97.

[32] *History of AFOAT-1*, 1955, 57.

[33] Ibid., 58-61.

[34] Codenamed shots One Tree, Marcoo, Kite, and Breakaway.

[35] Codenamed G1 and G2.

[36] *History of AFOAT-1*, 1956, 46.

[37] Ibid., 48-50:

[38] Ibid., 54-55.

[39] Technical memo #86, Northrup to Hooks, Subject: "AFOAT-1
Capability for Recording Nuclear Explosions," October 19, 1956,
AFTAC archives.

[40] *History of AFOAT-1*, 1956, 58.

[41] *History of AFOAT-1*, 1957, 28.

[42] The shots were codenamed Tadje, Biak, and Taranki.

[43] Kevin Ruane, *Churchill and the Bomb: In War and Cold War*
(London: Bloomsbury Publishing, 2016), 264. *January - 31 December
1957*. Washington, D.C.: The Air Force Office of Atomic Energy-
One, Headquarters, United States Air Force, 1957, 23-25.

[44] *History of AFOAT-1*, 1957, 1.

[45] Ibid.

[46] U.S. Department of State Bulletin, Vol. XXXVII, No. 959
(November 11, 1957), p. 740, as quoted in Jacobson and Eric Stein,
Diplomats, Scientists, and Politicians, 31.

[47] *History of AFOAT-1*, 1958, 35.

[48] Ibid., 45.

[49] *History of AFOAT-1,* 1959, 38.

ANNEX 2

The AEDS as a Wartime System

AFOAT-1's initial and extensive R&D plans implied there would be new fields of LRD capabilities in the future. Indeed, many believed in the early 1950s that LRD could perform a wartime mission, which might be more important than its peacetime mission. Consequently, the Special Projects Branch of AFOAT-1 undertook the preparation of a preliminary study for the employment of the AEDS during a period of open hostilities.[1]

Branch members completed their study, codenamed CODY, on August 8, 1950. CODY sought to determine the type and degree of participation the AEDS could contribute to combat operations in wartime. CODY suggested that many of the technical surveillance capabilities under development for the peacetime objectives of the AEDS could also be used in time of war to inform or augment conventional wartime intelligence. There were several basic assumptions of the study:

> 1. The USSR would initiate hostilities in conjunction with some or all of its allies at a time of Russian convenience.[2]

> 2. Nuclear weapons would be a major factor in the capabilities of the USSR and would be under American surveillance.

> 3. The AEDS would possess the capability to detect and locate nuclear explosions to a fair degree of accuracy, specifically:

>> a. To detect with high accuracy contaminated air masses.

>> b. To possess equipment adaptable to strategic reconnaissance.[3]

Initially, planners believed that, with proper priority and financial support, they could convert LRD equipment or

413

develop new technologies to meet additional wartime operational requirements.

By late 1950, SAC planners envisioned contingencies in which the USAF would drop a large number of atomic bombs on Soviet targets soon after the beginning of hostilities. SAC looked to LRD to possibly provide timely bomb damage assessments (BDA) for accuracy and efficiency of strikes to aid follow-on operational planning. SAC also saw tremendous value in AFOAT-1's experience with tracking detonation air masses. In warfare involving the use of large numbers of nuclear weapons, such air masses could contaminate operational Air Force equipment and endanger areas of operation (the LRD system had already demonstrated this capability in Operations CROSSROADS and SANDSTONE, and during the sampling operations of Joe-1). The CODY Study recommended the USAF should leverage the AEDS as a wartime system to:

1. Increase efforts to improve the existing available detection system

2. Increase emphasis on basic studies in order to determine new approaches to the detection problem

3. Plan for an expanded AEDS for combat operations

4. Assign specialized staff to perform certain operational radiological functions for the AEDS; and

5. Coordinate efforts between the AEDS and other USAF units in operations and R&D functions.[4]

In working on the CODY Study, the AFOAT-1 physicists were skeptical about the AEDS's wartime use. Although the AFOAT-1 Operations Branch had requested the Technical Branch to prepare the technical aspects of the study, the physicists failed to do so, and the report went to the Air Staff without their input.[5]

Similarly, the reviewers of CODY at Headquarters, USAF, viewed the study with some reservation. Some believed the recommendations were essentially sound but the entire study should go to an impartial agency such as the Scientific

Advisory Board (SAB) for further examination. Others, more reserved, argued the utilization of an expanded AEDS in combat operations would require a clearer definition of LRD responsibilities in time of war. They believed it was premature to present the study to the SAB for evaluation. Instead, they wanted the study returned to AFOAT-1 with instructions to re-orient the study to include technical details which, in turn, would enable the Air Staff to develop formal requirements. After that, the study could go to the SAB. AFOAT-1 largely accepted the latter view and embarked upon a revision of CODY, which was completed almost a year later in June 1951. Meanwhile, as AFOAT-1 was revising CODY, several important developments influenced the revision.[6]

In January 1951, SAC approached AFOAT-1 to determine how the AEDS could meet SAC wartime mission requirements. For example, on January 12, 1951, representatives of SAC and AFOAT-1 met at Offutt Air Force Base, where SAC presented several requirements for some of the services the AEDS could potentially provide in wartime. Five days later, on January 17th, Major General August W. Kissner, SAC Commander General Curtis LeMay's chief of staff, specified to AFOAT-1 that the AEDS could contribute to the strategic air offensive by supplying:

1. Information on the location of each atomic bomb burst occurring over the Eurasian continent, particularly over USSR;

2. Information on the time of each burst;

3. Information on the efficiency of each burst;

4. Information on the height of each burst;

5. Pre-strike reconnaissance to aid in pinpointing Soviet atomic energy targets;

6. Strike reconnaissance as an aid to target identification; and

7. Information on the location and intensity of airborne bomb debris generated by US atomic weapons[7]

Importantly, those specific categories of information were not intended as an overall statement of SAC's total requirements for AEDS data. Both SAC and AFOAT-1 fully expected that, in the near future, LRD R&D would result in additional capabilities that could contribute to combat operations.

Ten days after the SAC-AFOAT-1 meeting at Offutt, Operation RANGER commenced, followed by Operation GREENHOUSE in early April. As discussed in Chapter 5, the cloud-tracking activities in the five RANGER shots highlighted the feasibility of using manned aircraft for debris sampling (rather than drones). SAC observers at RANGER believed modified B-29s could meet the requirements outlined by General Kissner. After the GREENHOUSE shots, they were even more convinced, as those sorties demonstrated acceptable safety measures for the crews. Those measures included known time parameters for decay and dispersion (to reduce gamma exposure), and the use of sealed aircraft and closed breathing systems. While they acknowledged a real need for the development of adequate radiac equipment for the missions, the economy of manned aircraft rather than drones was indisputable. Five B-29's could collect 40 papers, while 20 B-17's could collect only 32.[8]

Several weeks after the completion of Operation GREENHOUSE, AFOAT-1 completed the revision of the CODY study. In the conclusions of the 33-page report, the authors stated the AEDS showed a potential to meet the short-range requirements previously identified for a wartime mission. AEDS equipment already fielded or under development could aid the strategic air offensive by locating Soviet bomb drop areas. They also believed AEDS instruments could supply information relative to the functioning and efficiencies of weapons, which in turn, could be used for indirect BDA. It was thought the extension of some AEDS techniques could also meet such strategic reconnaissance requirements as target location, target identification, and post-strike evaluations. Other techniques could be converted to aid air defense, and in the field of tactical air operations there was promise of assistance in the tactical employment of atomic bombs, radiological warfare agents, and in the support of ground forces. To move the effort forward, the revised CODY Study recommended AFOAT-1 receive increased resources to advance techniques and equipment pertinent to Air Force wartime requirements.

Interestingly, what AFOAT-1 really wanted was more detailed direction and guidance from the Air Staff. The report stated that AFOAT-1 should "be directed to proceed along such lines as Headquarters, United States Air Force, believes desirable at this time," and that Headquarters, United States Air Force, specify the war time mission of the Atomic Energy Detection System "to include delineation of responsibilities and degree of effort, based on . . . a complete review by operational and technical authorities to evaluate the potential of Atomic Energy Detection System type operations in an Air Force wartime role."[9]

Because LRD data collections would feed intelligence work during wartime, the director of Air Force Intelligence, Lieutenant General Charles P. Cabell, commented that the Air Staff should provide AFOAT-1 with guidance as to what areas they should investigate to make the AEDS a feasible wartime system, such as air defense, reconnaissance, and other areas of combat operations. After a follow-on meeting with AFOAT-1 on October 21, 1951, Cabell issued a statement covering his concept of the requirements for information prior to and during hostilities. To assist AFOAT-1 planning, he stated several requirements Air Force intelligence had identified. In general, they were:

> 1. To determine the details of the type and power of current Soviet weapons;

> 2. To determine the specific objectives of Soviet nuclear R&D; and

3. To assist in the determination of targeting information in regard to the location of air fields, depots, and other installations engaged in nuclear activities.[10]

Cabell also emphasized that AFOAT-1's short-range detection techniques could contribute to defensive requirements by determining the type and power of Soviet bomb drops in the continental United States in order to ascertain the nature of the weapons, and the extent of destruction caused by the Soviet attacks.

Despite the thoroughness of the 11-month effort to revise CODY and its optimistic assessments and conclusion, the Commander of AFOAT-1, Major General Raymond C. Maude, did not forward the study to the Air Staff. Indeed, Maude appeared much more reserved and cautious about the feasibility of using the AEDS in a wartime capacity. Instead, Maude sent a memo to the Air Staff on September 14, 1951, listing six abstract capabilities that had to be considered in arriving at a decision for an expanded LRD mission. Those were:

1. The detection of atomic weapons detonated by friendly powers;

2. The detection of atomic weapons detonated by enemy powers;

3. The performance of strike operations consisting of the location of targets prior to strike, target identification during strike, and post-strike reconnaissance;

4. The acquisition of information as to the location of foreign stockpile depots and airfield operational atomic stockpiles;

5. The creation of defensive measures to avoid atomic clouds, to survey sites contaminated by foreign atomic weapons, the analysis of suspect substances, the detection of toxic agents, and the establishment of an information center for the effective application of information on atomic operations; and

6. The derivation of a guidance system for missiles.[11]

Importantly, Maude stressed that these abstract capabilities were mere possibilities of what AFOAT-1 *might* attain.

There were many examples of the tentative nature of AEDS capabilities and reasons for Maude's restraint. For example, in the summer of 1951, none of the existing AEDS techniques appeared to have the capability of locating ground zero with the precision required for an accurate assessment of BDA. There was no way of determining the height of the burst. It was possible to locate a single atomic burst with some precision, but this capability would vanish in the case of multiple detonations. AFOAT-1's radiochemists could use techniques to determine the yield of a detonation within an accuracy of plus or minus ten percent, given a good sample. However, in wartime, it was necessary to gather the sample over the target area—most likely a severe hazard—and then require two or three weeks for analysis. A more promising technique was an adaptation of the Bhang meter principle by determining yield from optical measurements of variations in light intensity.[12] However, analysts faced inaccuracies ranging from plus or minus 20 percent to plus or minus 50 percent. Still, AFOAT-1 scientists felt the Bhangmeter principle might be important in getting an immediate estimate of bombs dropped on enemy targets. To measure the radiological contamination of terrain immediately under the burst, it seemed probable they could develop existing airborne atmospheric conductivity equipment and airborne scintillation counters for operational use in a relatively short period of time. Lastly, AEDS techniques did not appear adaptable to the problems of post-strike reconnaissance.[13]

Defensively, use of the AEDS seemed more feasible. AFOAT-1 could easily determine the efficiency and yield for detonations over the United States, given the current capabilities of operational sampling aircraft. Also, atmospheric conductivity and scintillation counter equipment, tested at RANGER and GREENHOUSE, gave some promise of detecting the orientation and proximity of atomic clouds. Aircraft equipped with current instrumentation could be used to conduct terrain surveys contaminated by foreign atomic weapons detonated in the U.S. But there were no capabilities for the detection of carriers of atomic weapons.[14]

Essentially, Maude provided the Air Staff a more scaled-down assessment of the AEDS, especially in regard to its current limitations. In retrospect, he did AFOAT-1 a huge favor by withholding the CODY study. LRD was in its infancy, and the AEDS had yet to establish itself in critical areas of the world. Maude did, however, concur with his staff's conclusions that the Air Staff should provide AFOAT-1 more detailed guidance.[15] In his September 14, 1951 memo, he recommended the Air Staff:

1. Analyze the requirements for AEDS wartime activities to determine their validity;

2. Determine which of the abstract capabilities were sufficiently promising to warrant development; and

3. Determine the Air Force agency or agencies to be charged with the development of the wartime capabilities[16]

At this point, enthusiasm for utilizing the AEDS in wartime seemed to wane. The Air Staff's reply to Maude's assessment was not enthusiastic. In their view, they had already determined that some capabilities were feasible and others less so. However, even in those areas in which the Air Staff exhibited some interest, there was no follow-up action, despite AFOAT-1's repeated request for clarification and guidance. By the fall of 1951, it was clear to all parties that the original question remained unanswered. Still, after discussions within AFOAT-1, the Air Staff assigned a project officer to continue to pursue the matter. Pending resolution, the new commander of AFOAT-1, Major General Donald J. Keirn, recognized an obligation to use AEDS capabilities in the event of war. The war plans program would continue but would be limited to emergency planning for the wartime employment of existing capabilities; in case of war, formal long-range planning involving R&D programs would be temporarily suspended.[17]

In the summer of 1952, AFOAT-1 completed its Emergency War Plan. The plan envisioned a coverage of intensive atomic strikes in addition to the current emphasis on sporadic Soviet atomic tests in the USSR. It outlined the technical limitations imposed by large-scale nuclear warfare, and evaluated the impact of foreseeable conventional military situations in regard to the world-wide deployment of the AEDS. In short, the plan played on LRD's *existing* strengths. The planners emphasized that the AEDS, already possessing strong detection capabilities, would go to war "as is."

On August 25, 1952, Major General William M. Canterbury, having just assumed command from Keirn, returned the AFOAT-1 Emergency War Plan 1-52 to the Air Staff. Notably, Canterbury requested comments from the staff concerning the validity of the promise that AEDS's requirements would continue during hostilities. Two weeks later, on September 5, 1952, the Director of Intelligence on the Air Staff, Major General John A. Samford, replied that the AEDS might well provide the best and possibly only means of measuring the overall damage to some Soviet targets after an outbreak of hostilities. He added the AEDS might also penetrate hostile territory to assess the Soviets' nuclear weapons program in order to help predict American requirements for countermeasures. In short, despite the reservations of some staff members, the senior members of the Air Staff, especially Samford, believed that even with serious limitations, the AEDS could provide invaluable information during wartime.[18]

As Samford passed the AFOAT-1 plan to the Director of Operations with the recommendation for final approval, Canterbury had no desire to obtain approval of a plan that did not have wide-spread support. After voicing his view to

Samford and the Director of Plans, Canterbury succeeded in getting the draft plan returned to AFOAT-1 without action, pending a detailed review of wartime AEDS information requirements and AFOAT-1's capabilities to satisfy them.[19]

General Samford wasted little time in providing Canterbury those requirements. On December 23, 1952, he sent Canterbury a memo articulating AFOAT-1's wartime requirements. As Director of Intelligence, Samford certainly had serious priority wartime requirements for atomic energy information, especially given the state of the Cold War escalation at that time. Samford understood that only the AEDS could collect and analyze nuclear scientific data. In his view, the requirements simply constituted the direct extension of the AFOAT-1 mission as a surveillance organization. He was primarily interested in the AEDS determining the type and power of enemy nuclear weapons. More specifically, he hoped the AEDS could identify and locate enemy carriers of atomic weapons (i.e., aircraft, guided missiles, and ships). His other requirements, such as the identification and location of enemy atomic energy installations for targeting purposes, resided outside the capabilities of the AEDS.[19]

Between December 1952 and May 1953, Canterbury and his staff held several meetings and conferences to consider Samford's numerous requirements for the AEDS. Simultaneously, the AFOAT-1 staff was also busy revising the AFOAT-1 war plan as it kept informed of the Canterbury-Samford exchanges. On April 13, 1953, the staff finalized AFOAT-1 Emergency War Plan (EWP) 1-53. In preparing the plan, the planners incorporated guidance from the latest Air Intelligence Estimate, studies by the Joint Strategic Plans Committee, and the larger Air Force Emergency War-Mobilization Plan 1-53. As the basis of its thinking in developing the atomic aspects of a wartime environment, AFOAT-1 accepted the following four JCS assumptions regarding the conduct of a Third World War:

1. Both sides would deploy large numbers of nuclear weapons early on, most likely on D-Day. Primary targets would be the Soviet and Allied zones in Germany, the British Isles, the United States, southern Canada, and U.S. nuclear strike and post-strike bases located outside the U.S.

2. The USSR would have the capability to attack any important installation in the continental United States.

3. Even the most elaborate air defense mechanism would not prevent bombers from penetrating the screen and delivering their bombs.

4. Critical areas in the quadrangle Duluth-St. Louis-Washington, DC-Boston, Seattle, San Francisco, Los Angeles, and locations such as SAC bases and atomic energy installations, would have top priority in the Soviet targeting system.[20]

The JCS assumptions were important to AFOAT-1 planners because the identified and prioritized geographical cities and regions offered an estimation of how much of the AEDS could be degraded in combat, initially and subsequently.

Having considered Canterbury's views, Samford recommended AFOAT-1 resubmit its Emergency War Plan and that Canterbury maintain direct liaison with the Office of the Director of Plans. Drawing on their revised war plan, AFOAT-1 declared on May 11, 1953, that it could determine the power of enemy nuclear weapons if the WB-29s could collect samples of the debris unique to each detonated bomb, and then subject the samples to radiochemical analysis. Although thoroughly aware of the significance of identifying and locating enemy carriers of atomic weapons, as of the spring of 1953, AFOAT-1 had been unable to develop any technique to meet that requirement. Also, there was no capability for in-flight targeting of known installations, or the establishment of the identity and location of suspected installations, either at short or long range. In addition, LRD techniques, as applied to the strategic reconnaissance problem, were inadequate to materially assist in the BDA program or in strategic offensive planning. However, AEDS techniques appeared to have some application in regard to locating Soviet guided missile development and testing programs.[21]

AFOAT-1 planners built their plan based on the capabilities of the AEDS as of July 1, 1953. At that time, the AEDS consisted of 67 stations that, collectively, were responsible for airborne and ground-based sampling, seismic and acoustic detection, and radiochemical analysis. In every case, AFOAT-1 possessed the best and most advanced equipment possible. In short, what AFOAT-1 commanders and scientists feared was that their critical peacetime operations would be changed in wartime by several other activities. Envisioned were planned redeployment of U.S. Forces, forced evacuation of existing sites, AEDS losses due to enemy action, the augmentation of the system as a result of new developments, and any new information requirements generated by hostilities. Internally, the AFOAT-1 staff developed a "concept of operations" that envisioned serious losses of personnel and AEDS stations. They recognized the vulnerability of overseas locations and how the loss of AFOAT-1 command and control facilities (especially the regional field offices) would impact AEDS operations.[22]

On June 4, 1953, Canterbury re-submitted EWP 1-53 to the Air Staff, and on July 22nd, with strong backing from Samford, the plan was approved. It is important to note that EWP 1-53 was not an affirmation of the AEDS capabilities highlighted in the first and second CODY studies, nor was it an acknowledgment of Samford's intelligence requirements issued the previous December. Rather, EWP 1-53 basically stated that existing LRD systems would continue their peacetime functions in the event of hostilities but could perform such additional duties as the situation might require. In reality, the existence of EWP 1-53 did not imply that the JCS or the Air Force Chief of Staff had assigned the AEDS a formal wartime mission. Even so, at the time, there was some consternation within AFOAT-1 that EWP 1-53, when viewed against the background of the strong 1950-1953 views of the Air Staff on wartime contingencies, might be seen as an unrealistic commitment of the AEDS. Subsequently, AFOAT-1 planners immediately reevaluated the AEDS's capabilities to determine a practical and defensible wartime mission for AFOAT-1.[23]

AFOAT-1 completed that reevaluation on July 15, 1953, when Northrup issued Technical Memorandum No. 80 (TM-80). Closely mirroring Canterbury's May 11, 1953 reply to Samford, TM-80 was a study of AEDS capabilities in meeting the wartime atomic energy intelligence requirements of the Air Force. The document specifically addressed the requirement for the AEDS to determine the power of "enemy" (i.e., Soviet) atomic weapons. It did so under the assumptions that Air Force war plans were referring to two types of weapons: weapons tested within the USSR as part of ongoing Soviet development programs, and those dropped during combat operations. Northrup reported that the AEDS could compile information of atomic bombs tested in the USSR if airborne samplers could collect adequate debris. However, to avoid cross contamination of such debris from other wartime detonations, sampling aircraft would have to fly relatively close to the enemy test site. The same principle applied to the work of compiling information of Soviet weapons dropped within the United States. If 50 or 100 bombs were dropped in quick succession, contamination would likely occur. Also, in multiple drops, acoustic, seismic, and electromagnetic signals would be confused, and the early warning features of those techniques would be lost. However, Bhangmeters estimating yields by light measurements could monitor enemy detonations over targeted areas. Northrup confirmed no AEDS technique could identify carriers of atomic weapons, as Samford had hoped. Likewise there was no capability for in-flight targeting of known installations, or the identification and location of suspected installations either by long-range or short-range methods.[24]

TM-80 specifically stated that the SAC plan for photography, radar observation, optical measurements, and electromagnetic signal observation was beyond the use of the AEDS for on-site BDA. The AEDS techniques, as strategic reconnaissance, did not possess sufficient accuracy. However, the techniques might possibly have some application to assessing the location or existence of Soviet guided missile programs. This was especially true of the seismic and acoustic techniques, since it was conceivable those techniques could detect sounds of rocket launches at long distances. It was also possible the new EMP technique could detect the ionization generated by a rocket's re-entry into the atmosphere.[25]

Prior to 1954, SAC and AFOAT-1 had yet to reach an agreement on the use of the AEDS in wartime. However, during 1954, the AFOAT-1 staff studied the problem of providing a realistic technical basis for an emergency war plan that could provide physical information and data on foreign nuclear operations. Other agencies contributed to the assessment, including various intelligence organizations and the Civil Defense Administration. The Joint Intelligence Community (JIC) also provided an estimate of Soviet capabilities.[26]

In June 1954, as a follow-up to TM-80, Northrup forwarded a memorandum to the new AFOAT-1 commander, Major General Daniel E. Hooks, identifying the collection means and types of physical data targeted during wartime. He emphasized the detonation of thousands of nuclear weapons in the first 60 days would contaminate the atmosphere in the northern hemisphere to a degree that would impede standard AEDS filtering operations. Therefore, AFOAT-1 would need a close-in sampling capability that would require sampling aircraft to be based within short flying distances of the various SAC bases targeted by the Soviets within the U.S. and overseas. Northrup also noted existing laboratory facilities could not accommodate the large number of simultaneous analyses, which would determine the type and size of Soviet atomic bombs. Multiple detonations of nuclear weapons would overwhelm acoustic stations, preventing them from discriminating between signals from individual bombs. The "B" technique would function well in the northern hemisphere because seismic records could be transmitted by mail or courier if communications should fail. The "Q" technique (electromagnetic) would be able to determine the yield, location, and time of detonation of each Soviet weapon on U.S. targets.[27] In finally drafting a revised war plan, the AFOAT-1 planners recommended the following:

1. Maintain the ground-based sampling programs at the current level of activity until the detonation of a large number of thermonuclear weapons rendered these systems useless.

2. Augment the electromagnetic and seismic systems in the European and American Theaters for the purpose of providing timely information for vectoring sampling missions

3. Suspend operation of the acoustic system on D-Day.

4. Establish a control center in conjunction with the electro-magnetic and seismic analysis centers for the ultimate purpose of vectoring air sampling missions in the U.S.

5. Establish a second center for controlling sampling operations in Europe. Concurrently, arrange with the Air Defense Command and US Air Forces, Europe, for air filtering capabilities.

6. Make arrangements with pertinent defense and other government agencies to report to AFOAT-1, the time, place and estimated yield of Soviet detonations.

7. Discontinue all routine aerial filtering operations on D-Day and return all personnel from the labs in Japan and Alaska to the McClellan Central Laboratory where

all radiochemical analyses would be accomplished for both the American and European Centers.

8. Utilize existing U.S. laboratory facilities and augment personnel with returnees from Japan and Alaska.

Although these recommendations became the basis for rewriting the AFOAT-1 Emergency War Plan, Northrup further recommended an exhaustive study to address the problem of obtaining physical data during wartime from qualified representatives of interested agencies.[28]

As a result of conferences and informal discussions between SAC and AFOAT-1, both parties agreed that the AEDS was sufficient in reporting SAC-delivered weapons results. Therefore, SAC requested Headquarters, U.S. Air Force, amend the AFOAT-1 wartime mission to include rapid notification to SAC of nuclear activity related to bomb strikes. In early November 1954, AFOAT-I submitted its war plan, aimed to reduce restrike and reconnaissance efforts, to SAC. The plan emphasized the capabilities of the seismic network to best detect explosions of weapons in the higher yield ranges, perhaps within a radius of 50 miles. On November 10th, SAC accepted AFOAT-1's basic plan and incorporated it into the SAC war plan as Annex E "SAC Support."[29]

The 1954 efforts to produce a realistic war plan for the AEDS underscored Canterbury's and Hooks' extensive efforts to offer up AEDS (despite its limitations) to the Air Staff and to SAC as an important component of their war plans. Still, they faced many frustrations in educating those organizations on the capabilities and, more importantly, the limitations of the adolescent AEDS. To be fair, in the era of massive retaliation, the Air Staff and SAC naturally viewed the AEDS as a vital system. Investments in AEDS R&D were huge and expectations high. However, AFOAT-1 commanders were simply not willing, quite responsibly, to fully endorse a system for national security that was not yet ready. LRD researchers were making great progress; with each U.S. test, their innovations were resulting in the development or advancement of higher fidelity instruments. AFOAT-1 commanders simply needed more time and were concerned that the Air Staff and SAC wartime requirements would divert the current LRD R&D path. Still, throughout 1954, Canterbury and Hooks succeeded in placating the Air Staff, who appeared satisfied that AFOAT-1 now had its own war plan tied into SAC planning. SAC, on the other hand, would continue to pressure AFOAT-1 in the foreseeable future to further commit the growing capabilities of the AEDS.

Such pressure became apparent in June 1955, when Northrup forwarded Technical Requirement Memorandum No. 12 to General Hooks, which outlined AFOAT- 1's wartime technical information objectives. He did so because AFOAT-1 had yet to receive any specific requirements for technical assessments from the Air Staff, SAC, or other partner agencies. In taking the initiative, Hooks and his staff produced the 1009th SWS War Plan 1-55 (WP 1-55) based primarily on Northrup's memo. WP 1-55 identified three specific missions:

1. Provide technical information concerning the enemy's nuclear weapons capability.

2. Provide technical information in support of the evaluation of the U.S. Strategic Air Offensive.

3. Be prepared to provide information on the degree of atmospheric radioactive contamination within the U.S.

WP 1-55 focused on accomplishing these three missions utilizing the existing scientific means within the entire AEDS. However, the war plan essentially extended, albeit in greater detail, the November 1954 plan AFOAT-1 had provided SAC. Specifically, it expanded on the seismic net's dual role in detecting and counting Soviet bombs in the various yield categories, and the seismic data used in the evaluation of U.S. strike missions.

The seismic net would also forward data to SAC to help determine which aircraft had detonated bombs in the targeted area. While the acoustic net would retain its current capabilities, Northrup anticipated the "Q" technique to be fully functional on D-Day. He did note, however, that ground-based particulate samplers would lose their efficiency almost immediately. Daily standard airborne collection missions, such as the Loon Charlie flights (i.e., the Alaska-Japan routes), would be terminated on D-Day and re-prioritized to collect against strikes on U.S. territory. Those collections, though, would be limited, since not every detonation would require radiochemical analysis. Given the anticipated high-priority target of Washington D.C., the plan envisioned the immediate loss of AFOAT-1 headquarters and the transfer of operational control to the Western Field Office. In short, WP 1-55 was clear the AEDS would go to war "as is" with all of its capabilities and limitations. WP 1-55 asserted that AFOAT-1 was combat ready because the AEDS had already been "at war" with the enemy for several years.[30]

As the year 1956 began, AFOAT-1 planners worked to extend and add detail to WP 1-55. The revisions built upon the fact that the AEDS was a "come as you are" system already operationally capable of meeting some of SAC's wartime requirements. Refinements generally aimed at determining the time and place of SAC's nuclear bomb explosions in enemy territory, the operational details of conducting air sampling in the United States and the Far East, and the subsequent radiochemical analysis of those collected samples.[31]

Airborne nuclear debris sampling missions in a nuclear war would require many more sorties. To fulfill that requirement, AFOAT-1 partnered with Air Defense Command (ADC) and the Air National Guard (ANG) to incorporate extensive augmentation from 14 ANG squadrons designated to conduct close-in sampling missions. ADC Operations Plan 7-56A (codenamed BIT BITE) outlined those tasks and procedures. From January 16 through 20, 1956, the 4926th Test Squadron (Sampling) trained 26 officers from the designated ANG squadrons at Kirtland Air Force Base, New Mexico. Those officers, in turn, returned to their squadrons to conduct training programs for the pilots and other personnel concerned with loading, recovery, monitoring, and decontamination procedures. AFOAT-1 and ADC included the ANG units in the U.S. nuclear test series codenamed Operation PLUMBBOB conducted in May 1957 (see Chapter 9).[32]

During the year, AFOAT-1 expanded WP 1-55 to determine how best to support SAC's long-range BDA assessment program. Now included as Annex E in SAC's war plan, AFOAT-1 personnel articulated how they could process SAC's request for information more rapidly, and reduce volume and length of data message traffic. Annex E also offered SAC reassurance that the AEDS would continue to function independently should AFOAT-1 Headquarters be neutralized. On May 8th, AFOAT-1 published its Emergency Relocation Plan, which detailed evacuation procedures to Camp Barrett at the Quantico Marine Base.[34]

In late 1956, AFOAT-1 began to exercise the war plan with SAC. In a command post exercise codenamed HARMONY HOUSE, AFOAT-1 demonstrated the capability of the Laramie Analysis Center (LAC) in Wyoming to process and evaluate the volume of seismic data resulting from multiple drops closely related in time. AFOAT-1 also successfully transferred command and control to the WFO in a scenario that saw the destruction of AFOAT-1 headquarters in Washington, D.C. On average, the reaction time between detection of an event and SAC receiving the AFOAT-1 report was approximately 28 hours. However, within AFOAT-1, the analysis (data processing and reporting) averaged six hours and 45 minutes. The unacceptable time lag of 28 hours was due to operational and administrative processing within SAC.

Obviously, communications needed improvement if the AEDS were to ever function in a viable BDA role.[33]

During the first half of 1957, SAC and AFOAT-1 addressed the shortcomings of HARMONY HOUSE. To implement improvements, SAC conducted another exercise codenamed GAME TIME from August 5 to 10, 1957. This exercise tested new communications procedures, which called for the LAC to function as AFOAT-1's command and control headquarters. The LAC's Team 139 controlled the AEDS data flow. With the Q systems now fully operational within the AEDS, exercise data included both seismic and electromagnetic feeds. SAC sent queries to the LAC, which were then forwarded to participating U.S. AEDS stations where records were screened. The location of strike data was then returned to SAC through the LAC. The average time for a reply was approximately 10 hours from electromagnetic teams and about 9 hours from the seismic teams. During this exercise, it became apparent SAC was most concerned about the survivability of Eielson AFB in Alaska and Thule AFB in Greenland. SAC requested AFOAT-1 to make these two locations a top priority in the AEDS. AFOAT-1 replied it would take two years to mature the AEDS to the point where AFOAT-1 could immediately provide SAC with reliable data on atomic explosions of at least 1MT at Thule AB or Eielson AFB. In providing SAC an honest assessment, AFOAT-1 stated the AEDS was incapable of doing so at the present time. SAC also recommended AFOAT-1 establish an electromagnetic station at each missile complex. Again, AFOAT-1 replied that an AEDS electromagnetic facility at each missile complex would not provide conclusive data.[34]

During the first quarter of 1958, SAC continued to press AFOAT-1 on using and improving the AEDS for timely assessments of strikes against Eielson and Thule. On March 11, 1958, SAC recommended that the AEDS be expanded to provide a general worldwide interim bomb alarm system reporting on enemy nuclear explosions to SAC. SAC requested information on the existing capability to provide data; future improvements in the detection capability; and funds, facilities, and personnel required to implement such a bomb alarm system.[35]

Under extensive pressure from SAC, which wanted an operational system in place by November 1958, AFOAT-1 completed a lengthy study of the entire problem on April 16, 1958.[36] In short, Northrup and his staff concluded it was "very optimistic" to expect the proposed bomb alarm system to be operational at the two bases by that time. In addition, the SAC commander, General Thomas S. Power, had set a requirement that he was to be informed within 30 minutes of any nuclear detonation detected by the AEDS occurring *at any SAC base in the world.*[37] While the report noted such a system was technologically feasible, several obstacles stood in the way. First, Canadian participation was essential, but those arrangements were still in flux. Second, advanced technical instruments (largely for the "Q" technique and a new computer system), while in production, had not been fully tested in the AEDS. Finally, Northrup informed his boss that SAC was now asking for confirmations, which the AEDS previously relied on from radiochemical analysis. This meant that AFOAT-1 had to confirm a detonation by immediately combining incoming seismic and electromagnetic data from more than one AEDS station. If both techniques' detonation times matched, then AFOAT-1 should "confirm" the event.[38]

Northrup's report was impressive given the time constraints involved. The report was lengthy due to the careful regional analyses which the staff prepared in assessing how the AEDS could conceivably report a detonation anywhere in the world within 30 minutes. The AFOAT-1 planners projected that, by January 1961, the AEDS could provide similar coverage for all overseas SAC bases if a reliable communications system were installed between AFOAT-1's analysis center and each AEDS station. This expansion would cost an estimated $5.4 million (2021 = $51 million), and the annual AFOAT-1 operating costs would be increased by more than $1,555,000 (2021 = $14.6 million). Also, AFOAT-1 manpower would increase by an additional 102 personnel. Northrup had his doubts. "It is with these reservations in mind and with the sense of urgency to get this report to the Strategic Air Command that I sign it and forward it for your consideration." By the summer of 1958, all parties agreed the expense was too great for a "bomb alarm system." However, the special alarm requirement for Eielson and Thule continued.[39]

By early 1959, AFOAT-1 was still attempting to provide SAC with a special bomb alarm capability on enemy nuclear explosions at Eielson and Thule. While the technological solution was feasible, AFOAT-1 notified the Air Staff that, in case of war, AFOAT-1 would require current intelligence on enemy targeting of those locations because the Thule-Eielson areas were not normally under surveillance. Refinements to the plan in 1959 included enhanced capabilities to provide SAC with post-strike information of explosions in enemy territory.[40]

When the Air Force inactivated the 1009th SWS/AFOAT-1 on July 6, 1959, the integration of wartime plans between SAC and AFOAT-1 had advanced about as far as they could go. With the exception of faster communication lines to improve data and reporting dissemination, AFOAT-1 had essentially put the Air Force on notice that the AEDS was a "go to war as is" system. In retrospect, SAC's extreme interest in utilizing the AEDS was no different from the general perceptions so apparent in the political debates over a test ban—that the AEDS was capable of much more than was really possible.

Notes for Annex 2

[1] *History of Long Range Detection: 1947-1953,* Chapter VII. Note that this annex draws heavily from that 25 page chapter due to the CODY Study missing from the AFTAC archives.

[2] The Warsaw Pact would form in May 1955 to formalize the alliance between the Soviet Union and its eastern European allies.

[3] *History of the Atomic Energy Detection System: 1947-1953,* 293.

[4] Ibid., 295.

[5] Ibid.

[6] Ibid., 296.

[7] Ibid., 300.

[8] Ibid., 301.

[9] Ibid., 302. In June 1951, AFOAT-1 briefers presented the revised CODY study to the assembled Air Staff and to SAC leaders. Operationally, SAC was interested in the use of the AEDS because the radarscope cameras on SAC bombers could not positively prove the detonation of their own bombs and because the crew might fail to return from such missions. The AFOAT-1 briefing to SAC occurred at a fortuitous time as SAC was planning to drop an atomic bomb on Bikini Atoll in an operation codenamed LAYOFF. In light of CODY, the planners decided to allow the AEDS to collect the debris rather than drone aircraft as used in the recent SANDSTONE shots. This meant that manned aircraft would make the collections that Major Fackler had advocated in his ad hoc experiment at SANDSTONE. Unfortunately, operational requirements in the Korean War precipitated the cancellation of Operation LAYOFF.

[10] Ibid., 297.

[11] Ibid., 303.

[12] The Bhangmeter was a non-imaging radiometer that was used to detect the distinctive bright double pulse of visible light emitted from a nuclear detonation. Bhangmeters were basically sensors that were deployed on satellites beginning in the 1960s when satellites became the "T" technique in the AEDS.

[13] *History of Long Range Detection: 1947-1953,* 304-305.

[14] Ibid., 306.

[15] Ibid., 303.

[16] Ibid., 306.

[17] Most likely, intense operations in the Korean War delayed their attention.

[18] *History of Long Range Detection: 1947-1953,* 307-308.

[19] Ibid., 309.

[20] Ibid., 312.

[21] Ibid., 310.

[22] Ibid., 313.

[23] Ibid., 314.

[24] Technical memo #80, Northrup to Canterbury, Subject: "Air Force Atomic Energy Intelligence Requirements," July 15, 1953, AFTAC archives.

[25] Ibid.

[26] *History of the Air Force Office for Atomic Energy-One (AFOAT-1): 1 January- 31 December 1954,* 46.

[27] *History of the Atomic Energy Detection System: 1947-1953,* 315.

[28] Ibid., 313.

[29] Letter from AFOAT-1 to SAC, Subject: "Transmittal of SAC Support Plan", dated October 20, 1954, and 1st Endorsement from SAC dated November 10, 1954.

[30] *History of the Air Force Office for Atomic Energy-One (AFOAT-1): 1 January- 31 December 1954,* 89-92.

[31] Glenn B. Fatzinger, *History of the Air Force Office for Atomic Energy- One (AFOAT-1): 1 January- 31 December 1957* (Washington, D.C.: The Air Force Office of Atomic Energy-One, Headquarters, United States Air Force, 1957), 122-127.

[32] Ibid.

[33] Ibid.

[34] Ibid., 38-39.

[35] Glenn B. Fatzinger, *History of the Air Force Office for Atomic Energy-One (AFOAT-1): 1 January - 31 December 1958* (Washington, D.C.: The Air Force Office of Atomic Energy-One, Headquarters, United States Air Force, 1958), 56-57.

[36] Technical memo #96, Northrup to Hooks, "Proposed System for Rapid Reporting of Nuclear Detonations," April 16, 1958, AFTAC archives. Throughout this report, it is clear that AFOAT-1 is not enthusiastic about adopting the AEDS as a bomb alert system. Up front, Northrup stated: 'The attached Technical Memorandum 96 has been prepared by my staff on terms of reference quite different from those upon which we usually base

technical reports. These terms were dictated by the urgency of the requirement for bomb alarm system by the Strategic Air Command."

[37] Emphasis added. Power had succeeded LeMay as SAC commander on July 1, 1957.

[38] Technical memo #96. Northrup explained how this was possible: 'The system proposed depends primarily on the interplay of data relating to the seismic and electromagnetic effects which are produced by large nuclear weapons detonated between the surface and 4,000 feet altitude. These are the only effects known to be detectable at great distances in sufficient time to meet SAC's requirements. An advantage in combining data from the seismic and electromagnetic techniques resides in the fact that, while neither technique alone is able to establish unequivocally that a nuclear burst has occurred, a nuclear burst is the only physical phenomenon known to produce both seismic waves and high-energy single pulse electromagnetic signals at the same time."

[39] Technical memo #96, Northrup to Hooks, "Proposed System for Rapid Reporting of Nuclear Detonations," April 16, 1958, AFTAC archives. Also Ibid.

[40] Glenn B. Fatzinger, *History of the Air Force Office for Atomic Energy- One (AFOAT-1): 1 January- 31 December 1959* (Washington, D.C.: The Air Force Office of Atomic Energy-One, Headquarters, United States Air Force, 1959), 48.

ANNEX 3

Chronology of LRD and the AEDS

1945

16 July: The first U.S. nuclear test, codenamed TRINITY, is conducted.

6 August: The U.S. drops an atomic bomb on the Japanese city of Hiroshima.

9 August   The U.S. drops an atomic bomb on the Japanese city of Nagasaki.

1946

14 June: The Baruch Plan that proposed a United Nations controlled program of oversight that would govern the development and use of atomic energy is presented to the UN's Atomic Energy Commission.

30 June: The nuclear test series—Operation CROSSROADS begin.

July: The Atomic Energy Commission (AEC) gains stewardship over all nuclear matters.

October: The Director of the Central Intelligence Group (CIG), Lieutenant General Hoyt S. Vandenberg, writes to Major General Leslie Groves seeking recommendations on how to set up a program to obtain information concerning the Soviet Union's possible possession of a nuclear bomb.

1947

1 January: The Atomic Energy Act of 1946 takes effect.

14 March: One week after the launching of the Truman Doctrine, Hoyt Vandenberg writes a letter to the Departments of the

War and Navy, the AEC, and the Joint Research and Development Board (JRDB) to express his concern for the need of an LRD capability.

14 May: Secretary of War Robert P. Patterson approves the Central Intelligence Group's (GIG) sponsorship of an LRD committee.

21 May: Representatives from the War Department, the Navy Department, AEC, and the CIG hold their first meeting to steer the birth of LRD.

6 June: A subcommittee of the LRD Committee concluded its work on identifying requirements and specifications for a LRD program. Their report advocated for an immediate implementation of a worldwide "system of systems" that could "determine the time and place of all large explosions which occur anywhere on the earth."

16 September: Eisenhower signs a directive charging the Commanding General of the AAF, with "overall responsibility for detecting atomic explosions anywhere in the world."

18 September: The Army Air Force becomes The United States Air Force, a separate uniformed service.

27 October: The Air Staff delegates responsibility for preparing the overall LRD plan to the Deputy Chief of Staff for Operations (DCS/O), General Lauris Norstad.

13 November: General Norstad submits a plan to Vandenberg that the Special Weapons Group (SWG) within the Deputy Chief of

Staff for Materiel (DCS/M) gain responsibility for LRD.

10 December: Lieutenant General Howard A. Craig, the DCS/M, organizes a LRD division within Kepner's SWG which is designated Air Force Materiel Special Weapons One (AFMSW-1). Its mission was to conduct R&D in order to design a functional LRD system.

14 December: Major General Alfred F. Hegenberger is appointed chief of the LRD division.

1948

January: AFOAT-1 Technical Director Ellis Johnson prepares a broad outline of a two-year program of intensive R&D in the seismic, acoustic, and nuclear fields.

January: The CAE unanimously pronounces that "we, on the Research and Development Board (RDB), deem it very unlikely that Russia will be in a position to test an atomic bomb as early as 1950, or within several years of that date."

14 January: AFOAT-1 Technical Director Ellis Johnson issues Technical Directive No.1, which was a preliminary estimate of the overall LRD problem.

February: AFOAT-1 first contracts with Tracerlab Inc., for radiochemistry laboratory services.

1 April: HQ USAF activates the 51st Air Force Base Unit as a field extension of the USAF Chief of Staff. Thus, the 51st became the military support organization for AFMSW-1 (the initial LRD organization).

14 April:	The nuclear test series Operation SANDSTONE begins.
28 April:	AFOAT-1 Technical Director Ellis Johnson reports that, at the time, the airborne collection method was the only LRD technique that could ensure a positive detection of a Russian test.
13 May:	Major General Albert Hegenberger, Commander of the 1009th SWS/AFOAT-1 requests a joint meeting of American, Canadian and British nuclear physicists to bring the best minds together in order to design an "interim network of long range detection stations."
1 June:	The AEDS first begins operations as the Air Weather Service initiates sorties of filter-equipped WB-29s as routine missions between Alaska and Japan.
24 June:	Russia initiates a blockade of Berlin to force the western allied powers to abandon their sectors of the divided city.
1 July:	The Air Staff transfers the SWG from the DCS/M to the DCS/O, and re-designates the SWG as the "Office of the Assistant Deputy Chief of Staff, Operations for Atomic Energy" (abbreviated "Office for Atomic Energy" or AFOAT).
1 July:	Hegenberger informs the AEC chairman (Lilienthal) that the initial challenges to establishing the AEDS could be more rapidly overcome by including the two allies (Canada and the UK) in its development.
2 July:	AFOAT-1's Technical Director, Ellis Johnson, submits a scathing 29-page memo to Hegenberger that condemns

	AMC for not supporting the LRD program.
7 July:	Johnson resigns over the issue of R&D budget restrictions (effective 4 August).
20 July:	Hegenberger submits a memo to Hoyt Vandenberg that stresses the need for Secretary of Defense James Forrestal to broaden the AFOAT-1 mission and to emphasize the priority of LRD.
28 August:	The 51st AF Base Unit is re-designated the 1009th Special Weapons Squadron (SWS) with the assumed office symbol of AFOAT-1. On that date, Hegenberger becomes the unit's first commander.
27 October:	AFOAT-1 presents a modified LRD plan to the Loomis Panel of the RDB. The RDB soon recommends that the Air Force begin the implementation of a comprehensive R&D program to determine the feasibility of utilizing seismic and acoustic techniques in an LRD surveillance system.
4 December:	The Loomis Panel formally submits its report to the CAE, noting that the exact date of a Soviet detonation is less important than gaining knowledge from nuclear research, thus reducing the sense of urgency to accelerate the LRD program.
17 December:	The CAE passes a resolution accepting the Loomis Panel's report with the provision that the panel could reconvene based on the impending JCS decision about the priority of LRD.
31 December:	Vandenberg informs the JCS that the RDB's decisions made it impossible for

AFOAT-1 to effect a fully operational AEDS by mid-1950.

<u>**1949**</u>

25 January: AFOAT-1 transfers 16 airmen to AWS to operate the Filter Center to support airborne debris collections. There are now 20 stations and 55 filter-equipped RB-29 AWS aircraft flying routine sampling missions from Guam to the North Pole. In total, the Interim Net (i.e. the AEDS) now employs 1,342 personnel in direct and secondary support roles.

7 March: The Air Force succeeds in persuading the RDB to convene a new joint committee of experts to review the LRD problem in the light of the strategic importance of LRD, as well as its technical aspects.

April: The Naval Research Laboratory (NRL) begins Project RAIN BARREL by constructing a nuclear debris collection station on Kodiak Island, Alaska.

May: The AWS establishes a routine weather reconnaissance/LRD track code-named "Loon Charlie" from Eielson AFB, Alaska, to Yokota, Airbase, Japan.

29 August: The Soviets detonate their first nuclear device—which the Americans nicknamed Joe-1—much earlier than anyone had expected. It is detected by AFOAT-1 personnel flying from Japan to Alaska.

31 August: Upon his retirement from the USAF, Hegenberger is replaced by his Deputy, Major General Morris Nelson, as Commander of the 1009th SWS/AFOAT-1.

August:	AFOAT-1 procures the first 10 Ground Filter Units (GFUs)
3 September:	A WB-29 flying from Japan to Alaska collects the first nuclear debris of the Russian Joe-1 detonation.
7 September:	AFOAT-1 receives the initial Tracerlab report that positively determines the presence of fission products from Joe-1.
19 September:	Vandenberg holds a meeting of American and British scientists to validate the Joe-1 analysis.
23 September:	President Truman announces to the world that the USSR had exploded its first nuclear device.

<u>1950</u>

20 January:	The JCS revalidates the LRD program as a top priority for the nation.
31 January:	President Truman authorizes research into the development of a hydrogen weapon.
2 February:	British authorities place Klaus Fuchs under arrest on charges of espionage; thus exacerbating restrictions on the sharing of information between AFOAT-1 and the British.
March:	R&D work begins on constructing a ground-based air sampler—the B/20.
20 March:	The Boner Panel completes its reassessment of the LRD R&D program and recommends that R&D should immediately develop LRD techniques that would assess and determine the detonation of a thermonuclear weapon.
5 April:	The RDB directs that a request be made to the Secretary of Defense for additional monies to immediately permit full

implementation of AFOAT-1's R&D program.

7 April: The National Security Council policy document NSC-68 is enacted. It dramatically increases the military budget in order to build up both conventional and nuclear arms. Importantly, it calls for the enhancement of American technical superiority through "an accelerated exploitation of [its] scientific potential."

1 June: Major General Morris Nelson announces that AFOAT-1 would complete its plans for a fully operational AEDS by 1 June 1951.

25 June: The Korean War begins.

8 August: AFOAT-1 completes a study, codenamed CODY, which determines the type and degree of participation that the AEDS could contribute to combat operations in wartime.

October: General Nelson and Northrup meet with members of the British embassy staff in Washington to explore the possibility of resuming LRD scientific conferences.

1951

January: The Strategic Air Command approaches AFOAT-1 to determine how the AEDS could meet SAC wartime mission requirements.

4 January: Brigadier General Raymond Coleman Maude assumes command of the 1009th SWS/AFOAT-1 from Nelson.

27 January: The nuclear test series Operation RANGER begins.

30 March:	Northrup sends a sharply worded six-page memo to General Maude to inform' him that a personnel crisis at Tracerlab threatens the operations of the AEDS and important LRD research
7 April:	The nuclear test series Operation GREENHOUSE begins.
2 July:	The Chairman of the Military Liaison Committee (MLC) is briefed that expansion of the AEDS could not achieve full functionality without the assistance of the British and the Canadians.
2 August:	A joint U.S.-British agreement is signed that will allow more LRD information sharing within the restrictions of the Atomic Energy Act.
16 September:	Major General Donald J. Keirn assumes command of the 1009th SWS/AFOAT-1 from Maude.
22 October:	The nuclear test series Operation BUSTER-JANGLE begins.

<u>1952</u>

4 March:	Northrup receives notification that the ARDC will now manage some of the AFOAT-1 R&D budget.
1 April:	The nuclear test series Operation TUMBLER-SNAPPER begins.
15 April:	A joint U.S.-Canadian agreement is signed that will allow more LRD information sharing within the restrictions of the Atomic Energy Act.
15 April:	The Air Force flies its first B-52 bomber; an airframe that will later allow AFOAT-1 to collect nuclear debris at high altitudes.

15 August:	Brigadier General William M. Canterbury assumes command of the 1009th SWS/AFOAT-1 from Keirn.
3 October:	The British begin nuclear test series Operation HURRICANE.
31 October:	The nuclear test series Operation IVY begins.
1 November:	The United States successfully tests the world's first thermonuclear weapon. Codenamed shot Mike, the world's first hydrogen bomb produced a yield of 10.4 megatons.
4 November:	Eisenhower wins the presidential election with a landslide victory of 442 electoral votes.

1953

4 March:	Soviet Premier Joseph Stalin dies, thus installing new leaders who would be amiable to future test ban negotiations.
17 March:	The nuclear test series Operation UPSHOT-KNOTHOLE begins.
13 April:	AFOAT-1 finalizes Emergency War Plan (EWP) 1-53.
May:	President Eisenhower implies that he is prepared to use tactical nuclear weapons to end the war in Korea if North Korea continues to stall the peace negotiations.
22 July:	The Air Staff approves AFOAT-1's Emergency War Plan.
27 July:	The Korean War ends.
12 August:	The Soviets detonate their first thermonuclear device (Joe-4) that produced a yield of 400kt.
1 September:	The electromagnetic ("Q") technique becomes operational in the AEDS.

15 October:	The British begin nuclear test series Operation TOTEM.
1 December:	Doyle Northrup provides General Canterbury with an assessment on Soviet capabilities to detect U.S. nuclear detonations.
8 December:	President Eisenhower delivers his famous "Atoms for Peace" proposal before the United Nations General Assembly.

<u>1954</u>

28 February:	The nuclear test series Operation CASTLE begins with the 15MT explosion of shot Bravo, the largest nuclear device ever detonated by the U.S.
17 March:	The Russian government directs its armed forces to create an LRD capability equivalent to the AEDS.
7 April:	The public views the freshly released film footage of the 10MT IVY Mike shot from October 1952, thus creating domestic and world-wide furor over the dangers of the H-bomb.
June:	Major General Daniel E. Hooks assumes command of the 1009th SWS/AFOAT-1 from Canterbury.
10 September:	In a memo to Hooks, Northrup documents his concerns about how extreme security restrictions in sharing information with the British are impeding progress.
10 November:	The Strategic Air Command incorporates AFOAT-1's basic war plan as Annex E "SAC Support" in the SAC war plan.

1955

18 February:	The nuclear test series Operation TEAPOT begins.
March:	Eisenhower appoints former Minnesota governor Harold Stassen as Assistant to the President for Disarmament.
April:	Northrup directs the removal of all ground-based particulate samplers from the AEDS.
15 June:	The U.S. and the UK sign a new agreement that reduces some of the restrictions on sharing nuclear information. Northrup is able to add several paragraphs to the document that relax some of the Area 5 security constraints.
18 July:	President Eisenhower announces his "Open Skies" proposal at the Geneva Four Powers Summit.
August:	AWS begins transitioning from the WB-29 to the WB-50 as the primary airborne collection aircraft.
16 September:	Stassen visits AFOAT-1 headquarters and receives a briefing on LRD techniques
November:	The Air Force conducts the initial test of a modified B-36 for use in collecting nuclear debris in the AEDS.

1956

January:	Harold Stassen testifies before a disarmament subcommittee of the Senate Foreign Relations Committee that was created to explore the feasibility of a nuclear test ban.
4 May:	The nuclear test series Operation REDWING begins.

8 May:	AFOAT-1 publishes its Emergency Relocation Plan that detailed evacuation procedures to re-locate to Camp Barrett at the Quantico Marine Base in case of a nuclear war with the Soviet Union.
16 May:	The British begin nuclear test series Operation MOSAIC.
May:	Stassen submits his first report to Eisenhower which concludes that a total elimination of all nuclear weapons is simply not realistic or feasible, and advocates for other forms of arms control.
4 July:	The CIA conducts the first U-2 reconnaissance flight over the Soviet Union.
27 September:	The British begin nuclear test series Operation BUFFALO.
8 October:	AFOAT-1 establishes Detachment 313 at Sonseca, Spain.
19 October:	Northrup releases an AEDS capabilities assessment that notes the difficulties in extending the AEDS into the southern Hemisphere.
6 November:	Eisenhower is reelected to a second term, sweeping 41 states and receiving 58 percent of the popular vote.

1957

21 January:	Dwight Eisenhower is sworn-in for a second term as President of the United States.
18 March:	The United Nations Disarmament Subcommittee (i.e., The London Disarmament Conference) begins with representatives from the U.S., the USSR, the UK, France, and Canada.

15 April:	The seismic station at Sonseca, Spain, becomes fully operational and is the 9th site to enter the AEDS.
6 May:	AFOAT-1 begins a major upgrade to the seismic network.
15 May:	The British begin nuclear test series Operation GRAPPLE.
28 May:	The nuclear test series Operation PLUMBBOB begins.
14 June:	At the London Conference on Disarmament, the Soviets suddenly agree to allow monitoring stations within the Soviet Union as part of a larger system of inspections within the nuclear nations. This constitutes a radical change in their position that appears to acquiesce to American demands. They also agree to suspend nuclear testing for two to three years.
21 June:	Northrup submits an assessment to General Hooks that succinctly underscores the limited capabilities of the AEDS.
21 August:	The Soviets conduct the first successful launch of an ICBM.
6 September:	The London Disarmament Conference ends with no consensus achieved.
14 September:	The British begin nuclear test series Operation ANTLER.
19 September:	The first underground nuclear test ever conducted occurs as shot Ranier in Operation PLUMBBOB
4 October:	The Soviets launch the world's first artificial satellite into space, called *Sputnik I*.

10 October:	The Soviets conduct their first underground nuclear detonation.
3 November:	The Soviets launch the *Sputnik II* satellite.
7 November:	The Gaither Report is released and fuels the widespread belief that the Soviets have achieved significant technological superiority.
7 November:	During a scheduled nationwide public address, Eisenhower announces the creation of the President's Science Advisory Committee (PSAC), and names James Killian, Jr. as Chairman and the first Special Assistant to the President for Science and Technology.
10 December:	The Soviet Premier recommends a summit to discuss an immediate moratorium on testing that would last for two to three years.

<u>1958</u>

9 January:	Killian, as directed by the NSC, appoints Hans Bethe to head up a working group to examine the complexities of monitoring compliance of a test ban.
12 January:	President Eisenhower writes to Soviet Premier Bulganin to insist on a conference to deal with the issue of a surprise attack (i.e., how to address such concerns).
5 February:	Northrup delivers a lengthy report entitled "Present and Potential Capabilities and Limitations of the AFOAT-1 Long Range Detection System" to Hans Bethe for the National Security Council.
11 March:	The Strategic Air Command (SAC) recommends that the AEDS be expanded

	to provide a general world-wide interim bomb alarm system that will report on enemy nuclear explosions to SAC.
17 March:	The United States launches the *Vanguard I* satellite.
28 March:	Hans Bethe's *ad hoc* working group delivers the NSC-directed top secret report to the PSAC (Killian).
31 March:	The Soviets announce that they will discontinue the testing of all types of atomic and hydrogen weapons in the Soviet Union.
28 April:	The nuclear test series Operation HARDTACK I begins.
9 May:	Northrup writes a letter to Killian expressing his concerns about widespread misperceptions that the AEDS is fully capable of monitoring a test ban agreement.
9 May:	The Soviets agree to Eisenhower's request for a Conference of Experts.
13 May:	The Special Monitoring Service (SMS), the Soviet equivalent to AFOAT-1, is formally established within the Ministry of Defense.
July:	The Conference of Experts begins in Geneva.
July:	Congress amends the Atomic Energy Act of 1954 to permit greater information sharing with U.S. allies.
1 August:	Brigadier General Jermain F. Rodenhauser assumes command of the 1009th SWS/ AFOAT-1.
19 August:	The Conference of Experts ends.

22 August:	The President informs the American people that the experts have all agreed that a system of international control to monitor a test ban is feasible.
27 August:	The nuclear test series Operation ARGUS begins.
12 September:	The nuclear test series Operation HARDTACK II begins.
25 September:	In anticipation of a comprehensive test ban, AFOAT-1 completes a study entitled "Incorporation of the Long Range Detection Capability of the AEDS into an Internationally Controlled Organization."
31 October:	The first session of the Geneva Conference begins.
10 November:	The Surprise Attack Conference begins in Geneva.
18 December:	The Surprise Attack Conference ends.
19 December:	The first session of the Geneva Conference recesses.

<u>1959</u>

5 January:	The second session of the Geneva Conference begins.
22 January:	The PSAC visits the University of California Radiation Laboratory. Dr. Edward Teller presents the topic of "decoupling" as a major problem for LRD.
19 March:	The second session of the Geneva Conference ends.
13 April:	The third session of the Geneva Conference reconvenes.
25 June:	Northrup presents his study "Improvement and Expansion of the A.E.D.S." that underscores the inability of the AEDS to adequately monitor a test ban.

6 July: The Air Force inactivates the 1009th
             SWS/AFOAT-1.

BIBLIOGRAPHY

Archives

Air Force Technical Applications Center (AFTAC)
Cornell University, Rare and Manuscript Collections
Dwight D. Eisenhower Presidential Library and Museum
Harry S. Truman Presidential Library and Museum
Los Alamos National Laboratory archives

Articles

"Accident Revealed after 29 Years: H-Bomb Fell Near Albuquerque in 1957."). *Los Angeles Times* (August 27, 1986).

Barth, Kai-Henrik. "The Politics of Seismology: Nuclear Testing, Arms Control, and the Transformation of a Discipline." *Social Studies of Science* 33, No. 5 (October 2003): 743-781.

Christofilos, Nicholas C. "The ARGUS Experiment." *Geophysics* 45 (1959): 1144-1152, http://www.pnas.org.

Damms, Richard V. "James Killian, the Technological Capabilities Panel, and the Emergence of President Eisenhower's 'Scientific-Technological Elite.'" *Diplomatic History* 24, Issue 1 (January 2000): 57-78.

Friedman, Herbert, Luther B. Lockhart, and Irving H. Blifford. "Detecting the Soviet Bomb: Joe-1 in a Rain Barrel." *Physics Today* (November 1996).

Goodby, James E. "The Limited Test Ban Negotiations, 1954-63: How a Negotiator Viewed the Proceedings." *International Negotiation* 10 (2005): 381-404.

Hegenberger, Robert F. "The Bird of Paradise: The Significance of the 1927 Hawaiian Flight." *Air Power History* (Summer 1991).

Latter, A.L., R.E. LeLevier, E.A. Martinelli, and W.G. McMillan. "A Method of Concealing Underground Nuclear Explosions." *The Journal of Geophysical Research* 66, No. 3 (March 1961): 943-946.

MacDonald, Julia M. "Eisenhower's Scientists: Policy Entrepreneurs and the Test-Ban Debate 1954-1958." *Foreign Policy Analysis* 11, (2015): 1-21.

Morland, Howard. "Born Secret." *Cardozo Law Review* 26, No.4 (March 2005).

Murphy, C.J.V. "Nuclear Inspections: A Near Miss." *Fortune* LIX, No. 3 (March 1959).

Norris, Robert S. and Hans M. Kristensen. "Global Nuclear Weapons Inventories, 1945- 2010." *Bulletin of the Atomic Scientists* (July/August 2010).

"Nuclear Test Debate: A Foolproof System Needs a Rogueproof Agreement." Editorial. *Time Magazine* (2 February 1959): 17.

Payne, Rodger A. "Public Opinion and Foreign Threats: Eisenhower's Response to Sputnik." *Armed Forces & Society 21*, No.1 (Fall 1994): 89-112.

Powell, Alvin. "How Sputnik Changed U.S. Education." *The Harvard Gazette,* (October 11, 2007), https://news.harvard.edu/gazette/story/2007/10/how-sputnik- changedu-s-education.

Robinson, Paul. "'Crucified on a Cross of Atoms': Scientists, Politics, and the Test Ban Treaty." *Diplomatic History* 35, No. 2 (April 2011): 283-319.

Shreve, Bradley G. 'The US, the USSR, and Space Exploration, 1957-1963." *International Journal on World Peace,* Vol. 20, No.2 (June 2003): 67-83.

Smith-Norris, Martha. "The Eisenhower Administration and the Nuclear Test Ban Talks, 1958-1960: Another Challenge to Revisionism." *Diplomatic History* 27, No. 4 (September 2003): 503-541.

Suri, Jeremi. "America's Search for a Technological Solution to the Arms Race: The Surprise Attack Conference of 1958 and a Challenge for 'Eisenhower Revisionists.'" *Diplomatic History 21,* No.3 (Summer 1997): 417-451.

Tal, David. "From the Open Skies Proposal of 1955 to the Norstad Plan of 1960: A Plan Too Far." *Journal of Cold War Studies* 10, No. 4 (Fall 2008): 66-93.

Volmar, Axel. "Listening to the Cold War: The Nuclear Test Ban Negotiations, Seismology, and Psychoacoustics, 1958-1963." *The History of Science Society 28* (2013): 80-102.

York, Herbert F. "Making Weapons, Talking Peace." *Physics Today* (April 1988): 40-52.

Ziegler, Charles. "Waiting for Joe-1: Decisions Leading to the Detection of Russia's First Atomic Bomb Test." *Social Studies of Science* 18 (1988): 197-229.

Books

Ambrose, Stephen E. *Eisenhower: The President.* New York, NY: Simon and Schuster, 1984.

Bolt, Bruce A *Nuclear Explosions and Earthquakes.* San Francisco, CA: W.H. Freeman, 1976.

Bowe, Robert R. and Richard H. Immerman. *Waging Peace: How Eisenhower Shaped an Enduring Cold War Strategy.* New York, NY: Oxford University Press, 1998.

Buck, Alice L. *A History of the Atomic Energy Commission.* Washington DC: U.S. Department of Energy, 1983.

Cantelon, Philip L., Richard G. Hewlett, and Robert Williams. *The American Atom: A Documentary History of Nuclear Policies from the Discovery of Fission to Present.* Philadelphia, PA: University of Pennsylvania Press, 1991.

Cherepanov, Y. V. and Alexey Pavlovich Vasiliev, Ed. *Born by the Atomic Age: 1958-1998.* Moscow: 2002.

Clemens, Walter C. and Franklyn Griffiths. *The Soviet Position on Arms Control and Disarmament- Negotiations and Propaganda, 1954-1964.* Cambridge, MA: Center for International Studies, 1965.

Divine, Robert A. *Blowing on the Wind: The Nuclear Test Ban Debate, 1954-1960.* New York, NY: Oxford University Press, 1978.

_____ . *Eisenhower and the Cold War.* New York, NY: Oxford University Press, 1981.

Gaddis, John Lewis. *We Now Know: Rethinking Cold War History.* New York, NY: Oxford University Press, 1998.

Goodman, Michael E. *Spying on the Nuclear Bear: Anglo- American Intelligence and the Soviet Bomb.* Stanford, CA: Stanford University Press, 2007.

Gordin, Michael D. *Red Cloud at Dawn: Truman, Stalin, and the End of the Atomic Monopoly.* New York, NY: Picador, 2010.

Gorn, Michael H. *Harnessing the Genie: Science and Technology Forecasting for the Air Force, 1944-1986.* Washington, DC: Office of Air Force History, 1989.

Goudsmit, Samuel A. *Alsos.* Newbury, NY: AIP Press, 1996.

Groves, Leslie M. *Now It Can Be Told.* NY: Harper, 1962.

Halberstam, David. *The Fifties.* New York, NY: Fawcett, 1994.

Herken, Gregg. *Brotherhood of the Bomb: The Tangled Lives and Loyalties of Robert Oppenheimer, Ernest Lawrence, and Edward Teller.* New York, NY: John Macrae/Owl Book, Henry Holt and Co, 2003.

Herken, Gregg. *Cardinal Choices: Presidential Science Advising from the Atomic Bomb to SDI.* Stanford, CA: Stanford University Press, 2000.

Jacobson, Harold Karan and Eric Stein. *Diplomats, Scientists, and Politicians.* Ann Arbor, Ml: University of Michigan Press, 1966.

Larson, Deborah Welch. *Anatomy of Mistrust: U.S.-Soviet Relations during the Cold War.* Ithaca, NY: Cornell University Press, 2000.

Lawren, William. *The General and the Bomb: A Biography of General Leslie R. Groves, Director of the Manhattan Project.* New York, NY: Dodd, Mead, 1988.

LeMay, Curtis E., and MacKinlay Kantor. *Mission with LeMay.* Garden City, NJ: Doubleday and Company, Inc., 1965.

Meilinger, Phillip S. *Hoyt S. Vandenberg, the Life of a General.* Bloomington, IN: Indiana University Press, 1989.

Miller, Richard L. *Under the Cloud: The Decades of Nuclear Testing.* Woodlands, TX: Two-Sixty Press, 1991.

Neff, Donald. *Warriors at Suez: Eisenhower Takes America into the Middle East.* New York, NY: Linden Press, Simon and Schuster, 1981.

Peierls, Rudolf Ernst. *Atomic Histories.* Woodbury, NY: American Institute of Physics, 1997.

Reed, Thomas C., and David B. Stillman. *The Nuclear Express: A Political History of the Bomb and Its Proliferation.* Minneapolis, MN: Zenith Press, 2010.

Rhodes, Richard. *The Making of the atomic bomb.* New York, NY: Touchstone Book, 1986.

Rhodes, Richard. *Dark Sun: The Making of the Hydrogen Bomb.* New York, NY: Touchstone, 1996.

Romney, Carl. *Detecting the Bomb: The Role of Seismology in the Cold War.* Washington, DC: New Academia Publishing, 2009.

———. *Reflections.* Bloomington, IN: AuthorHouse, 2012.

Ruane, Kevin. *Churchill and the Bomb: In War and Cold War.* London: Bloomsbury Publishing, 2016.

Schwerber, Silvan S. *In the Shadow of the Bomb: Bethe, Oppenheimer, and the Moral Responsibility of the Scientist.* Princeton, NJ: Princeton University Press, 2000.

Sherwin, Martin J. *A World Destroyed: The Atomic Bomb and the Grand Alliance.* New York, NY: Vintage Books, 1977.

Shrader, Charles R. *History of Operations Research in the United States Army: Volume 1: 1942-1946.* Washington, DC: Office of the Deputy Under Secretary of the Army for Operations Research, 2000.

Siddiqi, Asif A. *Sputnik and the Soviet Space Challenge.* Gainsville, FL: University of Florida Press, 2003.

Stassen, Harold Edward, and Marshall Houts. *Eisenhower: Turning the World toward Peace.* St. Paul, MN: Merrill-Magnus, 1990.

Steil, Ben. *The Marshall Plan: Dawn of the Cold War.* New York, NY: Simon and Schuster, 2018.

Taylor, Leland B. *History of Air Force Atomic Cloud Sampling.* Andrews AFB, MD: Air Force Systems Command, 1963.

Teller, Edward and Albert Latter. *Our Nuclear Future: Facts, Dangers, and Opportunities.* New York, NY: Criterion Books, 1958.

Walker, Martin J. *The Cold War: A History.* New York, NY: H. Holt, 1994.

Wang, Jessica. *American Science in an Age of Anxiety: Scientists, Anticommunism, and the Cold War.* Chapel Hill, NC: University of North Carolina Press, 1999.

Wadsworth, James J. *The Price of Peace,* New York, NY: Frederick A. Praeger, 1962.

Wang, Zuoyue. *In Sputnik's Shadow: The President's Science Advisory Committee and Cold War America.* New Brunswick, NJ: Rutgers University Press, 2009.

U.S. Department of State. *Documents on Disarmament, 1945-1959.* Washington, DC: Government Printing Office, 1960.

Williams, Robert Chadwell. *Klaus Fuchs, Atom Spy.* Cambridge, MA: Harvard University Press, 1987.

Zoppo, Giro E. *The Issue of Nuclear Test Cessation at the London Disarmament Conference of 1957: A Study in East-West Negotiation.* Santa Monica, CA: RAND Corporation, September 1961.

Dissertations

Barth, Kai-Henrik. "Detecting the Cold War: Seismology and Nuclear Weapons Testing, 1945-1970." PhD dissertation. University of Minnesota, July 2000.

Fitzpatrick, Anne. "Igniting the Light Elements: The Los Alamos Thermonuclear Weapon Project, 1942-1952." PhD dissertation. George Washington University, 1999.

Reports

Berkhouse, L.H., S.E. Davis, F.R. Gladeck, J.H. Hallowell, C.B. Jones, E.J. Martin, F.W. McMullan, and M.J.

Osborn. "Operation GREENHOUSE: 1951,"Technical Report. Washington, DC: Defense Nuclear Agency, June 1983.

_____ . *Final Report of Operation Crossroads-1946.* Washington, DC: Defense Nuclear Agency, May 1984.

Berkhouse, L.H., J.H. Hallowell, F.W. McMullan, S.E. Davis, C.B. Jones, M.J. Osborn, F.R. Gladeck, EJ. Martin, and W.E. Rogers. "Operation SANDSTONE: 1948," Technical Report. Washington, DC: Defense Nuclear Agency, December 1983.

Blumenson, Martin and Hugh D. Hexamer. "A History of Operation Redwing: The Atomic Weapons Teats in the Pacific, 1956." Washington, DC: Headquarters, Joint Task Force Seven, December 1956.

Clemens, Walter C., and Franklyn Griffiths. "The Soviet Position on Arms Control and Disarmament-Negotiations and Propaganda, 1954-1964: An Annex to Report on Soviet Interests in Arms Control and Disarmament." Center for International Studies. Cambridge, MA: Massachusetts Institute of Technology, February 1965.

Gladeck, F.R., J.H. Hallowell, E.J. Martin, F.W. McMullan, R.J. Miller, R. Pozega, W.E. Rogers, R.H. Roland, C.F. Shelton, and L. Berkhouse. *OPERATION IVY: 1952.* Washington, DC: Defense Nuclear Agency, December 1982.

Harris, P.S., C. Lowery, A Nelson, S. Obermiller, W.J. Ozeroff, and S.E. Weary. *PLUMBBOB Series 1957.* Alexandria, VA: Defense Nuclear Agency, September, 1981.

Hull, J.E. Report to the Joint Chiefs of Staff. "Atomic Weapons Tests, Enewetak Atoll, Operation

464

Sandstone, 1948." San Francisco, CA: Headquarters, Joint Task Force Seven, 16 June 1948.

Ponton, Jean, Carl Maag, Mary Francis Barrett, and Robert Shepanek. "Operation TUMBLER-SNAPPER, 1952." Washington D.C: Defense Nuclear Agency, June 1982.

U.S. Air Force. *The Roswell Report: Fact Versus Fiction in the New Mexico Desert.* Washington, DC: Headquarters, United States Air Force, 1995.

_____. "4950th Test Group (N) Final Report, Operation Plumbbob." Kirtland, AFB, NM: Headquarters, Air Force Special Weapons Center, 20 November 1957.

_____. *Report of Operation Fitzwilliam.* Washington, D.C.: Headquarters, United States Air Force, October 1952. U.S. Department of Defense.

_____. *Defense's Nuclear Agency: 1947-1997.* Defense Threat Reduction Agency. Washington, DC: Government Printing Office, 2002.

U.S. Department of State. Conference of Experts to Study the Methods of Detecting Violations of a Possible Agreement of the Suspension of Nuclear Tests. Verbatim Record of the Eighth Meeting, 10 July 1958. Washington, DC: Government Printing Office, 1958.

U.S. Navy. "Operation WIGWAM." Operation Plan 1-55. Task Group 7.3, 25 March 1955. DTIC ADA078579.

U.S. Senate. Hearing before a Subcommittee of the Committee on Foreign Relations. Control and Reduction of Armaments. 84th Congress, 2nd Session, January 25, 1956. Washington, DC: Government Printing Office, 1957.

_____ . Hearings: Disarmament and Foreign Policy, 86th Congress, 1st Session, January 28, 1958. Government Printing Office, 1959.

_____ . Hearings: Disarmament and Foreign Policy, 86th Congress, 1st Session, February 2, 1958. Government Printing Office, 1959.

GLOSSARY

A

AAF	Army Air Force
ADC	Air Defense Command
AEC	Atomic Energy Commission
AEDS	Atomic Energy Detection System
AEIU	Atomic Energy Intelligence Unit
AFCRC	Air Force Cambridge Research Center
AFMSW-1	Air Force Materiel Special Weapons-One
AFOAT-1	Office of the Assistant Deputy Chief of Staff, Operations for Atomic Energy
AFSWP	Armed Forces Special Weapons Project
AFTAC	Air Force Technical Applications Center
AMC	Air Materiel Command
ANG	Air National Guard
ARDC	Air Research & Development Command
ARPA	Advanced Research Projects Agency
AWRE	Atomic Weapons Research Establishment
AWS	Air Weather Service

B

B&H	Beers & Heroy
BDA	Bomb Damage Assessment

C

CAE	Committee on Atomic Energy
CFO	Central Field Office
CIA	Central Intelligence Agency
GIG	Central Intelligence Group
CPM	Counts per Minute

D

DCS/M   Deputy Chief of Staff for Materiel
DCS/O   Deputy Chief of Staff for Operations
DoD   Department of Defense
DRL   Defense Research Laboratory
DSI   Directorate of Scientific Intelligence

E

EDC   European Defense Community
EFO   Eastern Field Office
EG&G   Edgerton, Germeshausen, and Grier, Inc.
EGMG   Experimental Guided Missiles Group
EMP   Electromagnetic Pulse
EWP   Emergency War Plan

F

FAC   Forward Air Controller
FAG   Field Activities Group
FBI   Federal Bureau of Investigations
FEAF   Far East Air Force
FY   Fiscal Year

G

GAG   General Advisory Committee
GFU   Ground Filter Units
GMT   Greenwich Mean Time

H

HUAC   Committee on Un-American Activities

I

ICBM   Intercontinental Ballistic Missile
IMS   International Monitoring System

| IR | Infrared |

J

JAEIC	Joint Atomic Energy Intelligence Committee
JCAE	Joint Committee on Atomic Energy
JCS	Joint Chiefs of Staff
JIC	Joint Intelligence Committee
JRDB	Joint Research and Development Board
JTF	Joint Task Force

K

| Kt | Kiloton |

L

LAC	Laramie Analysis Center
LASL	Los Alamos Scientific Laboratory
LRD	Long Range Detection

M

MATS	Military Air Transportation Service
MCL	McClellan Central Laboratory
MEW	Methane End Window
MLC	Military Liaison Committee
MT	Megaton
MoD	Ministry of Defence

N

NATO	North Atlantic Treaty Organization
NBS	National Bureau of Standards
NEL	Navy Electronics Laboratory
NOL	Naval Ordnance Laboratory
NPG	Nevada Proving Grounds
NSC	National Security Council

NRL	Naval Research Laboratory

O

ODM	Office of Defense Mobilization
ONR	Office of Naval Research

P

PM	Program Manager
PPG	Pacific Proving Ground
PRC	Peoples' Republic of China
PSAC	President's Scientific Advisory Committee

R

R/hr	Roentgens per hour
R&D	Research and Development
RAF	Royal Air Force
RD	Restricted Data
RDB	Research and Development Board
RS	Reconnaissance Squadron

S

SAC	Science Advisory Committee
SAC	Strategic Air Command
SEATO	Southeast Asia Treaty Organization
SFO	Southern Field Office
SMS	Special Monitoring Service
SOFAR	Sound Fixing and Ranging
SWG	Special Weapons Group
SWS	Special Weapons Squadron

T

TD	Technical Directorate

TOG	Technical Operations Group
TWG	Technical Working Group

U

UNAEC	United Nations Atomic Energy Commission
USC&GS	U.S. Coast and Geodetic Society
USAEDS	United States Atomic Energy Detection System

V

VLF	Very Low Frequency
VLR	Very Long Range

W

WFO	Western Field Office
WRS	Weather Reconnaissance Squadron

Made in the USA
Columbia, SC
02 January 2025

51047651R00267